KAPLAN

MCAT®

Critical Analysis and Reasoning Skills Review

2020–2021

Edited by Alexander Stone Macnow, MD

ONLINE + BOOK

ACKNOWLEDGMENTS

Editor-in-Chief, 2020–2021 Edition
Elizabeth Flagge

Contributing Editors, 2020–2021 Edition
Brandon Deason, MD, Dom Eggert, Tyler Fara, Lauren White

Prior Edition Editorial Staff: Brandon Deason, Christopher Durland, Dom Eggert, Tyler Fara, Elizabeth Flagge, Adam Grey, Lauren White

Published by Kaplan Publishing, a division of Kaplan, Inc.
750 Third Avenue
New York, NY 10017

ISBN: 978-1-5062-4871-4

10 9 8 7 6 5 4 3 2 1

TABLE OF CONTENTS

GO ONLINE

www.kaptest.com/booksonline

THE *KAPLAN MCAT REVIEW* TEAM

Alexander Stone Macnow, MD
Editor-in-Chief

Áine Lorié, PhD
Editor

Kristen L. Russell, ME
Editor

Derek Rusnak, MA
Editor

Pamela Willingham, MSW
Editor

Mikhail Alexeeff
Kaplan MCAT Faculty

Melinda Contreras, MS
Kaplan MCAT Faculty

Laura L. Ambler
Kaplan MCAT Faculty

Samantha Fallon
Kaplan MCAT Faculty

Krista L. Buckley, MD
Kaplan MCAT Faculty

Jason R. Pfleiger
Kaplan MCAT Faculty

Faculty Reviewers and Editors: Elmar R. Aliyev; James Burns; Jonathan Cornfield; Alisha Maureen Crowley; Nikolai Dorofeev, MD; Benjamin Downer, MS; Colin Doyle; Christopher Durland; M. Dominic Eggert; Marilyn Engle; Eleni M. Eren; Raef Ali Fadel; Elizabeth Flagge; Adam Grey; Tyra Hall-Pogar, PhD; Scott Huff; Samer T. Ismail; Elizabeth A. Kudlaty; Kelly Kyker-Snowman, MS; Ningfei Li; John P. Mahon; Matthew A. Meier; Nainika Nanda; Caroline Nkemdilim Opene; Kaitlyn E. Prenger; Uneeb Qureshi; Bela G. Starkman, PhD; Michael Paul Tomani, MS; Nicholas M. White; Allison Ann Wilkes, MS; Kerranna Williamson, MBA; and Tony Yu.

Thanks to Kim Bowers; Eric Chiu; Tim Eich; Tyler Fara; Owen Farcy; Dan Frey; Robin Garmise; Rita Garthaffner; Joanna Graham; Allison Harm; Beth Hoffberg; Aaron Lemon-Strauss; Keith Lubeley; Diane McGarvey; Petros Minasi; John Polstein; Deeangelee Pooran-Kublall, MD, MPH; Rochelle Rothstein, MD; Larry Rudman; Sylvia Tidwell Scheuring; Carly Schnur; Karin Tucker; Lee Weiss; and the countless others who made this project possible.

GETTING STARTED CHECKLIST

Getting Started Checklist

- ☐ Register for your free online assets—including three full-length MCATs, Science Review Videos, and additional practice materials—at kaptest.com/booksonline
- ☐ Download the Kaplan app to access the digital materials on your mobile device.
- ☐ Create a study calendar that ensures you complete content review and sufficient practice by Test Day!
- ☐ As you finish a chapter (including the practice problems!), check it off on the table of contents.
- ☐ Register to take the MCAT at aamc.org/mcat
- ☐ Set aside time during your prep to make sure the rest of your application—personal statement, recommendations, and other materials—is ready to go!
- ☐ Take a moment to admire your completed checklist, then get back to the business of prepping for this exam!

PREFACE

And now it starts: your long, yet fruitful journey toward wearing a white coat. Proudly wearing that white coat, though, is hopefully only part of your motivation. You are reading this book because you want to be a healer.

If you're serious about going to medical school, then you are likely already familiar with the importance of the MCAT in medical school admissions. While the holistic review process puts additional weight on your experiences, extracurricular activities, and personal attributes, the fact remains: along with your GPA, your MCAT score remains one of the two most important components of your application portfolio—at least early in the admissions process. Each additional point you score on the MCAT pushes you in front of thousands of other students and makes you an even more attractive applicant. But the MCAT is not simply an obstacle to overcome; it is an opportunity to show schools that you will be a strong student and a future leader in medicine.

We at Kaplan take our jobs very seriously and aim to help students see success not only on the MCAT, but as future physicians. We work with our learning science experts to ensure that we're using the most up-to-date teaching techniques in our resources. Multiple members of our team hold advanced degrees in medicine or associated biomedical sciences and are committed to the highest level of medical education. Kaplan has been working with the MCAT for over 50 years and our commitment to premed students is unflagging; in fact, Stanley Kaplan created this company when he had difficulty being accepted to medical school due to unfair quota systems that existed at the time.

We stand now at the beginning of a new era in medical education. As citizens of this 21st-century world of healthcare, we are charged with creating a patient-oriented, culturally competent, cost-conscious, universally available, technically advanced, and research-focused healthcare system, run by compassionate providers. Suffice it to say, this is no easy task. Problem-based learning, integrated curricula, and classes in interpersonal skills are some of the responses to this demand for an excellent workforce—a workforce of which you'll soon be a part.

We're thrilled that you've chosen us to help you on this journey. Please reach out to us to share your challenges, concerns, and successes. Together, we will shape the future of medicine in the United States and abroad; we look forward to helping you become the doctor you deserve to be.

Good luck!

Alexander Stone Macnow, MD
Editor-in-Chief
Department of Pathology and Laboratory Medicine
Hospital of the University of Pennsylvania

BA, Musicology—Boston University, 2008
MD—Perelman School of Medicine at the University of Pennsylvania, 2013

ABOUT *SCIENTIFIC AMERICAN*

As the world's premier science and technology magazine, and the oldest continuously published magazine in the United States, *Scientific American* is committed to bringing the most important developments in modern science, medicine, and technology to our worldwide audience in an understandable, credible, and provocative format.

Founded in 1845 and on the "cutting edge" ever since, *Scientific American* boasts over 150 Nobel laureate authors including Albert Einstein, Francis Crick, Stanley Prusiner, and Richard Axel. *Scientific American* is a forum where scientific theories and discoveries are explained to a broader audience.

Scientific American published its first foreign edition in 1890, and in 1979 was the first Western magazine published in the People's Republic of China. Today, *Scientific American* is published in 14 foreign language editions. *Scientific American* is also a leading online destination (**www.ScientificAmerican.com**), providing the latest science news and exclusive features to millions of visitors each month.

The knowledge that fills our pages has the power to spark new ideas, paradigms, and visions for the future. As science races forward, *Scientific American* continues to cover the promising strides, inevitable setbacks and challenges, and new medical discoveries as they unfold.

ABOUT THE MCAT

Anatomy of the MCAT

Here is a general overview of the structure of Test Day:

Section	Number of Questions	Time Allotted
Test-Day Certification		4 minutes
Tutorial (optional)		10 minutes
Chemical and Physical Foundations of Biological Systems	59	95 minutes
Break (optional)		10 minutes
Critical Analysis and Reasoning Skills (CARS)	53	90 minutes
Lunch Break (optional)		30 minutes
Biological and Biochemical Foundations of Living Systems	59	95 minutes
Break (optional)		10 minutes
Psychological, Social, and Biological Foundations of Behavior	59	95 minutes
Void Question		3 minutes
Satisfaction Survey (optional)		5 minutes

The structure of the four sections of the MCAT is shown below.

Chemical and Physical Foundations of Biological Systems	
Time	95 minutes
Format	• 59 questions • 10 passages • 44 questions are passage-based, and 15 are discrete (stand-alone) questions. • Score between 118 and 132
What It Tests	• Biochemistry: 25% • Biology: 5% • General Chemistry: 30% • Organic Chemistry: 15% • Physics: 25%

Critical Analysis and Reasoning Skills (CARS)	
Time	90 minutes
Format	• 53 questions • 9 passages • All questions are passage-based. There are no discrete (stand-alone) questions. • Score between 118 and 132
What It Tests	Disciplines: • Humanities: 50% • Social Sciences: 50% Skills: • *Foundations of Comprehension*: 30% • *Reasoning Within the Text*: 30% • *Reasoning Beyond the Text*: 40%
Biological and Biochemical Foundations of Living Systems	
Time	95 minutes
Format	• 59 questions • 10 passages • 44 questions are passage-based, and 15 are discrete (stand-alone) questions. • Score between 118 and 132
What It Tests	• Biochemistry: 25% • Biology: 65% • General Chemistry: 5% • Organic Chemistry: 5%
Psychological, Social, and Biological Foundations of Behavior	
Time	95 minutes
Format	• 59 questions • 10 passages • 44 questions are passage-based, and 15 are discrete (stand-alone) questions. • Score between 118 and 132
What It Tests	• Biology: 5% • Psychology: 65% • Sociology: 30%
Total	
Testing Time	375 minutes (6 hours, 15 minutes)
Total Seat Time	447 minutes (7 hours, 27 minutes)
Questions	230
Score	472 to 528

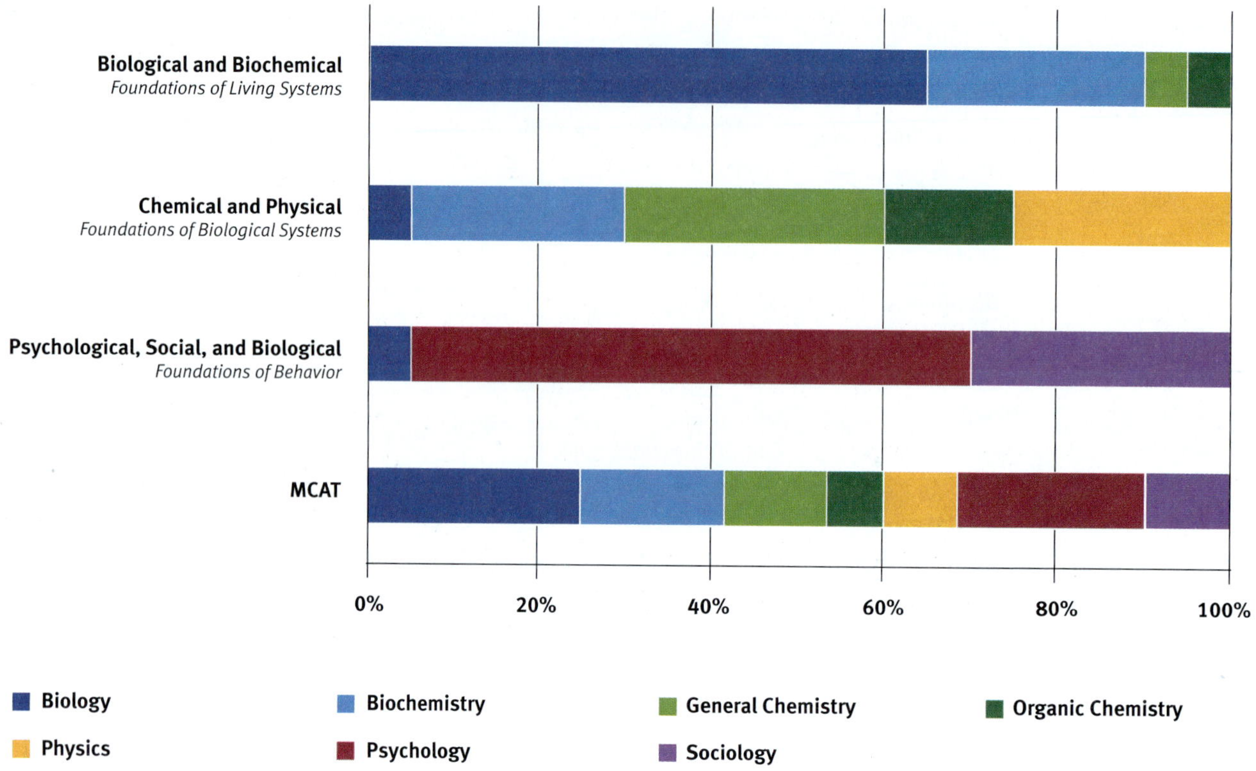

Scientific Inquiry and Reasoning Skills (SIRS)

The AAMC has defined four *Scientific Inquiry and Reasoning Skills* (SIRS) that will be tested in the three science sections of the MCAT:

1. *Knowledge of Scientific Concepts and Principles* (35% of questions)
2. *Scientific Reasoning and Problem-Solving* (45% of questions)
3. *Reasoning About the Design and Execution of Research* (10% of questions)
4. *Data-Based and Statistical Reasoning* (10% of questions)

Let's see how each one breaks down into more specific Test Day behaviors. Note that the bullet points of specific objectives for each of the SIRS are taken directly from the *Official Guide to the MCAT Exam*; the descriptions of what these behaviors mean and sample question stems, however, are written by Kaplan.

Skill 1: *Knowledge of Scientific Concepts and Principles*

This is probably the least surprising of the four SIRS; the testing of science knowledge is, after all, one of the signature qualities of the MCAT. Skill 1 questions will require you to do the following:

- Recognize correct scientific principles
- Identify the relationships among closely related concepts
- Identify the relationships between different representations of concepts (verbal, symbolic, graphic)
- Identify examples of observations that illustrate scientific principles
- Use mathematical equations to solve problems

At Kaplan, we simply call these Science Knowledge or Skill 1 questions. Another way to think of Skill 1 questions is as "one-step" problems. The single step is either to realize which scientific concept the question stem is suggesting or to take the concept stated in the question stem and identify which answer choice is an accurate application of it. Skill 1 questions are particularly prominent among discrete questions (those not associated with a passage). These questions are an opportunity to gain quick points on Test Day—if you know the science concept attached to the question, then that's it! On Test Day, 35% of the questions in each science section will be Skill 1 questions.

Here are some sample Skill 1 question stems:

- How would a proponent of the James–Lange theory of emotion interpret the findings of the study cited in the passage?
- Which of the following most accurately describes the function of FSH in the human female menstrual cycle?
- If the products of Reaction 1 and Reaction 2 were combined in solution, the resulting reaction would form:
- Ionic bonds are maintained by which of the following forces?

Skill 2: *Scientific Reasoning and Problem-Solving*

The MCAT science sections do, of course, move beyond testing straightforward science knowledge; Skill 2 questions are the most common way in which they do so. At Kaplan, we also call these Critical Thinking questions. Skill 2 questions will require you to do the following:

- Reason about scientific principles, theories, and models
- Analyze and evaluate scientific explanations and predictions
- Evaluate arguments about causes and consequences
- Bring together theory, observations, and evidence to draw conclusions
- Recognize scientific findings that challenge or invalidate a scientific theory or model
- Determine and use scientific formulas to solve problems

Just as Skill 1 questions can be thought of as "one-step" problems, many Skill 2 questions are "two-step" problems, and more difficult Skill 2 questions may require three or more steps. These questions can require a wide spectrum of reasoning skills, including integration of multiple facts from a passage, combination of multiple science content areas, and prediction of an experiment's results. Skill 2 questions also tend to ask about science content without actually mentioning it by name. For example, a question might describe the results of one experiment and ask you to predict the results of a second experiment without actually telling you what underlying scientific principles are at work—part of the question's difficulty will be figuring out which principles to apply in order to get the correct answer. On Test Day, 45 percent of the questions in each science section will be Skill 2 questions.

Here are some sample Skill 2 question stems:

- Which of the following experimental conditions would most likely yield results similar to those in Figure 2?
- All of the following conclusions are supported by the information in the passage EXCEPT:
- The most likely cause of the anomalous results found by the experimenter is:
- An impact to a man's chest quickly reduces the volume of one of his lungs to 70% of its initial value while not allowing any air to escape from the man's mouth. By what percentage is the force of outward air pressure increased on a 2 cm^2 portion of the inner surface of the compressed lung?

Skill 3: *Reasoning About the Design and Execution of Research*

The MCAT is interested in your ability to critically appraise and analyze research, as this is an important day-to-day task of a physician. We call these questions Skill 3 or Experimental and Research Design questions for short. Skill 3 questions will require you to do the following:

- Identify the role of theory, past findings, and observations in scientific questioning
- Identify testable research questions and hypotheses
- Distinguish between samples and populations and distinguish results that support generalizations about populations
- Identify independent and dependent variables
- Reason about the features of research studies that suggest associations between variables or causal relationships between them (such as temporality and random assignment)
- Identify conclusions that are supported by research results
- Determine the implications of results for real-world situations
- Reason about ethical issues in scientific research

Over the years, the AAMC has received input from medical schools to require more practical research skills of MCAT test takers, and Skill 3 questions are the response to these demands. This skill is unique in that the outside knowledge you need to answer Skill 3 questions is not taught in any one undergraduate course; instead, the research design principles needed to answer these questions are learned gradually throughout your science classes and especially through any laboratory work you have completed. It should be noted that Skill 3 comprises 10 percent of the questions in each science section on Test Day.

Here are some sample Skill 3 question stems:

- What is the dependent variable in the study described in the passage?
- The major flaw in the method used to measure disease susceptibility in Experiment 1 is:
- Which of the following procedures is most important for the experimenters to follow in order for their study to maintain a proper randomized sample of research subjects?
- A researcher would like to test the hypothesis that individuals who move to an urban area during adulthood are more likely to own a car than are those who have lived in an urban area since birth. Which of the following studies would best test this hypothesis?

Skill 4: *Data-Based and Statistical Reasoning*

Lastly, the science sections of the MCAT test your ability to analyze the visual and numerical results of experiments and studies. We call these Data and Statistical Analysis questions. Skill 4 questions will require you to do the following:

- Use, analyze, and interpret data in figures, graphs, and tables
- Evaluate whether representations make sense for particular scientific observations and data
- Use measures of central tendency (mean, median, and mode) and measures of dispersion (range, interquartile range, and standard deviation) to describe data
- Reason about random and systematic error

- Reason about statistical significance and uncertainty (interpreting statistical significance levels and interpreting a confidence interval)
- Use data to explain relationships between variables or make predictions
- Use data to answer research questions and draw conclusions

Skill 4 is included in the MCAT because physicians and researchers spend much of their time examining the results of their own studies and the studies of others, and it's very important for them to make legitimate conclusions and sound judgments based on that data. The MCAT tests Skill 4 on all three science sections with graphical representations of data (charts and bar graphs), as well as numerical ones (tables, lists, and results summarized in sentence or paragraph form). On Test Day, 10 percent of the questions in each science section will be Skill 4 questions.

Here are some sample Skill 4 question stems:

- According to the information in the passage, there is an inverse correlation between:
- What conclusion is best supported by the findings displayed in Figure 2?
- A medical test for a rare type of heavy metal poisoning returns a positive result for 98% of affected individuals and 13% of unaffected individuals. Which of the following types of error is most prevalent in this test?
- If a fourth trial of Experiment 1 was run and yielded a result of 54% compliance, which of the following would be true?

SIRS Summary

Discussing the SIRS tested on the MCAT is a daunting prospect given that the very nature of the skills tends to make the conversation rather abstract. Nevertheless, with enough practice you'll be able to identify each of the four skills quickly, and you'll also be able to apply the proper strategies to solve those problems on Test Day. If you need a quick reference to remind you of the four SIRS, these guidelines may help:

Skill 1 (**Science Knowledge**) questions ask:

- Do you remember this science content?

Skill 2 (**Critical Thinking**) questions ask:

- Do you remember this science content? And if you do, could you please apply it to this novel situation?
- Could you answer this question that cleverly combines multiple content areas at the same time?

Skill 3 (**Experimental and Research Design**) questions ask:

- Let's forget about the science content for a while. Could you give some insight into the experimental or research methods involved in this situation?

Skill 4 (**Data and Statistical Analysis**) questions ask:

- Let's forget about the science content for a while. Could you accurately read some graphs and tables for a moment? Could you make some conclusions or extrapolations based on the information presented?

Critical Analysis and Reasoning Skills (CARS)

The *Critical Analysis and Reasoning Skills* (CARS) section of the MCAT tests three discrete families of textual reasoning skills; each of these families requires a higher level of reasoning than the last. Those three skills are as follows:

1. *Foundations of Comprehension* (30 percent of questions)
2. *Reasoning Within the Text* (30 percent of questions)
3. *Reasoning Beyond the Text* (40 percent of questions)

These three skills are tested through nine humanities- and social sciences–themed passages, with approximately 5 to 7 questions per passage. Let's take a more in-depth look into these three skills. Again, the bullet points of specific objectives for each of the CARS are taken directly from the *Official Guide to the MCAT Exam*; the descriptions of what these behaviors mean and sample question stems, however, are written by Kaplan.

Foundations of Comprehension

Questions in this skill will ask for basic facts and simple inferences about the passage; the questions themselves will be similar to those seen on reading comprehension sections of other standardized exams like the SAT® and ACT®. *Foundations of Comprehension* questions will require you to do the following:

- Understand the basic components of the text
- Infer meaning from rhetorical devices, word choice, and text structure

This admittedly covers a wide range of potential question types including Main Idea, Detail, Inference, and Definition-in-Context questions, but finding the correct answer to all *Foundations of Comprehension* questions will follow from a basic understanding of the passage and the point of view of its author (and occasionally that of other voices in the passage).

Here are some sample *Foundations of Comprehension* question stems:

- **Main Idea**—The author's primary purpose in this passage is:
- **Detail**—Based on the information in the second paragraph, which of the following is the most accurate summary of the opinion held by Schubert's critics?
- **(Scattered) Detail**—According to the passage, which of the following is FALSE about literary reviews in the 1920s?
- **Inference (Implication)**—Which of the following phrases, as used in the passage, is most suggestive that the author has a personal bias toward narrative records of history?
- **Inference (Assumption)**—In putting together her argument in the passage, the author most likely assumes:
- **Definition-in-Context**—The word "obscure" (paragraph 3), when used in reference to the historian's actions, most nearly means:

Reasoning Within the Text

While *Foundations of Comprehension* questions will usually depend on interpreting a single piece of information in the passage or understanding the passage as a whole, *Reasoning Within the Text* require more thought because they will ask you to identify the purpose of a particular piece of information in the context of the passage, or ask how one piece of information relates to another. *Reasoning Within the Text* questions will require you to:

- Integrate different components of the text to draw relevant conclusions

The CARS section will also ask you to judge certain parts of the passage or even judge the author. These questions, which fall under the *Reasoning Within the Text* skill, can ask you to identify authorial bias, evaluate the credibility of cited sources, determine the logical soundness of an argument, identify the importance of a particular fact or statement in the context of the passage, or to search for relevant evidence in the passage to support a given conclusion. In all, this category includes Function and Strengthen–Weaken (Within the Passage) questions, as well as a smattering of related—but rare—question types.

Here are some sample *Reasoning Within the Text* question stems:

- **Function**—The author's discussion of the effect of socioeconomic status on social mobility primarily serves which of the following functions?
- **Strengthen–Weaken (Within the Passage)**—Which of the following facts is used in the passage as the most prominent piece of evidence in favor of the author's conclusions?
- **Strengthen–Weaken (Within the Passage)**—Based on the role it plays in the author's argument, *The Possessed* can be considered:

Reasoning Beyond the Text

The distinguishing factor of *Reasoning Beyond the Text* questions is in the title of the skill: the word *Beyond*. Questions that test this skill, which make up a larger share of the CARS section than questions from either of the other two skills, will always introduce a completely new situation that was not present in the passage itself; these questions will ask you to determine how one influences the other. *Reasoning Beyond the Text* questions will require you to:

- Apply or extrapolate ideas from the passage to new contexts
- Assess the impact of introducing new factors, information, or conditions to ideas from the passage

The *Reasoning Beyond the Text* skill is further divided into Apply and Strengthen–Weaken (Beyond the Passage) questions, and a few other rarely appearing question types.

Here are some sample *Reasoning Beyond the Text* question stems:

- **Apply**—If a document were located that demonstrated Berlioz intended to include a chorus of at least 700 in his *Grande Messe des Morts*, how would the author likely respond?
- **Apply**—Which of the following is the best example of a "virtuous rebellion," as it is defined in the passage?
- **Strengthen–Weaken (Beyond the Text)**—Suppose Jane Austen had written in a letter to her sister, "My strongest characters were those forced by circumstance to confront basic questions about the society in which they lived." What relevance would this have to the passage?
- **Strengthen–Weaken (Beyond the Text)**—Which of the following sentences, if added to the end of the passage, would most WEAKEN the author's conclusions in the last paragraph?

CARS Summary

Through the *Foundations of Comprehension* skill, the CARS section tests many of the reading skills you have been building on since grade school, albeit in the context of very challenging doctorate-level passages. But through the two other skills (*Reasoning Within the Text* and *Reasoning Beyond the Text*), the MCAT demands that you understand the deep structure of passages and the arguments within them at a very advanced level. And, of course, all of this is tested under very tight timing restrictions: only 102 seconds per question—and that doesn't even include the time spent reading the passages.

Here's a quick reference guide to the three CARS skills:

Foundations of Comprehension questions ask:

- Did you understand the passage and its main ideas?
- What does the passage have to say about this particular detail?
- What must be true that the author did not say?

Reasoning Within the Text questions ask:

- What's the logical relationship between these two ideas from the passage?
- How well argued is the author's thesis?

Reasoning Beyond the Text questions ask:

- How does this principle from the passage apply to this new situation?
- How does this new piece of information influence the arguments in the passage?

Scoring

Each of the four sections of the MCAT is scored between 118 and 132, with the median at 125. This means the total score ranges from 472 to 528, with the median at 500. Why such peculiar numbers? The AAMC stresses that this scale emphasizes the importance of the central portion of the score distribution, where most students score (around 125 per section, or 500 total), rather than putting undue focus on the high end of the scale.

Note that there is no wrong answer penalty on the MCAT, so you should select an answer for every question—even if it is only a guess.

The AAMC has released the 2018–2019 correlation between scaled score and percentile, as shown on the following page. It should be noted that the percentile scale is adjusted and renormalized over time and thus can shift slightly from year to year.

Total Score	Percentile	Total Score	Percentile
528	100	499	46
527	100	498	42
526	100	497	39
525	100	496	35
524	100	495	32
523	99	494	29
522	99	493	26
521	99	492	24
520	98	491	21
519	97	490	18
518	96	489	16
517	95	488	14
516	94	487	12
515	92	486	10
514	91	485	9
513	89	484	7
512	86	483	6
511	84	482	5
510	81	481	4
509	79	480	3
508	76	479	2
507	73	478	2
506	69	477	1
505	66	476	1
504	63	475	<1
503	59	474	<1
502	56	473	<1
501	52	472	<1
500	49		

Source: AAMC. 2018. *Summary of MCAT Total and Section Scores.* Accessed November 2018. **https://students-residents.aamc.org/advisors/article/percentile-ranks-for-the-mcat-exam/.**

Further information on score reporting is included at the end of the next section (see *After Your Test*).

MCAT Policies and Procedures

We strongly encourage you to download the latest copy of *MCAT® Essentials*, available on the AAMC's website, to ensure that you have the latest information about registration and Test Day policies and procedures; this document is updated annually. A brief summary of some of the most important rules is provided here.

MCAT Registration

The only way to register for the MCAT is online. You can access AAMC's registration system at: **www.aamc.org/mcat**.

You will be able to access the site approximately six months before Test Day. The AAMC designates three registration "Zones"—Gold, Silver, and Bronze. Registering during the Gold Zone (from the opening of registration until approximately one month before Test Day) provides the most flexibility and lowest test fees. The Silver Zone runs until approximately two to three weeks before Test Day and has less flexibility and higher fees; the Bronze Zone runs until approximately one to two weeks before Test Day and has the least flexibility and highest fees.

Fees and the Fee Assistance Program (FAP)

Payment for test registration must be made by MasterCard or VISA. As described earlier, the fees for registering for the MCAT—as well as rescheduling the exam or changing your testing center—increase as one approaches Test Day. In addition, it is not uncommon for test centers to fill up well in advance of the registration deadline. For these reasons, we recommend identifying your preferred Test Day as soon as possible and registering. There are ancillary benefits to having a set Test Day, as well: when you know the date you're working toward, you'll study harder and are less likely to keep pushing back the exam. The AAMC offers a Fee Assistance Program (FAP) for students with financial hardship to help reduce the cost of taking the MCAT, as well as for the American Medical College Application Service (AMCAS®) application. Further information on the FAP can be found at: **www.aamc.org/students/applying/fap**.

Testing Security

On Test Day, you will be required to present a qualifying form of ID. Generally, a current driver's license or United States passport will be sufficient (consult the AAMC website for the full list of qualifying criteria). When registering, take care to spell your first and last names (middle names, suffixes, and prefixes are not required and will not be verified on Test Day) precisely the same as they appear on this ID; failure to provide this ID at the test center or differences in spelling between your registration and ID will be considered a "no-show," and you will not receive a refund for the exam.

During Test Day registration other identity data collected may include: a digital palm vein scan, a Test Day photo, a digitization of your valid ID, and signatures. Some testing centers may use a metal detection wand to ensure that no prohibited items are brought into the testing room. Prohibited items include all electronic devices, including watches and timers, calculators, cell phones, and any and all forms of recording equipment; food, drinks (including water), and cigarettes or other smoking paraphernalia; hats and scarves (except for religious purposes); and books, notes, or other study materials. If you require a medical device, such as an insulin pump or pacemaker, you must apply for accommodated testing. During breaks, you are allowed to access food and drink, but not electronic devices, including cell phones.

Testing centers are under video surveillance and the AAMC does not take potential violations of testing security lightly. The bottom line: *know the rules and don't break them.*

Accommodations

Students with disabilities or medical conditions can apply for accommodated testing. Documentation of the disability or condition is required, and requests may take two months—or more—to be approved. For this reason, it is recommended that you begin the process of applying for accommodated testing as early as possible. More information on applying for accommodated testing can be found at: **www.aamc.org/students/applying/mcat/accommodations**.

After Your Test

When your MCAT is all over, no matter how you feel you did, be good to yourself when you leave the test center. Celebrate! Take a nap. Watch a movie. Ride your bike. Plan a trip. Call up all of your neglected friends or stalk them on Facebook. Totally consume a cheesesteak and drink dirty martinis at night (assuming you're over 21). Whatever you do, make sure that it has absolutely nothing to do with thinking too hard—you deserve some rest and relaxation.

Perhaps most importantly, do not discuss specific details about the test with anyone. For one, it is important to let go of the stress of Test Day, and reliving your exam only inhibits you from being able to do so. But more significantly, the Examinee Agreement you sign at the beginning of your exam specifically prohibits you from discussing or disclosing exam content. The AAMC is known to seek out individuals who violate this agreement and retains the right to prosecute these individuals at their discretion. This means that you should not, under any circumstances, discuss the exam in person or over the phone with other individuals—including us at Kaplan—or post information or questions about exam content to Facebook, Student Doctor Network, or other online social media. You are permitted to comment on your "general exam experience," including how you felt about the exam overall or an individual section, but this is a fine line. In summary: *if you're not certain whether you can discuss an aspect of the test or not, just don't do it!* Do not let a silly Facebook post stop you from becoming the doctor you deserve to be.

Scores are released approximately one month after Test Day. The release is staggered during the afternoon and evening, ending at 5 p.m. Eastern Standard Time. This means that not all examinees receive their scores at exactly the same time. Your score report will include a scaled score for each section between 118 and 132, as well as your total combined score between 472 and 528. These scores are given as confidence intervals. For each section, the confidence interval is approximately the given score ± 1; for the total score, it is approximately the given score ± 2. You will also be given the corresponding percentile rank for each of these section scores and the total score.

AAMC Contact Information

For further questions, contact the MCAT team at the Association of American Medical Colleges:

MCAT Resource Center
Association of American Medical Colleges
www.aamc.org/mcat
(202) 828-0600
aamc.org/contactmcat

HOW THIS BOOK WAS CREATED

The *Kaplan MCAT Review* project began shortly after the release of the *Preview Guide for the MCAT 2015 Exam*, 2nd edition. Through thorough analysis by our staff psychometricians, we were able to analyze the relative yield of the different topics on the MCAT, and we began constructing tables of contents for the books of the *Kaplan MCAT Review* series. A dedicated staff of 30 writers, 7 editors, and 32 proofreaders worked over 5,000 combined hours to produce these books. The format of the books was heavily influenced by weekly meetings with Kaplan's learning-science team.

In the years since this book was created, a number of opportunities for expansion and improvement have occurred. The current edition represents the culmination of the wisdom accumulated during that time frame, and it also includes several new features designed to improve the reading and learning experience in these texts.

These books were submitted for publication in April 2018. For any updates after this date, please visit www.kaptest.com/pages/retail-book-corrections-and-updates.

If you have any questions about the content presented here, email KaplanMCATfeedback@kaplan.com. For other questions not related to content, email booksupport@kaplan.com.

Each book has been vetted through at least ten rounds of review. To that end, the information presented in these books is true and accurate to the best of our knowledge. Still, your feedback helps us improve our prep materials. Please notify us of any inaccuracies or errors in the books by sending an email to KaplanMCATfeedback@kaplan.com.

USING THIS BOOK

Kaplan MCAT Critical Analysis and Reasoning Skills Review, and the other six books in the *Kaplan MCAT Review* series, bring the Kaplan classroom experience to you—right in your home, at your convenience. This book offers the same Kaplan content review, strategies, and practice that make Kaplan the #1 choice for MCAT prep.

This book is designed to help you review the *Critical Analysis and Reasoning Skills* section of the MCAT. Unlike other books in this MCAT series, there is no content to review for the *Critical Analysis and Reasoning Skills* section. The questions are written in such a way that they do not presume any prior fund of knowledge. In other words, all the support that is needed to answer the questions correctly is found in the corresponding pages.

Learning Objectives

At the beginning of each chapter, you'll find a short list of objectives describing the skills covered within that chapter. Learning objectives for these texts were developed in conjunction with Kaplan's learning science team. They have been designed specifically to focus your attention on tasks and concepts that are likely to be relevant to your MCAT testing experience. These learning objectives will function as a means to guide your review of the chapter and indicate what information and relationships you should be focused on within each chapter. Before starting each chapter, read these learning objectives carefully. They will not only allow you to assess your existing familiarity with the content of the chapter, but also provide a goal-oriented focus for your studying experience.

Sidebars

The following is a guide to the five types of sidebars you'll find in *Kaplan MCAT Critical Analysis and Reasoning Skills Review*:

- **Bridge:** These sidebars create connections between science topics that appear in multiple chapters throughout the *Kaplan MCAT Review* series.
- **Key Concept:** These sidebars draw attention to the most important takeaways in a given topic, and they sometimes offer synopses or overviews of complex information. If you understand nothing else, make sure you grasp the Key Concepts for any given subject.
- **MCAT Expertise:** These sidebars point out how information may be tested on the MCAT or offer key strategy points and test-taking tips that you should apply on Test Day.
- **Mnemonic:** These sidebars present memory devices to help recall certain facts.
- **Real World:** These sidebars illustrate how a concept in the text relates to the practice of medicine or the world at large. While this is not information you need to know for Test Day, many of the topics in Real World sidebars are excellent examples of how a concept may appear in a passage or discrete (stand-alone) question on the MCAT.

In the end, this is your book, so write in the margins, draw diagrams, highlight the key points—do whatever is necessary to help you get that higher score. We look forward to working with you as you achieve your dreams and become the doctor you deserve to be!

Online Resources

In addition to the resources located within this text, you also have additional online resources awaiting you at **www.kaptest.com/booksonline**. Make sure to log on and take advantage of free practice and access to online versions of the book!

Please note that access to the online resources is limited to the original owner of this book.

STUDYING FOR THE MCAT

The first year of medical school is a frenzied experience for most students. To meet the requirements of a rigorous work schedule, students either learn to prioritize their time or else fall hopelessly behind. It's no surprise, then, that the MCAT, the test specifically designed to predict success in medical school, is a high-speed, time-intensive test. The MCAT demands excellent time-management skills, endurance, as well as grace under pressure both during the test as well as while preparing for it. Having a solid plan of attack and sticking with it are key to giving you the confidence and structure you need to succeed.

Creating a Study Plan

The best time to create a study plan is at the beginning of your MCAT preparation. If you don't already use a calendar, you will want to start. You can purchase a planner, print out a free calendar from the Internet, use a built-in calendar or app on one of your smart devices, or keep track using an interactive online calendar. Pick the option that is most practical for you and that you are most likely to use consistently.

Once you have a calendar you'll be able to start planning your study schedule with the following steps:

1. **Fill in your obligations and choose a day off.**

   Write in all your school, extracurricular, and work obligations first: class sessions, work shifts, and meetings that you must attend. Then add in your personal obligations: appointments, lunch dates, family and social time, etc. Making an appointment in your calendar for hanging out with friends or going to the movies may seem strange at first, but planning social activities in advance will help you achieve a balance between personal and professional obligations even as life gets busy. Having a happy balance allows you to be more focused and productive when it comes time to study, so stay well-rounded and don't neglect anything that is important to you.

   In addition to scheduling your personal and professional obligations, you should also plan your time off. Taking some time off is just as important as studying. Kaplan recommends taking at least one full day off per week, ideally from all your study obligations but at minimum from studying for the MCAT.

2. **Add in study blocks around your obligations.**

   Once you have established your calendar's framework, add in study blocks around your obligations, keeping your study schedule as consistent as possible across days and across weeks. Studying at the same time of day as your official test is ideal for promoting recall, but if that's not possible, then fit in study blocks wherever you can.

   To make your studying as efficient as possible, block out short, frequent periods of study time throughout the week. From a learning perspective, studying one hour per day for six days per week is much more valuable than studying for six hours all at once one day per week. Specifically, Kaplan recommends studying for no longer than three hours in one sitting. Within those three-hour blocks, also plan to take ten-minute breaks every hour. Use these breaks to get up from your seat, do some quick stretches, get a snack and drink, and clear your mind. Although ten minutes of break for every 50 minutes of studying may sound like a lot, these breaks will allow you to deal with distractions and rest your brain so that, during the 50-minute study blocks, you can remain fully engaged and completely focused.

3. **Add in your full-length practice tests.**

   Next, you'll want to add in full-length practice tests. You'll want to take one test very early in your prep and then spread your remaining full-length practice tests evenly between now and your test date. Staggering tests in this way allows you to form a baseline for comparison and to determine which areas to focus on right away, while also providing realistic feedback throughout your prep as to how you will perform on Test Day.

When planning your calendar, aim to finish your full-length practice tests and the majority of your studying by one week before Test Day, which will allow you to spend that final week completing a final review of what you already know. In your online resources, you'll find sample study calendars for several different Test Day timelines to use as a starting point. The sample calendars may include more focus than you in need in some areas, and less in others, and it may not fit your timeline to Test Day. You will need to customize your study calendar to your needs using the steps above.

The total amount of time you spend studying each week will depend on your schedule, your personal prep needs, and your time to Test Day, but it is recommended that you spend somewhere in the range of 300–350 hours preparing before taking the official MCAT. One way you could break this down is to study for three hours per day, six days per week, for four months, but this is just one approach. You might study six days per week for more than three hours per day. You might study over a longer period of time if you don't have much time to study each week. No matter what your plan is, ensure you complete enough practice to feel completely comfortable with the MCAT and its content. A good sign you're ready for Test Day is when you begin to earn your goal score consistently in practice.

How to Study

The MCAT covers a large amount of material and studying for Test Day can initially seem daunting. To combat this, we have some tips for how to take control of your studying and make the most of your time.

Goal Setting

To take control of the amount of content and practice required to do well on the MCAT, break the content down into specific goals for each week instead of attempting to approach the test as a whole. A goal of "I want to increase my overall score by 5 points" is too big, abstract, and difficult to measure on the small scale. More reasonable goals are "I will read two chapters each day this week." Goals like this are much less overwhelming and help break studying into manageable pieces.

Active Reading

As you go through this book, much of the information will be familiar to you. After all, you have probably seen most of the content before. However, be very careful: Familiarity with a subject does not necessarily translate to knowledge or mastery of that subject. Do not assume that if you recognize a concept you actually know it and and can apply it quickly at an appropriate level. Don't just passively read this book. Instead, read actively: Use the free margin space to jot down important ideas, draw diagrams, and make charts as you read. Highlighting can be an excellent tool, but use it sparingly: highlighting every sentence isn't active reading, it's coloring. Frequently stop and ask yourself questions while you read (e.g., *What is the main point? How does this fit into the overall scheme of things? Could I thoroughly explain this to someone else?). By making connections and focusing on the grander scheme, not only will you ensure you know the essential content, but you also prepare yourself for the level of critical thinking required by the MCAT.*

Focus on Areas of Greatest Opportunity

If you are limited by only having a minimal amount of time to prepare before Test Day, focus on your biggest areas of opportunity first. Areas of opportunity are topic areas that are highly tested and that you have not yet mastered. You likely won't have time to take detailed notes for every page of these books; instead, use your results from practice materials to determine which areas are your biggest opportunities and seek those out. After you've taken a full-length test, make sure you are using Smart Reports to best identify areas of opportunity. Skim over content matter for which you are already demonstrating proficiency, pausing to read more thoroughly when something looks unfamiliar or particularly difficult. Consider starting with the Review Problems at the end of each chapter. If you can get all of those questions correct within a reasonable amount of time, you may be able to quickly skim through that chapter, but if the questions prove to be more difficult, then you may need to spend time reading the chapter or certain subsections of the chapter more thoroughly.

Practice, Review, and Tracking

Leave time to review your practice questions and full-length tests. You may be tempted, after practicing, to push ahead and cover new material as quickly as possible, but failing to schedule ample time for review will actually throw away your greatest opportunity to improve your performance. The brain rarely remembers anything it sees or does only once. When you carefully review the questions you've solved (and the explanations for them), the process of retrieving that information reopens and reinforces the connections you've built in your brain. This builds long-term retention and repeatable skill sets—exactly what you need to beat the MCAT!

While reviewing, take notes about the specific reasons why you missed questions you got wrong or had to guess on, perhaps by using a spreadsheet like the one below. Keep adding to the same Why I Missed It Sheet (WIMIS) as you complete more practice questions, and periodically review your WIMIS to identify any patterns you see, such as consistently missing questions in certain content areas or falling for the same test-maker traps. As you move through your MCAT program, adjust your study plan based on your available study time and the results of your review. Your strengths and weaknesses are likely to change over the course of your prep. Keep addressing the areas that are most important to your score, shifting your focus as those areas change. For more help with reviewing and making the most of your full-length tests, including a Why I Missed It Sheet template, make sure to check out the videos and resources in your online syllabus.

Section	Q #	Topic or Type	Wrong answer chosen	Why I missed it
Chem/Phys	*42*	*Nuclear chem.*	*Opposite*	*Confused electron absorption and emission*
Chem/Phys	*47*	K_{eq}	*Miscalculation*	*Need to memorize equation*
CARS	*2*	*Detail*	*Opposite*	*Didn't read "not" in answer choice; slow down!*
CARS	*4*	*Inference*	*Out of Scope*	*Forgot to research answer*

Where to Study

One often-overlooked aspect of studying is the environment where the learning actually occurs. Although studying at home is many students' first choice, several problems can arise in this environment, chief of which are distractions. Studying can be a mentally draining process, so as time passes, these distractions become ever more tempting as escape routes. Although you may have considerable willpower, there's no reason to make staying focused harder than it needs to be. Instead of studying at home, head to a library, quiet coffee shop, or another new location whenever possible. This will eliminate many of the usual distractions and also promote efficient studying; instead of studying off and on at home over the course of an entire day, you can stay at the library for three hours of effective studying and enjoy the rest of the day off from the MCAT.

No matter where you study, make your practice as much like Test Day as possible. Just as is required during the official test, don't have snacks or chew gum during your study blocks. Turn off your music, television, and phone. Practice on the computer with your online resources to simulate the computer-based test environment. When completing practice questions, do your work on scratch paper or noteboard sheets rather than writing directly on any printed materials since you won't have that option on Test Day. Because memory is tied to all of your senses, the more test-like you can make your studying environment, the easier it will be on Test Day to recall the information you're putting in so much work to learn.

CHAPTER 1

About CARS

CHAPTER 1

About CARS

In This Chapter

Introduction

LEARNING OBJECTIVES

After Chapter 1, you will be able to:

- Recite the major structural features of the CARS section of the MCAT
- Recall the two major passage topic categories
- Explain the major differences between *Foundations of Comprehension, Reasoning Within the Text,* and *Reasoning Beyond the Text* question categories

Congratulations! You are about to embark upon an exciting journey down the path to medical school to achieve your goal of becoming a doctor. As you might expect, this particular journey will require thorough preparation. Fortunately, you don't have to prepare on your own: we are here to help!

As a pre-medical student, you have already seen at least some of the wide variety of science topics that will be tested in the three science sections of the exam. In contrast, the *Critical Analysis and Reasoning Skills* (CARS) section will present you with a variety of passages from myriad disciplines to which you may have never been exposed. You could see a musicological analysis of Johannes Brahms's 1868 masterpiece *Ein Deutsches Requiem*, a philosophical diatribe criticizing Immanuel Kant's *Metaphysics*, or a dissection of the political underpinnings of the development of the Medicare system. For the CARS section, you will be expected to read, understand, and apply the knowledge you gain from these passages. Students often feel ill-equipped for the CARS section of the test, but Kaplan is here to help! This book will help you understand what is expected of you in CARS and will teach you the Kaplan strategies that have paved the way for many thousands of students to become the doctors they deserve to be.

In this chapter, we will go over the structure of the CARS section of the MCAT, as well as the diverse disciplines encountered in CARS passages. We'll provide a brief overview of the question categories identified by the Association of American Medical Colleges (AAMC). Finally, we'll discuss how to use this book and how it can guide you in preparing for your MCAT and the journey beyond. The journey to becoming a physician may be long, but it is extremely rewarding. Someday in the future, you'll find yourself putting on your white coat and changing patients' lives, and having the right plan for success is what will make that future possible.

1.1 The CARS Section

In some ways, the *Critical Analysis and Reasoning Skills* (CARS) section of the MCAT will be nothing new to you; it is similar to many of the standardized tests you have taken throughout your academic career, presenting you with passages to read and multiple-choice questions to gauge your understanding. In 90 minutes, you will be presented with 9 passages, each of which will be followed by approximately 5 to 7 questions, for a total of 53 questions. The passages you encounter will be relatively short (but lengthier than the science passages on the test), typically ranging from 500 to 600 words.

Unlike reading comprehension sections you have come across previously, such as those in the SAT® or ACT®, the CARS section of the MCAT has been designed to assess analytical and reasoning skills that are required in medical school. The passages you will face in CARS will be multifaceted, incorporating advanced vocabulary, presenting varied writing styles, and requiring higher-level thought. To answer the accompanying questions, you will have to go beyond merely comprehending the content of a CARS passage: you will need to analyze its rhetorical and logical structure and assess how it impacts (or is impacted by) outside information.

1.2 Passages

The types of passages chosen for CARS consist of multiple paragraphs that require active, critical reading to answer the questions that follow. The passages included in the section are from an array of disciplines in the social sciences and humanities, as listed in Table 1.1. Approximately half of the passages (and questions) that you encounter on Test Day will fall in the realm of the humanities, while the other half will be in the social sciences. All of the passages that appear in CARS are selected from books, journals, and other publications similar to those you have come across in academic settings.

Humanities	Social Sciences
Architecture	Anthropology
Art	Archaeology
Dance	Economics
Ethics	Education
Literature	Geography
Music	History
Philosophy	Linguistics
Popular Culture	Political Science
Religion	Population Health
Studies of Diverse Cultures*	Psychology
Theater	Sociology
	Studies of Diverse Cultures*

* Note: Studies of Diverse Cultures can be tested in both humanities and social sciences passages.

Table 1.1 Humanities and Social Sciences Disciplines in the CARS Section[1]

For students who have exclusively focused on the sciences, information for the fields used in the CARS section may be presented in a strikingly different way that can sometimes seem overwhelming. This book will review the writing styles used for the passages in CARS and explain how to read these passages with purpose, which will ultimately make them much less intimidating and significantly more manageable.

1.3 Question Categories

The AAMC has identified three categories of questions in CARS that will assess your critical thinking skills: *Foundations of Comprehension*, *Reasoning Within the Text*, and *Reasoning Beyond the Text.*

Foundations of Comprehension

These questions tend to be straightforward. They will ask about the main idea of a passage, specific details from within the passage, inferences that can be drawn from the passage, or the likely meaning of a word or phrase based on context. These questions are the most similar to those you have seen in previous standardized tests because they ask only for reading comprehension (understanding what you have read). Questions in *Foundations of Comprehension* will make up approximately 30 percent of the questions in CARS, or about 16 questions.

[1] Adapted from AAMC, *The Official Guide to the MCAT 2015 Exam* (Washington, D.C.: Association of American Medical Colleges, 2014), 311–322.

In Chapter 9 of *MCAT CARS Review*, we will further dissect the four question types within *Foundations of Comprehension*:

- Main Idea
- Detail
- Inference
- Definition-in-Context

Reasoning Within the Text

Reasoning Within the Text questions require greater thought than *Foundations of Comprehension* questions because they will ask you to identify the purpose of a particular piece of information in the context of the passage, or ask how one piece of information relates to another (as a piece of evidence that supports a conclusion, for example). Questions in *Reasoning Within the Text* will also make up approximately 30 percent of the questions in CARS, or 16 questions.

In Chapter 10 of *MCAT CARS Review*, we will further dissect the two main question types within *Reasoning Within the Text* and a few other, rare questions that fit into this category:

- Function
- Strengthen–Weaken (Within the Passage)

Reasoning Beyond the Text

Reasoning Beyond the Text questions focus on two specific skills: the capacity to extrapolate information from the passage and place it within new contexts, and the ability to ascertain how new information would relate to and affect the concepts in the passage. Questions in *Reasoning Beyond the Text* will make up approximately 40 percent of the questions in CARS, or 21 questions.

In Chapter 11 of *MCAT CARS Review*, we will further dissect the two main question types within *Reasoning Beyond the Text* and a few other, rare questions that fit into this category:

- Apply
- Strengthen–Weaken (Beyond the Passage)

Conclusion

This chapter is only the beginning. Now that we have covered the structure of the CARS section, we will dive into the tools and strategies that will help you score points on Test Day. In Chapter 2, we will begin with a thorough explanation of the tools (both physical and within the testing interface) that will be available to you on Test Day.

In Chapter 3, we begin with an examination of rhetoric and its application in the CARS section, then move into analysis of arguments. Chapter 4 explains the Kaplan CARS Passage Strategy, with guidance and tips on previewing CARS passages. Chapter 5 continues the strategy with a discussion of the use of keywords to extract valuable information as you read the passage. Chapter 6 describes the three ways in which you can choose to Distill information from the passage in order to best tackle questions. Chapter 7 integrates the contents of Chapters 4, 5, and 6 into a cohesive passage strategy you can use to attack any passage presented in the CARS section. Chapter 8 introduces the Kaplan Method for CARS Questions, and this method is then applied in the three following chapters to each of the different AAMC categories. Finally, we end with a look at how to review your practice tests to find your personal test-taking pathologies and keep improving that score.

Concept and Strategy Summary

The CARS Section

- The *Critical Analysis and Reasoning Skills* (CARS) section lasts 90 minutes and contains 53 questions, divided among 9 passages.
    - Passages range from 500 to 600 words.
    - Each passage has approximately 5 to 7 questions.
- CARS requires going beyond merely comprehending the content: you must analyze a passage's rhetorical and logical structure, and even be able to assess relationships between information given in the passage and new outside information.

Passages

- Half of the passages will be in the humanities (architecture, art, dance, ethics, literature, music, philosophy, popular culture, religion, studies of diverse cultures, and theater).
- Half of the passages will be in the social sciences (anthropology, archaeology, economics, education, geography, history, linguistics, political science, population health, psychology, sociology, and studies of diverse cultures).

Question Categories

- *Foundations of Comprehension* questions ask about the main idea of a passage, specific details from within the passage, inferences that must be true based on the passage content, or the likely meaning of a word or phrase based on context.
- *Reasoning Within the Text* questions ask you to identify the function of a particular piece of information within the context of the passage or ask how one piece of information relates to another (as a piece of evidence that supports a conclusion, for example).
- *Reasoning Beyond the Text* questions ask you to extrapolate information from the passage and place it within a new context or to ascertain how new information would relate to and affect the concepts in the passage.

CHAPTER 2

Using Your Test Day Tools

CHAPTER 2

Using Your Test Day Tools

In This Chapter

Introduction

LEARNING OBJECTIVES

After Chapter 2, you will be able to:

- Describe the computer-based nature of the MCAT and the writing tools available on Test Day
- Apply Test Day tools to expertly navigate and review an MCAT section
- Leverage Test Day tools while approaching MCAT passages and questions

"10 blade," "forceps," "suction"—if you've ever had the opportunity to observe a surgery or just marathon your favorite medical drama, you've likely heard these tools called for by a surgeon or an actor playing a surgeon. Even if Hollywood's portrayal of these events can be exaggerated, the truth remains that the tools of a surgeon are vital for a successful surgery. To this end, surgeons must not only be experts in their anatomical and procedural knowledge, but also experts on the tools at their disposal and their capabilities. In this way, the MCAT is much like surgery. Not only does the MCAT requires science knowledge and critical reasoning, but also a thorough understanding of the tools provided on Test Day. In this chapter, we will explain the tools available to you on Test Day and explore best practices for their usage.

2.1 Test Day Tools

The MCAT has been administered as a computer-based test (CBT) since 2007. In the years since, many adjustments have been made to the CBT interface, with the latest changes (as of the writing of this textbook) occurring in January 2018. For many students beginning their MCAT studies, learning that the MCAT is administered via computer can come with feelings of uncertainty or even anxiety. But there is good news: like all aspects of the MCAT, once the nature of the challenge is understood it can be prepared for and ultimately leveraged in your favor. In fact, compared to a standardized pencil-and-paper test, the tools offered on the CBT MCAT will allow your Test Day experience to go far more smoothly! All that's needed on your end is a little knowledge and some practice with the testing interface. Before jumping into the specifics, let's discuss a few key factors to keep in mind.

Interface Controls

Inputting commands on the MCAT can be done two ways. One input method is via on-screen buttons using the mouse, and the second is via keyboard based commands, also known as **hotkeys** (Table 2.1). The MCAT aims to be as versatile as possible, so while all functions on Test Day can be accomplished with the mouse and on-screen buttons, those functions have associated hotkeys as well. As a general rule, the use of hotkeys is more efficient than on-screen controls, thus it is highly recommended that test takers become comfortable with the hotkey system for the MCAT prior to Test Day. If you find yourself using a function only once or twice per section, then the keyboard shortcut may not be worth incorporating into your Test Day habits.

Shortcut	Function	Shortcut (Section Review)	Function (Section Review)
Alt + N	Advance to Next; Answer No	Alt + E	End Review of Section/Exam
Alt + P	Return to Previous	Alt + W	Return to Section Review
Alt + V	Open Navigation	Alt + A	Review All Questions
Alt + H	Highlight/Remove Highlight	Alt + I	Review Incomplete
Alt + S	Strikethrough/ Remove Strikethrough	Alt + R	Review Flagged
Alt + T	Open Periodic Table	Alt + Y	Answer Yes
Alt + C	Close Navigation/ Close Periodic Table	Alt + O	Answer OK
Alt + F	Flag for Review		

Table 2.1 Keyboard Shortcuts

Selecting an Answer

The most important feature on Test Day is the ability to select an answer. This can be done in two ways. First, you can select an answer by clicking on the answer choice (either the letter or the answer). Alternatively, pressing the A, B, C, or D key on the keyboard selects the corresponding answer choice. Upon selecting an answer, the circle beside the answer choice will fill in. To deselect an answer choice, select the same answer choice again or select a different answer choice.

Noteboard Booklet

Despite being a computer based test, on Test Day you will have the opportunity to use a test-center provided **noteboard booklet**. The noteboard booklet will be 8″ by 14″ in size, contain 9 white laminated pages, and come with a black fine point marker. The pages themselves are made of wet-erase material preventing any written work to be erased. If additional scratch-work space is needed, raise your hand and the proctor will exchange your booklet for a clean one. Strategically, it is best to exchange your noteboard booklet during the breaks in order to avoid losing your previously done scratch work and valuable test time.

The marker and noteboard booklet should be used just like a pen and paper booklet. Use of a similar noteboard booklet during practice, particularly a full-length simulation, will prevent any unforeseen frustrations on Test Day. As a general note, it is good practice to keep your written work relatively neat and organized in case you find yourself needing to revisit a question. In later chapters, we will go into greater detail about the types of things that are most useful to write down when attacking MCAT passages and questions.

2.2 Navigation Tools

Compared to a standard pencil-and-paper exam, where you can glance at upcoming questions or even flip pages back and forth, the computer based MCAT may seem like an unnecessary nuisance. In reality, the navigation tools available on the MCAT provide time-saving opportunities not possible with a standard pencil-and-paper exam. Let's discuss the tools that will allow you to expertly navigate each section of the MCAT.

Section Interface

The section interface is the main interface you will use on Test Day (Figure 2.1), displaying both a passage and one associated question. On Test Day you will use this interface both to analyze passages and to answer questions. The features of this interface allow efficient navigation between questions and passages as the you work through the section.

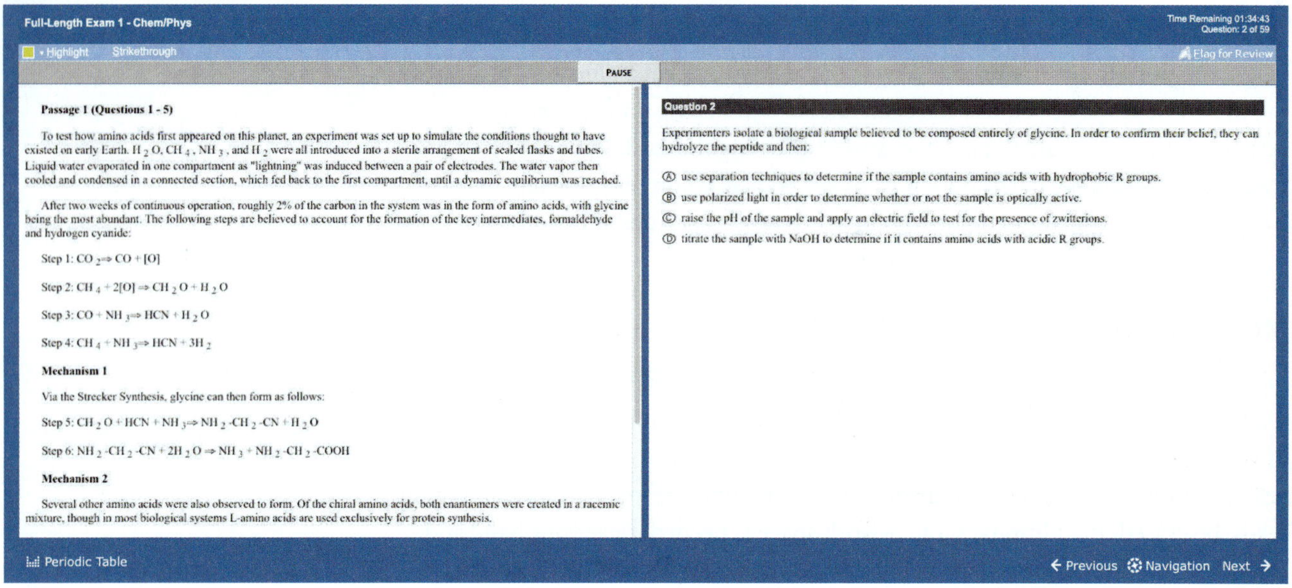

Figure 2.1 Section Interface

To navigate the section, use the three on-screen buttons at the bottom right of the screen, or their associated hotkeys:

- **Next** (Alt + N) brings up the next question.
- **Previous** (Alt + P) brings up the previous question.
- **Navigation** (Alt + V) brings up the Navigator window.

Navigator Window

The **Navigator window** lists the questions, displays question status, and displays flagged status (Figure 2.2). Understanding how to use this tool is vital for the passage strategies discussed in later chapters!

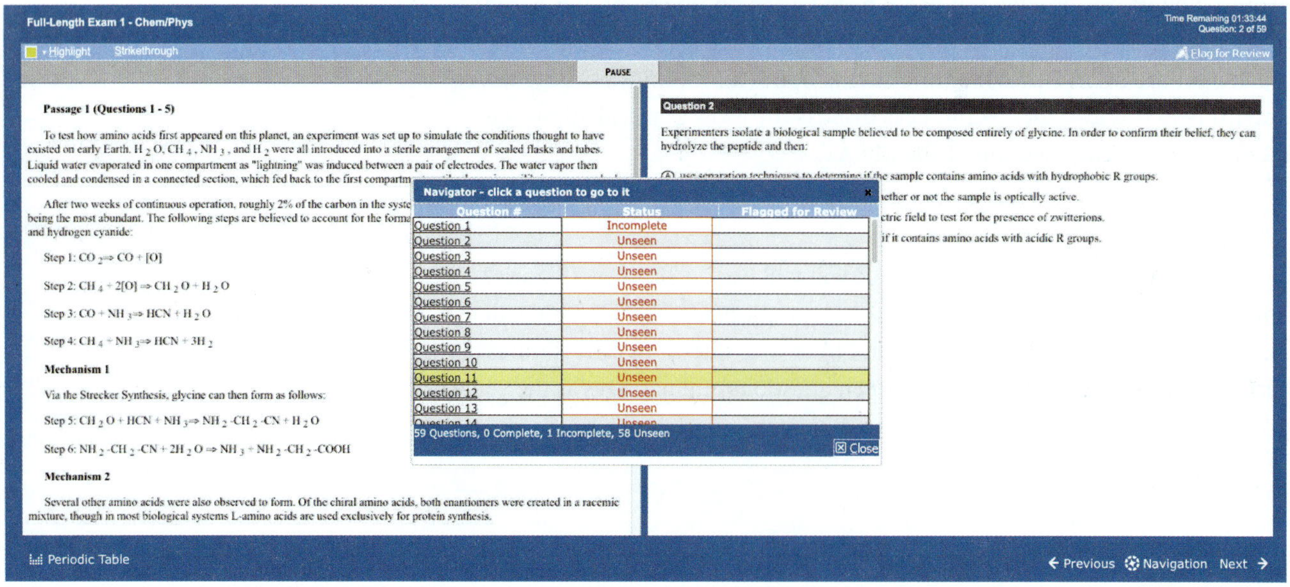

Figure 2.2 Navigator Window

The left column lists question numbers and can be used to navigate to specific questions. To do so, left-click on the question number. The center column displays question status, which updates automatically as you progress through the test:

- **Completed**—The question has an answer selected.
- **Incomplete**—The question has been seen but no answer has been selected.
- **Unseen**—The question has not been seen.

In the right column, the question's flagged status is shown. As you take the exam, you can flag a question to remind you to return to it at the end of the section. We'll expand upon this strategy later in the chapter.

The Navigator window itself can be moved by clicking and dragging the top bar and resized by clicking and dragging the borders. To close the navigation window without jumping to a different question, click **Close** or press Alt + C.

With the ability to jump to any question in the section, the Navigator window can allow a savvy MCAT tester to skip difficult passages (saving them for last) and quickly return to them near the end of the section. To skip a passage, note the title of a passage, such as "Passage 1 (Questions 1–5)," then use the Navigator window to bring up the first question of the next passage (e.g., Question 6). Returning to any previously seen passage, including passages skipped due to their difficulty, requires a bit of foresight. Reserve a space in your noteboard booklet for passage navigation, a place where you can jot down each passage number and the number of the first question associated with each passage (e.g., "P6: Q31"). When you wish to return to a skipped passage, simply bring up the Navigator window and click on the first question of the passage (e.g., Question 31). We will discuss these strategies further in Chapter 4: Previewing the Passage.

Section Review Tool

Transition from the general section interface (seen above) to the **Section Review** interface (Figure 2.3) occurs when the section is finished (more specifically, when the **Next** button is clicked on the last question of the section). The purpose of the Section Review interface is to allow the tester to return to any lingering questions and to ensure no questions remain unanswered. Toward this end, the Section Review provides three ways to review the section.

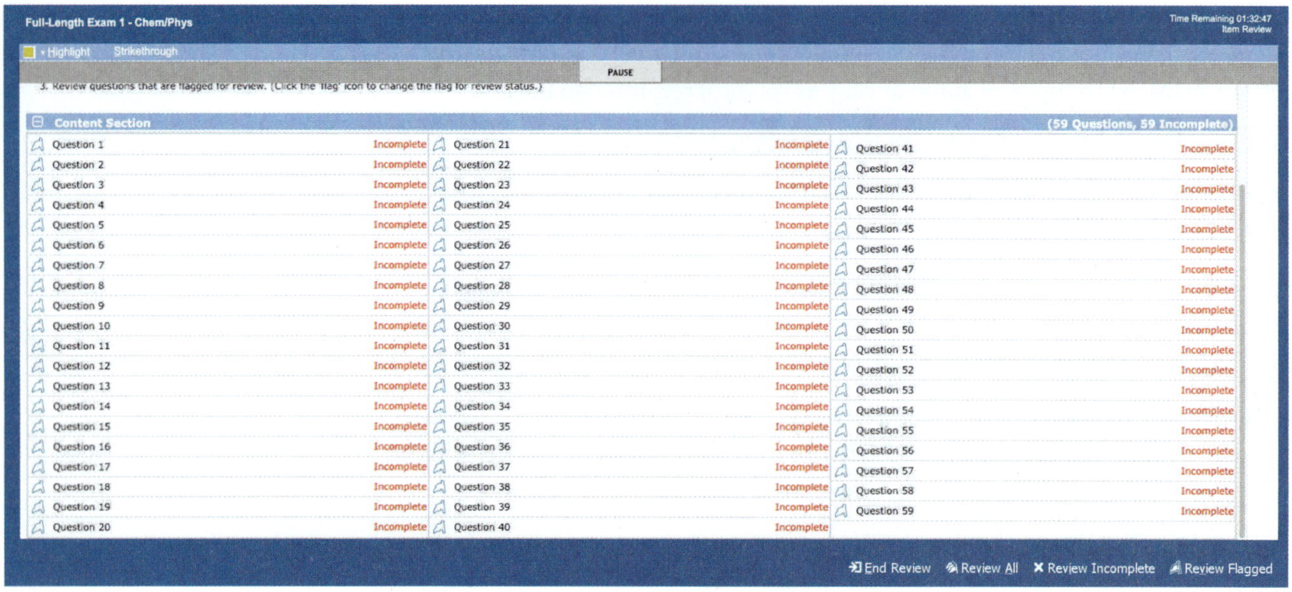

Figure 2.3 Section Review

1. **Review All (Alt + A)**—brings up Question 1 and allows for navigation to every question with the next/previous buttons.
2. **Review Incomplete (Alt + I)**—allows for review of incomplete questions only. The next and previous button only navigate to incomplete questions, skipping completed questions.
3. **Review Flagged (Alt + R)**—allows for review of flagged question only. The next/previous buttons will only navigate to flagged questions, skipping non-flagged questions.

Similar to the Navigator window, Section Review allows you to review individual questions by clicking the question number on the screen. When viewing questions through the Section Review interface, the **Navigation** button is replaced with the **Review Section** button (Alt + V), which returns to the Section Review screen.

Most testers will have limited time remaining when they enter the Section Review interface, so it's important to use its functions strategically. We recommend using the **Review Incomplete** function to ensure no questions remain unanswered and then **Review Flagged** to return to those questions you've marked for further review. To end the section, click the **End Review** (Alt + E) button in the bottom right of the screen.

Timer and Question Progress

The time remaining and question progress are displayed at the top right of all interfaces within a section of the MCAT. Although these items can be hidden by left-clicking on them, it is highly recommended to keep both visible—they are important tools for managing your section pacing, as discussed in Chapter 4: Previewing the Passage.

2.3 Interface Tools

In addition to the tools available to navigate throughout the sections of the MCAT, the test maker has made several other tools available via the CBT interface (Figure 2.4). Just as it takes practice to become comfortable adapting from reading books and articles on paper to reading text on a screen, so too do these interface tools require a bit of practice. But with a little dedication, using these tools will become second nature by Test Day. When you master these tools, you can efficiently highlight text, strike out text, flag questions, and call up the periodic table in the science sections.

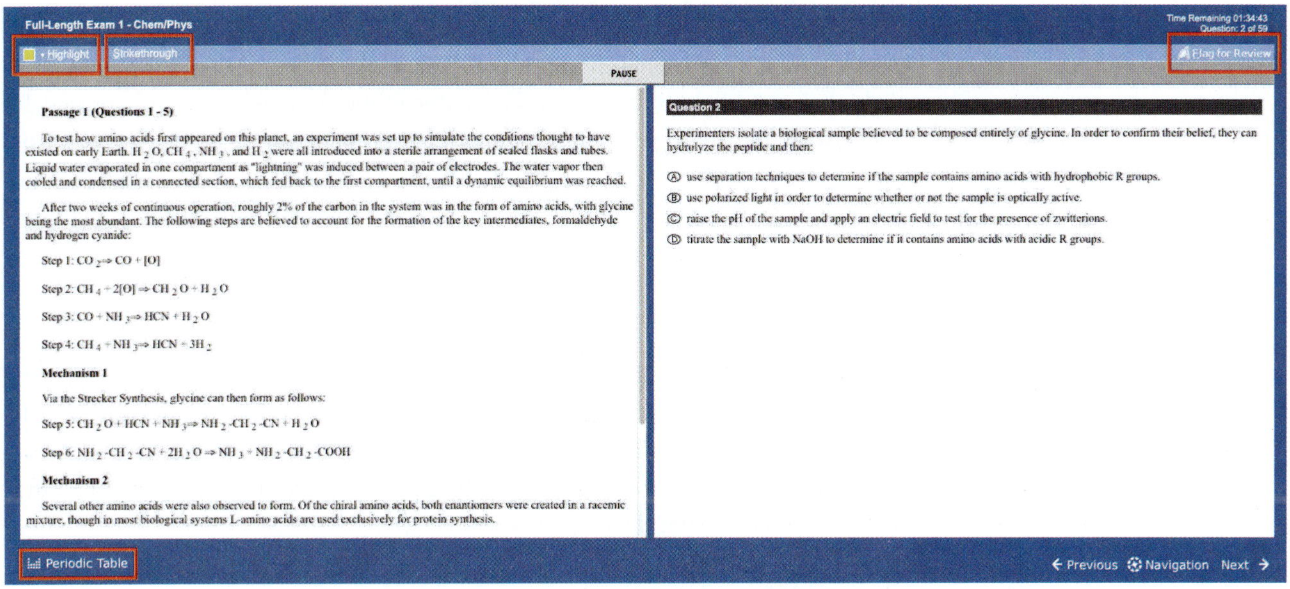

Figure 2.4 Interface Tools

Highlight Function

Much like its everyday counterpart, the highlight function highlights text in yellow. Strategically, highlighting text should stress the text's importance. To highlight text:

1. Ensure the **Highlight** option is selected in the top left corner, indicated by a yellow box (this is the default setting).
2. Use the cursor to left-click and hold while dragging over the desired text.
3. Press Alt + H or alternatively left-click the **Highlight** button in the top left of the screen.

MCAT highlighting also allows a feature that ordinary highlighting does not: erasing. Unfortunately, this process is somewhat cumbersome. To remove highlighting from text:

1. Select the **Remove Highlight** option by left-clicking the box beside the word "Highlight" in the top left corner. Alternatively, pressing Alt + H (with no text selected), then pressing the down arrow key twice, and then the space bar will also select the remove highlight option.
2. Left-click and hold while dragging over the highlighted text using the cursor.
3. Press Alt + H, alternatively left-click the **Highlight** button, ensuring that the remove highlight option is selected.

When highlighting, we highly recommend using the hotkey, Alt + H, rather than clicking the highlight button. In addition, due to the cumbersome nature of switching between the highlighting and remove highlighting modes, use the highlight tool carefully to avoid ever needing to use the remove highlight function. And if you do highlight something you didn't intend to, consider just leaving it on screen unless you're sure it would distract you.

Highlighting in Passages

If you choose to highlight within the passage, we recommend that you do so sparingly to avoid creating more work for yourself when referring back to the passage later. In general, "less is more" when it comes to highlighting. In the sciences, it can be useful to highlight key terms or given values within passages. We will discuss highlighting within CARS in greater detail in later chapters.

Highlighting in Questions

Highlighting in question stems should be used as needed. To clarify, many students will find highlighting question stems unnecessary, while others may find highlighting helps them avoid careless mistakes. For example, if you find yourself often overlooking phrases such as "is not to be expected" in question stems, spending a moment highlighting the phrase can help ensure it stays in your mind while you are answering the question. Highlighting in the answer choices, although possible, is not recommended.

Strikethrough Function

The **strikethrough** function crosses out selected text, ~~as seen here~~. Typically, this is used to indicate that the crossed out text is not important. To strikethrough text:

1. Left-click and hold while dragging over the desired text using the cursor.
2. Press Alt + S or alternatively left-click the **Strikethrough** button in the top left of the screen.

To remove **Strikethrough** from text, click and drag to select *only* the strikethrough text, then press Alt + S. If non-strikethrough text is accidentally selected along with the strikethrough text, then the Alt + S function will strikethrough *all* selected text.

Strikethrough in Passages

Generally speaking, a limited use of the Strikethrough tool during passage reading is best. Although it is no secret that each passage on the MCAT has its own amount of "fluff" (irrelevant) text, the Strikethrough tool should be used sparingly, since it's often more efficient to simply read past the fluff than to stop and strikethrough the text. It can be difficult to determine when text is actually irrelevant in CARS passages, so we recommend using the Strikethrough tool more on passages in the science sections (where it is easier to determine when material won't be targeted by questions) than in CARS.

Strikethrough in Questions

Like highlighting, the Strikethrough tool should be used as needed in questions. Most testers find little value using the Strikethrough tool in the question stem, but find it useful to eliminate answer choices. The key here is to keep use of the tool purpose driven; do not strikethrough answer choices simply because they are wrong, as these extra seconds add up over a section! Rather, aim to use the Strikethrough tool in a way that reduces your chance of error. For instance, let's say you begin working on a question, then ultimately choose to triage it. Before moving on to the next question, first use Strikethrough on any choices you managed to eliminate. When you return to the triaged question later, those eliminated choices will remain crossed out, even if you don't remember why!

Question Flagging

The **Flag** (Alt + F) function allows testers to flag a specific question, usually to remind them to return to it at the end of the section. Flagged questions can be seen at a glance in the Navigator window and the Section Review screen. To flag a specific question, click the flag icon in the top right corner or use the hotkey Alt + F. Once flagged, the flag icon will turn yellow. In order to make the most of this function, it's important to use it sparingly. Remember, the MCAT is a timed test: if there is extra time at the end of a section, it is best spent reviewing questions you could feasibly answer. In other words, do not flag every difficult question or every question that you guess on; rather, flag only those questions that you know you could answer correctly if you just devoted some extra time to them.

It is worth pointing out that the Flag function loses much of its usefulness in the CARS section. CARS questions are completely passage based and therefore best answered shortly after analyzing the passage. Thus, you should aim to answer every question in a CARS passage before moving on to the next one, and save the Flag function for the science sections.

The Periodic Table

The periodic table of elements is made available in both the Chem/Phys and Bio/Biochem sections of the MCAT and provides for each element its symbol, atomic number, and atomic weight. To bring up the periodic table, left-click the **Periodic Table** (Alt + T) button in the bottom left. This will bring up a centered window displaying the periodic table (Figure 2.5). Similar to the Navigator window discussed previously, the Periodic Table window can be manipulated by clicking and dragging the top bar and window borders. To close the periodic table left-click Close (Alt + C).

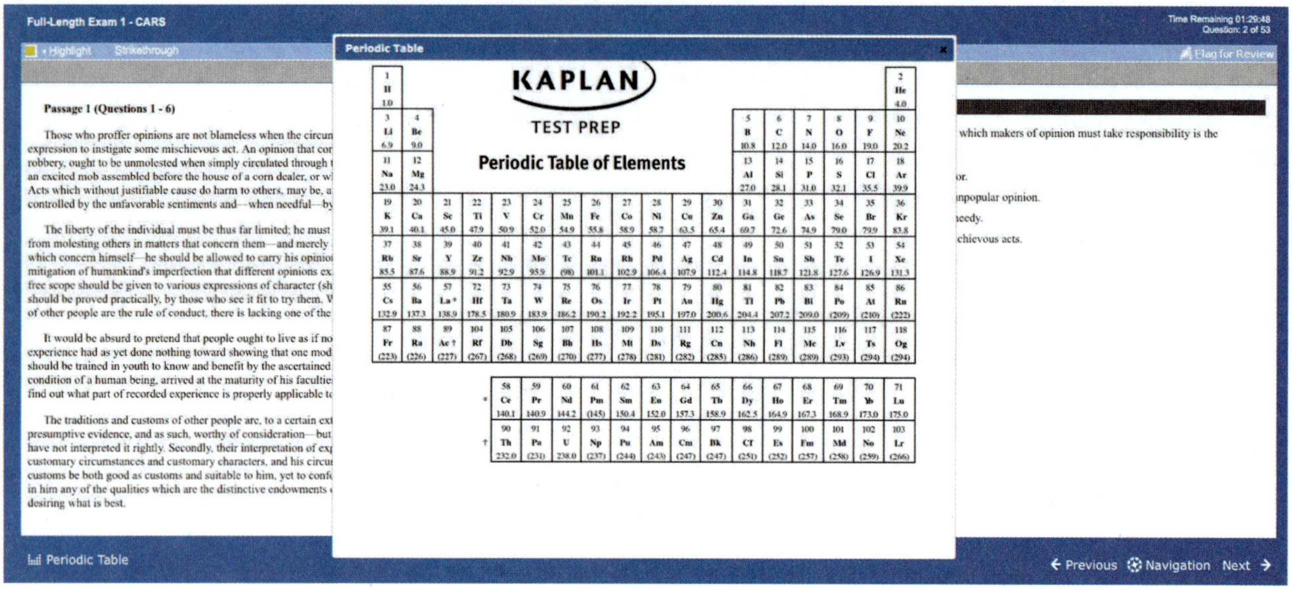

Figure 2.5 Periodic Table

Strategically, the periodic table should be used only when needed. Keep in mind that the periodic table always displays the same information and that bringing it up obscures the view of the question and passage. Thus, you should use the periodic table only when seeking a specific piece of information, e.g., an element's group or an element's atomic weight. In other words, do not bring up the periodic table simply to check if there is anything useful for a particular question. Instead, when opening up the periodic table, you should know exactly what you are looking for.

Conclusion

Well done! You've taken your first step toward mastering the computerized tools, that allow for smooth navigation of the interface on Test Day. Now you're ready to practice with them! If you're learning how to swim, it's not enough to read a book about different swim strokes; you're better off finding a shallow body of water where you can practice those strokes safely. So too with these CBT tools: to truly master them, practice using them whenever you use a test-like MCAT interface.

During your practice, you will likely find some tools are more useful to you than others, and you might even find yourself disregarding some of the recommendations in this chapter. This is perfectly acceptable: the tools are yours to make use of (or not) based on your preferences. The key is to ask yourself, before your test-taking habits become too entrenched, "Is my usage of the tools helping or hurting?" As long as it helps you, you're using the tools in the right way for you.

Concept and Strategy Summary

Test Day Tools

- The MCAT is a **computer-based test (CBT)** with unique tools to master.
- Interacting with the CBT MCAT can be done via on-screen buttons or keyboard shortcuts, listed in Table 2.2.

Shortcut	Function	Shortcut (Section Review)	Function (Section Review)
Alt + N	Advance to Next; Answer No	Alt + E	End Review of Section/Exam
Alt + P	Return to Previous	Alt + W	Return to Section Review
Alt + V	Open Navigation	Alt + A	Review All Questions
Alt + H	Highlight/Remove Highlight	Alt + I	Review Incomplete
Alt + S	Strikethrough/ Remove Strike-through	Alt + R	Review Flagged
Alt + T	Open Periodic Table	Alt + Y	Answer Yes
Alt + C	Close Navigation/ Close Periodic Table	Alt + O	Answer OK
Alt + F	Flag for Review		

Table 2.2 Keyboard Shortcuts

- On Test Day, you will be provided with a 9-page **noteboard booklet** and fine-point marker to use for scratchwork. If you need additional space, you may exchange your booklet for a clean one with the test administrator.

Navigation Tools

- **Section Interface** displays the passage and associated questions one at a time. Use the buttons listed below to navigate the section.
  - **Next** brings up next question.
  - **Previous** brings up previous question.
  - **Navigation** brings up the Navigator window which displays the status of all questions at a glance and allows the tester to jump to any specific question.

- Use the **Navigator window** like an expert.
  - Jot down in your noteboard booklet the passage number and first question number for each passage (e.g., "P6: Q31").
  - When you wish to return to a skipped passage, simply bring up the Navigator window and click on the first question of the passage (e.g., Question 31).
- The **Section Review** interface shares most of the functions of the Navigator interface, but allows for targeted review of questions, including:
  - **Review Incomplete** allows for review of incomplete questions only.
  - **Review Flagged** allows for review of flagged questions only.
  - Use the Review Incomplete function first, then the Review Flagged before ending your section (Alt + E).
- **Timer and Question Progression** functions are located at the top right of the screen and should be kept visible to assist in your pacing strategy.

Interface Tools

- **Highlight** keywords or phrases, as needed.
  - Use the cursor to select text and the keyboard shortcut, Alt + H, to highlight text.
  - Reserve highlighting for important terms or phrases, while avoiding the need to remove highlighting.
  - Highlight key phrases in question stems that have been missed on previous questions.
  - The less text that is highlighted, the more effective it is.
- **Strikethrough** irrelevant text, as needed.
  - Use the cursor to select text and then the keyboard shortcut, Alt + S, to strikethrough text.
  - Strikethrough should be used for wrong answer choices, in particular when a question may be returned to later.
  - Strikethrough with purpose! Do not simply strikethrough wrong answers because they are wrong; this will use too much time.
- **Flag** questions that you want to return to at the end of the section.
  - Be conservative in your question flagging because there will be limited time at the end of the section.
  - You can navigate through only flagged items from the Section Review interface.
- The **periodic table** is available in the Chem/Phys and Bio/Biochem sections.
  - The periodic table contains an element symbol, atomic number, and atomic weight for each element.
  - When using the periodic table, you should know what you are looking for. Do not use the periodic table for inspiration.

CHAPTER 3

Rhetoric and Arguments

CHAPTER 3

Rhetoric and Arguments

In This Chapter

Introduction

LEARNING OBJECTIVES

After Chapter 3, you will be able to:

- Describe the key components of rhetoric and their impact on passage reading on Test Day
- Describe the key components of an argument and their impact on passage reading on Test Day
- Use rhetorical and argument analysis to understand the central ideas of CARS passages

By now you've learned about the CARS section in outline and about the CBT tools at your disposal on Test Day, and perhaps you've even attacked a couple CARS passages yourself. During all this, you've seen that CARS passages can explore almost any topic, which may have led to the question, *"How can the CARS section be standardized if each passage discusses something unique? Isn't that the opposite of standardized?"* Well, not quite. The standardization of the CARS section is achieved not through a shared set of content topics, but rather by means of *how* the passages are written (as well as through the common types of questions accompanying each passage). Thus, rather than trying to familiarize yourself with all the topics that could show up in CARS passages on Test Day, your aim should be to recognize the common patterns authors use to construct their passages. By Test Day, whether you see passages discussing the workings of ancient Greek currency, the appeal of a mystery novel, or even the ramifications of an obscure philosophical theory, you will be comfortable analyzing these passages by identifying their basic rhetorical and argumentative components.

Toward that end, this chapter will focus on two nearly universal aspects of CARS passages, **rhetoric** and **arguments**, with the ultimate goal of introducing you to their most common (and most testable) patterns. A strong command of rhetoric and argumentation will allow you to navigate any passage on Test Day.

3.1 Key Components of Rhetoric

It is safe to assume that the authors of CARS passages are competent writers. In fact, we could even go so far as to say they are effective writers. Therefore, it makes sense to begin our discussion of *how* CARS authors construct their passages by examining rhetoric, the art of effective communication. The theory of rhetoric is a wide-ranging topic that can literally fill a university-level writing course but, fortunately, only a couple of key components of rhetoric are necessary to gain proficiency in CARS.

Rhetoric

Most of us are familiar with the device known as a **rhetorical question**. Although it ends in a question mark, a rhetorical question tends to have only one plausible and obvious answer. What makes it rhetorically effective is that it forces readers to reach the conclusion themselves, so that readers are more convinced of the intended conclusion than if the author had simply stated the conclusion. However, effective use of rhetoric is much more nuanced than simply asking questions that aren't really questions. Considered broadly, **rhetoric** is the art of effective communication, both in speech and in text. Because the MCAT is a written exam (as opposed to oral), we will discuss only the textual side of rhetoric throughout this chapter. While language may serve many purposes, the study of rhetoric tends to focus on persuasion—the attempt to influence others to adopt particular beliefs or to engage in certain behaviors. **Rhetorical analysis**, then, is an examination of speech or writing that goes beyond *what* the author is saying (the content) to consider *how* and *why* the author is saying it.

Many other standardized tests focus predominantly on understanding the details of what the author has said. However, the MCAT takes it a step further, sometimes asking you to use the text to infer characteristics about the author, the audience, and the goal of the passage. To that end, we will define these fundamental aspects of rhetoric (author, audience, and goal) and demonstrate how Distilling them will translate to points on Test Day.

REAL WORLD

Discussion of the art of rhetoric often begins by describing the **rhetorical situation**: an interplay between author, audience, message, purpose, genre, exigence, and context. However, discussion of the rhetorical situation as a whole is not needed for the MCAT.

Author and Tone

The **author**, in the most basic sense, is the individual or group that wrote the text. Authors can be distinguished by how much expertise they have on the topic at hand and by how passionate or vested in the topic they are.

Authors who are experts in a topic—and who know that their intended audience is also knowledgeable in the topic—tend to use a lot of jargon in their writing. **Jargon** refers to technical words and phrases that belong to a particular field. For example, *transcriptional repression*, *zwitterion*, and *anabolism* are all biochemical jargon; *homunculus*, *Gesellschaften*, and *negative symptoms* are all behavioral sciences jargon. Authors who are less expert, or who are writing to a less-informed audience, tend to use more common terminology and provide more explicitly detailed descriptions of their ideas. Authors who consider themselves less expert than their audiences may use an abundance of Moderating keywords, described in Chapter 5 of *MCAT CARS Review*.

MCAT EXPERTISE

The more knowledgeable authors are in a topic, the more jargon may appear in their writing. This may make for a challenging passage to read, but recognize that the MCAT does not expect you to know any field-specific terminology in CARS. Any important jargon will be defined in the passage—or the definition will be strongly implied.

Tone reflects the author's attitude toward the subject matter. When an author is passionate about a topic, this emotion often manifests as strong language. Extreme keywords, also described in Chapter 5 of *MCAT CARS Review*, may suggest that an author is emotionally invested in the piece. Less-invested authors may use more emotionally neutral words to describe the same ideas. In addition to an author's word choice, what the author chooses to discuss about a subject matter can also reveal the author's tone. For instance, an author may use neutral language, but use that neutral language to describe disadvantages of a particular activity and to list suitable alternatives. Despite the lack of Negative keywords (also discussed in Chapter 5), the author would still have a negative attitude toward the subject matter. On Test Day, aim to identify whether, and to what extent, the author's attitude toward the subject matter is positive or negative.

Determining Author Tone

Identifying the author's tone can help you keep track of the author's opinions while reading the passage, but more importantly it will pay off when answering questions. Not only will you be asked questions directly focused on the attitudes and opinions of the author, but you will also see questions where simply knowing the author's tone will allow you to narrow down the answer choices. To determine the author's tone in a passage, examine the words that are used while considering the question: What imagery or feelings do these words convey? Let's try an example!

MCAT EXPERTISE

On Test Day, be on the lookout for descriptive words, such as adjectives and adverbs. These words were specifically chosen by the author and can convey tone.

Consider this passage excerpt:

> One of the first examples of the ascendance of abstraction in 20th-century art is the Dada movement, which Lowenthal dubbed "the groundwork to abstract art and sound poetry, a starting point for performance art, a prelude to postmodernism, an influence on pop art . . . and the movement that laid the foundation for surrealism." Dadaism was ultimately premised on a philosophical rejection of the dominant culture, which is to say the dominating culture of colonialist Europe. Not content with the violent exploitation of other peoples, Europe's ruling factions once again turned inward, reigniting provincial disputes into the conflagration that came to be known by the Eurocentric epithet "World War I"—the European subcontinent apparently being the only part of the world that mattered.

It's clear that the author is discussing the Dada movement, but what is the author's view of the Dada movement? Notice how the author's chosen quotation describes the Dada movement: "...groundwork...starting point...prelude...foundation..." The quotation makes it clear that the Dada movement is one of the first instances of artistic abstraction, but why did the author choose a quotation that seems to be saying the same thing in four different ways? The answer is because the author wants to stress the influential nature of the Dada movement, implying that the author's opinion of Dadism is positive! In contrast, notice the author's word choice to describe the "dominating culture" of Europe: "violent exploitation." This is clearly negative. Furthermore, if you read the last sentence of the excerpt and thought it sounded sarcastic, you're right. This echoes the author's negative view of early twentieth-century European culture. All this provides further evidence of the author's positive view of Dadaism because the author claims Dadaism is based on a "philosophical rejection of the dominant culture."

In short, there are two opinions being put forward in this excerpt: the author's positive view of the Dada movement and his negative view of the European culture of the time. Using the author's tone, let's attack this Inference question (Inference questions will be further discussed in Chapter 9):

1. It is reasonable to infer that the author believes that:
   - (A) the rejection of Dadaism in Europe was a cause of World War I.
   - (B) the Dada movement was an insignificant blip in art history.
   - (C) the Dada movement was a step forward for artistic abstraction.
   - (D) Dadaism should not have gained popularity in Europe.

When answering an Inference question on Test Day, you can use the author's tone to set expectations before reviewing the answer choices. Keeping in mind the author's opinions toward Dadaism and European culture of the time, let's evaluate each of the answer choices. Choice (A) suggests something negative about European culture, that its rejection of Dadaism led to World War I. We know that this author has a negative opinion about European culture, so we can't immediately eliminate this answer. But, there's more we need to unpack with this choice to determine whether it's what we're looking for. So let's see if we can eliminate any of the other choices. Choice (B) portrays Dadaism negatively with the word "insignificant" and is thus wrong. Choice (C), however, paints Dadaism in a positive light with the phrase "a step forward." This is consistent with the author's expressed opinions. On the other hand, Choice (D), like (B), represents Dadaism negatively, making (D) incorrect as well.

Using the author's tone alone, we were able to get this Inference question down to a 50/50 choice between (A) and (C)! To decide between them, simply return to the passage and verify whether the facts match up as well as the opinions do. The author never suggests that rejecting Dadaism was a cause of World War I—rather, he blames the "inward" turn of "Europe's ruling factions." However, he does describe Dadaism as an important early example of abstract art. That means (A) can be ruled out and (C) is correct.

Audience and Genre

The **audience** is the person, or persons, for whom the text is intended. In daily life, the audience could be a single person with whom you have a dialogue, but publications typically have considerably larger audiences. Many CARS passages address an academic audience—perhaps other specialists in the author's field. Even when writing for the "general public," authors may draw upon idioms, clichés, symbols, and references that may have been recognizable to people of a particular time and place, but are less commonly used today.

In CARS, you will rarely be a member of the passage's intended audience, but you can still develop the ability to recognize the audience for whom a passage was originally written. One potential source of information about the author's intended audience is the genre of the passage. The **genre** is the particular category to which the written work belongs—for example, fiction, nonfiction, drama, poetry, and so on. Genre can also represent the more tangible form of the work: book, scholarly journal article, case study, essay, letter, email, and so forth.

Determining Audience and Genre

A passage's audience and genre are tested less frequently in the CARS section than are the author's opinions and tone. Nevertheless, being able to recognize audience and genre can sometimes yield points on Test Day. To determine the audience, begin by identifying the topic of the passage, then consider *how* the author presents the major ideas in the passage. Does she explicitly define key concepts and use simple terms that most readers would recognize? Or does she use jargon and assume her readers will already understand her meaning? Generally speaking, an author who writes more accessibly aims for a general audience, while an author who assumes reader knowledge is writing for an expert or specialized audience.

In terms of genre, most of the passages you'll encounter on Test Day will be nonfiction prose, excerpted from books or articles. A piece intended for a general audience is likely to be taken from a popular magazine or book, while journal articles and other scholarly publications are more often associated with specialized audiences. Let's try an example!

Consider this excerpt:

> The most prevalent argument against doctor-assisted suicide relies upon a distinction between *passive* and *active* euthanasia—in essence, the difference between killing someone and letting that person die. On this account, a physician is restricted by her Hippocratic oath to do no harm and thus cannot act in ways that would inflict the ultimate harm, death. In contrast, failing to resuscitate an individual who is dying is permitted because this would be only an instance of refraining from help and not a willful cause of harm. The common objection to this distinction, that it is vague and therefore difficult to apply, does not carry much weight. After all, applying ethical principles of *any sort* to the complexities of the world is an enterprise fraught with imprecision.

The MCAT might test your ability to recognize the passage's intended audience or genre with a question like the one below.

1. Which of the following titles would best be suited for this passage?
   (A) A Medical Student Handbook: Euthanasia
   (B) Exploring the Ethics of Doctor-Assisted Suicide
   (C) A Case Study on Euthanasia
   (D) Euthanasia, Society's Next Plight

To answer this question, consider the author's tone as well as the passage's intended audience. Unlike the previous example, the author's tone is fairly neutral, with few words suggesting strong authorial opinions. Despite this, the author does share one opinion in the next-to-last sentence: the common objection to the active vs. passive distinction is weak. This excerpt describes subject matter that might appeal to medical professionals, professors of ethics, or the general public. In this case, considering subject matter alone, it may be difficult to determine the intended audience. But by considering *how* the author constructs his passage, we can arrive at a more precise answer. The author explicitly defines concepts like active euthanasia, passive euthanasia, and the Hippocratic oath—terms that specialists such as doctors and ethics scholars would already know. Overall, the language is accessible and the amount of jargon minimal, which suggests that this article's audience is the segment of the general public that may have an interest in ethics. Choice (B), "Exploring the Ethics of Doctor-Assisted Suicide," matches the tone of the author, the topic of the passage, and the intended audience. In terms of genre, that title suggests an excerpt from a popular book or magazine, and this makes more sense than (A)'s "handbook" or (C)'s "case study," which are more specialized publications that don't fit the language of the passage.

REAL WORLD

Why do the test makers care so much about your ability to think beyond just understanding the content of an author's writing? This is a skill you'll use every day when talking with patients (to discern what they are really thinking from their body language and tone of voice) and when critically appraising research (to look for potential biases or conflicts of interest on the part of the researcher). To make good choices for your patients, you must be able to move beyond simply reading words on a page and consider the bigger picture, such as the author's goal.

Goal

The **goal** of a passage, sometimes called the purpose, is why the author wrote the passage. In other words, what was she hoping to accomplish? In some cases, the author's goal may be simply to inform her audience. Passages with an informative goal tend to read like textbooks or encyclopedia entries, providing detailed descriptions nearly devoid of the author's opinion. In the CARS section, however, more often the goal is persuasion, in which the author aims to influence the audience to adopt new beliefs. Persuasive passages are the most common on the MCAT; most passages on Test Day will contain at least one opinion the author tries to get the reader to endorse (with varying degrees of forcefulness). Besides altering beliefs, persuasion can also motivate individuals to take action. Such persuasion is often encountered in speeches, but it could appear on the MCAT as a set of recommendations for solving a particular problem.

On Test Day, you will encounter Main Idea questions, covered in Chapter 9: Foundations of Comprehension, which will directly test your ability to distill the author's goal.

BRIDGE

Impression management, discussed in Chapter 9 of *MCAT Behavioral Sciences Review*, focuses on how we present ourselves to accomplish specific goals. Some of the impression management strategies can be employed in writing as well, such as self-disclosure, ingratiation, and alter-casting.

Distilling the Goal of a Passage

To distill the goal of the passage, consider the major ideas of the passage while keeping the author's tone in mind. Ask yourself, *Why did the author write this passage?* We recommend beginning your answer an infinitive verb ("to X"), such as "to explain," "to argue," or "to compare." Let's try an example!

Consider this excerpt, which includes and continues the previous excerpt:

> The most prevalent argument against doctor-assisted suicide relies upon a distinction between *passive* and *active* euthanasia—in essence, the difference between killing someone and letting that person die. On this account, a physician is restricted by her Hippocratic oath to do no harm and thus cannot act in ways that would inflict the ultimate harm, death. In contrast, failing to resuscitate an individual who is dying is permitted because this would be only an instance of refraining from help and not a willful cause of harm. The common objection to this distinction, that it is vague and therefore difficult to apply, does not carry much weight. After all, applying ethical principles of *any sort* to the complexities of the world is an enterprise fraught with imprecision.
>
> Rather, the fundamental problem with the distinction is that it is not an ethically relevant one, readily apparent in the following thought experiment. Imagine a terminally ill patient hooked up to an unusual sort of life support device, one that only functioned to prevent a separate "suicide machine" from administering a lethal injection so long as the doctor pressed a button on it once per day. Would there be any relevant difference between using the suicide machine directly and not using the prevention device? The intention of the doctor would be the same (fulfilling the patient's wish to die), and the effect would be the same (an injection causing the patient's death). The only variance here is the means by which the effect comes about, and this is not an ethical difference but merely a technical one.

Consider the ideas discussed in both paragraphs and determine how they connect to serve the purpose of the passage. In the first paragraph, the author describes the distinction between active and passive euthanasia, providing examples. He then offers the "common objection" to this distinction, but immediately shoots this common objection down. In paragraph 2, the author provides *his own* objection ("fundamental problem") with the active and passive distinction: that the distinction is ethically irrelevant. The author then supports his objection with a thought experiment.

With the passage's structure laid out, the goal should be more visible. The author clearly disagrees with those who distinguish between active and passive euthanasia. After describing these terms in paragraph 1, he dedicates the entire second paragraph to debunking this distinction. But more specifically, what reason does this author give for disagreeing with this distinction? In paragraph 1, the author describes one reason why some people object to the distinction between passive and active euthanasia, which is that the distinction is vague. However, notice that the author distances himself from this point of view: he calls this "the common objection" and says it "does not carry much weight." The author's own objection comes in paragraph 2, where he claims that the distinction between passive and active euthanasia is not "ethically relevant." Thus, the author's goal is broadly persuasive and could be phrased more precisely as: *to argue that the distinction between passive and active euthanasia is not ethically relevant.* Armed with the goal of the passage, let's attack the following Main Idea question.

1. Which of the following best reflects the central purpose of the passage?
   (A) To explain how euthanasia is ethically unjustified in all circumstances
   (B) To suggest there is no distinction between active and passive euthanasia
   (C) To contend that active and passive are not morally significant categories for euthanasia
   (D) To argue that euthanasia is ethical only if it is neither active nor passive

Choice (C)'s phrase "morally significant" is just another way of saying "ethically relevant," so (C) best reflects the goal we distilled previously.

3.2 Key Components of Arguments

You may have noticed that we used the verb "to argue" in our formulation of the last passage's purpose. But contrary to popular usage, an "argument" in the CARS section is not a heated verbal dispute between people, but a specific kind of logical structure that authors use for persuasive purposes. In this section of the chapter, we will investigate the structure of arguments and prepare you to recognize their components in CARS passages.

Argument Structure

In the study of logic, an argument is the combination of one claim, known as the **conclusion**, and one or more other assertions, known as the **evidence**, explicitly used to support the conclusion. In this way, a mere claim becomes a supported conclusion as long as it's accompanied by supporting evidence. Evidence is like the friend who vouches for you to get into the VIP area at a nightclub (*conclusions only!*): it offers a reason to "trust" that the conclusion is what it claims to be. In other words, the evidence is an answer to the question, *"Why should I believe that conclusion?"* Together, the evidence, the conclusion, and the one-way connection between them constitute the simplest form of an **argument**. Consider the following statements:

- **Statement A:** Cats are great pets.
- **Statement B:** Cats are cute.

These two statements can create a plausible argument, but only if arranged the proper way. Specifically, you must correctly identify which of the statements is the conclusion and which is the evidence. To determine the relationship between two statements in an argument, we recommend using the One Sentence Test. Phrase a sentence in the following form: "<conclusion> because <evidence>." The order that makes more sense will reveal which statement is the evidence and which is the conclusion.

Let's try out the One Sentence Test on Statements A and B:

- Cats are great pets *because* cats are cute.
- Cats are cute *because* cats are great pets.

The first arrangement makes sense; it's logical to think that at least one reason why cats make great pets is their cuteness. That doesn't have to be the full explanation of why they're great pets, as long as it gives us some reason to believe that they are. The second arrangement, however, seems confused. An animal's appearance is based on genetics and environmental conditions, not on their relationships with humans (as pets). Notice that arguments have a one-way connection: evidence supports conclusion and not vice versa.

We can summarize the argument and its components as follows:

- **Evidence:** Cats are cute.
- **Conclusion:** Cats are great pets.
- **Argument:** Cats are great pets because they are cute. (Or, equivalently: Cats are cute and therefore make great pets.)

KEY CONCEPT

An argument is the combination of one claim, known as the conclusion, and one or more other assertions, known as the evidence, explicitly used to support it.

While one piece of evidence is the minimum required to support a conclusion, authors typically rely upon multiple sources of support. In fact, authors will often build layers of support for their arguments by providing additional reasons to believe the evidence. In other words, providing evidence for their evidence! So, with respect to our cat example, perhaps you're a dog person, someone who's skeptical that cats are so cute, especially compared to that new puppy your friend just adopted. You might ask, why should I believe that cats are cute? After all, that claim seems like a matter of opinion, one that you might not share. The cat advocate could attempt to address your concern by providing evidence for her evidence, perhaps by stating that cats are cute because they have large eyes and small mouths. More formally, this argument appears as follows:

- **Evidence:** Cats have large eyes and small mouths.
- **Conclusion:** Cats are cute.
- **Argument:** Cats are cute because they have large eyes and small mouths.

This may seem peculiar because what was before called the evidence is now the conclusion. Can the exact same statement be both evidence and conclusion? Yes, but only in relation to two different claims. This is nothing special: the same man can be both a son and a father, but only to two different individuals. Anytime a claim is used to *support* something else, the claim is acting as evidence; but whenever the claim in turn is *supported by* something else, the claim is acting as a conclusion. If we treated the second argument as a "subargument" within the first, we could represent it like this:

- **Subevidence:** Cats have large eyes and small mouths.
- **Subconclusion/evidence:** Cats are cute.
- **Conclusion:** Cats are great pets.

We can also use arrows to represent the relevant support relationships, simplifying the depiction of the argument further:

MCAT EXPERTISE

When authors employ especially difficult arguments, you may find it useful to sketch a diagram on your noteboard, using arrows to denote the support connections. Place the author's main conclusion, or thesis, either at the top with arrows pointing up toward it or, as we model here, at the bottom with arrows pointing down to it. Multiple pieces of evidence for the same conclusion would appear as branches extending from a common stem; thus, we sometimes call these drawings "argument trees."

Cats have large eyes and small mouths.

↓

Cats are cute.

↓

Cats are great pets.

When determining relationships such as these, you can use the textual clues discussed in Chapter 5 of *MCAT CARS Review*, especially Evidence and Conclusion keywords (the two most important subtypes of Logic keywords). However, as noted there, these keywords are less common than the other two types, and some authors use them quite sparingly. Be alert for hidden support relationships!

With an understanding of the basic components of an argument, let's take a moment to consider how this will affect your CARS reading on Test Day. As previously stated, most passages on Test Day will contain at least one argument, but is it necessary to identify its components clearly when you first analyze it? After all, we were able to analyze the euthanasia passage from earlier without applying knowledge of arguments. But even in that analysis, we identified the author's conclusion (that the active/passive distinction is morally irrelevant) and recognized that the purpose of the passage was to argue for that conclusion. In addition, we noticed that the author supported that conclusion with a thought experiment—in other words, the thought experiment was the evidence in that argument. Thus, just as a little rhetorical analysis can aid us in better understanding the passage (and in answering more questions correctly), so too can attention to argumentation. On Test Day, you'll typically find it most worthwhile to identify conclusions because they are more often featured in questions. Often, it's enough just to note the location of evidence without spelling out exactly what it says (until prompted by a question).

KEY CONCEPT

Unstated claims in arguments are known as inferences. Inferences are either assumptions (unstated evidence) or implications (unstated conclusions).

Inferences

While an argument could contain as little as one stated conclusion and one piece of stated evidence, most arguments contain unstated, or implicit, parts as well. The most commonly appearing terms for these are *implications*, *assumptions*, and *inferences*. For the sake of precision, whenever we use the word **implication**, we specifically will refer to an unstated conclusion, and **assumption** will be used only for unstated evidence. **Inference** will be used generally to cover any unstated part of an argument, whether an implication or an assumption.

It's important to note that inferences (whether assumptions or implications) are not simply claims that are *possibly* or *probably* true given what is said. Rather, inferences *must* be true or—at the very least—must be highly probable, the most likely option among the alternatives. One way to recognize an inference is by the negative effect it would have on the argument if the inference were denied. An example will help clarify what this means.

In our previous argument about the cuteness of cats, *cats are cute because they have large eyes and small mouths* the connection between the evidence and the conclusion might seem unclear. For that argument to be valid, we need another piece of information that connects those features (big eyes and tiny mouths) to that trait (cuteness). This additional claim is an assumption. In short, an assumption is an unstated piece of evidence that must be true in order for the conclusion to follow from the stated evidence. In this case, we can identify the assumption as follows:

- **Conclusion:** Cats are cute.
- **Evidence:** Cats have large eyes and small mouths.
- **Assumption:** Animals with large eyes and small mouths are cute.

We can be sure that we have correctly identified an assumption by considering what would happen to the conclusion if we negated that suspected assumption. So, for example, if we were to say instead that *animals with large eyes and small mouths are NOT cute*, then it would make no sense to conclude that cats are cute—because we've already stated as our evidence that they have those very features. This procedure for identifying inferences by negating them is called the Denial Test, and is explained further in Chapter 9.

So where do inferences fit into your Test Day experience? Well, it's worth mentioning that a common question type in the CARS section is the Inference question type, which will be discussed in detail in Chapter 9. So, you should have the ability to identify an argument's assumptions and implications as necessary. However, spelling out the inferences of every argument that appears in the passage is generally not a good use of your time. Instead, only consider the inferences of an argument if and when a question demands it of you.

Strengthening and Weakening Arguments

Although claims or statements can be called true or false, it is a misnomer to say that an argument is true or false. Rather, the **validity** or **strength** of arguments varies along a continuum with certain proof at one end and complete refutation at the other. Most arguments (virtually all of the ones you see on the MCAT) fall somewhere short of either extreme, with evidence making conclusions more probable and refutations making them less probable.

KEY CONCEPT

Arguments are evaluated on the basis of their strength or validity, which varies along a spectrum. An argument is strengthened whenever the truth of its conclusion is made more probable and weakened whenever it is made less probable.

CARS questions, particularly Strengthen–Weaken questions (discussed in Chapters 10 and 11), will often require you to imagine how an argument's validity might be modified. More supporting evidence will *strengthen*, *bolster*, or *further* an argument. This evidence can come in many forms, and often what counts as compelling evidence will depend on the topic of the CARS passage. For instance, quotations from the novel *Moby Dick* are excellent evidence when an author is advocating for his interpretation of *Moby Dick*. But those same quotes—coming from a work of fiction—would not be great evidence in a passage about the history of whaling in the Atlantic ocean.

On the other hand, a claim is said to *weaken*, *challenge*, or *undermine* an argument whenever the claim's truth would make the conclusion more likely to be false. Such claims serve as refutations or counterarguments (discussed more below). Note that most refutations do not come in the form of outright contradictions of the original conclusion or evidence. In other words, while directly attacking a conclusion will definitely weaken an argument, there are many ways to challenge an argument other than to simply state the opposite conclusion. For example, if an author draws a general conclusion about US citizens on the basis of results from a psychological survey, then evidence that the survey used a demographically unrepresentative sample would undermine the author's argument. In this way, it is not the author's conclusion that is directly undermined, but rather the evidence (that psychological survey) the conclusion stands on.

With an understanding of how claims can strengthen or weaken an argument, let's revisit our argument on cats as great pets.

- **Subevidence:** Cats have large eyes and small mouths.
- **Subconclusion/evidence:** Cats are cute.
- **Conclusion:** Cats are great pets.

This argument could be *weakened* by a claim that many of the cutest animals of the world make terrible pets, which would undermine the assumption that cute animals make for great pets. Similarly, we might challenge the assumption that large eyes and small mouths indicate cuteness by pointing to a study that identified other features (say, floppy ears or wagging tails) as being more salient indicators of cuteness. In contrast, this argument could be *strengthened* by a study that found that most people choose pets based on their apparent cuteness or by a survey showing that a majority of people describe cats as "adorable." Any argument can be strengthened *or* weakened with the appropriate evidence!

On Test Day, strengthening and weakening arguments should be treated just like making inferences. Don't bother trying to figure out all the ways an argument could be bolstered or hindered as you read through a passage—the number of possibilities is enormous, but your time is limited. Instead, you should only make such judgments when a particular Strengthen–Weaken question (see Chapters 10 and 11) requires it.

Counterarguments

A related concept is the counterargument. **Counterarguments**, also known as refutations, objections, or challenges, are simply claims that weaken a particular conclusion. Counterarguments are often worth noting when they appear in passages, because they're high yield for questions. Counterarguments are sometimes indicated through the use of Refutation keywords, discussed in Chapter 5 of *MCAT CARS Review*.

A counterargument can take aim at any component of an argument. Consider the original version of our cats as great pets argument:

- **Evidence:** Cats are cute.
- **Conclusion:** Cats are great pets.

Each of the following are possible counterarguments:

- **Refute Evidence:** Cats are not cute.
- **Refute Conclusion:** Cats are bad pets.
- **Refute Assumption:** Cute animals make for poor pets.

When you encounter a counterargument in a passage on Test Day, consider its relative strength compared to the original argument and the author's tone when presenting it then ask yourself, *Why did the author include this?* In most cases, an author includes counterarguments for one of two reasons: either she's constructing a counterargument in order to shoot down an argument she opposes, or she's describing someone else's counterargument (and probably then refuting it!) as an indirect way of supporting her own point of view.

3.3 Analyzing Passages with Rhetoric and Arguments

Now that you've seen the basic elements of rhetorical and argumentative structure, let's consider in more detail how to apply this knowledge when reading and analyzing CARS passages.

Using Rhetoric to Analyze Passages

While analyzing passages, always be on the lookout for clues that reveal the author's tone and attitudes toward the subject matter. Pay attention to opinions suggested by the author's use of adjectives and adverbs. At the start of your CARS journey, this might mean consciously stopping and asking yourself, *What is the author's feeling about this?* With practice, you will likely begin to identify author attitudes automatically, without the need for conscious effort. In addition to tone and author opinion, the author's goal is also prime material for questions. When you finish reading a CARS passage, you should always mentally answer the question, *Why did the author write this?* Other rhetorical elements like the intended audience and the genre are really only useful to the extent that they help you to identify the author's goal or to answer a particular question.

Using Arguments to Analyze Passages

As with rhetorical analysis, logical analysis is most effective on Test Day if you focus on only some of the elements of arguments. Conclusions, in particular, should be explicitly identified because they most often appear in questions. And whenever you identify a conclusion, you should also be able to locate the evidence that supports that conclusion, but you will typically not need to unpack that evidence further unless a question demands it. Counterarguments are less common but are likely to show up in questions when they do appear, so these are also worth locating. Be particularly careful when arguments are complex, with multiple layers of evidence or counterargument; try to keep clear which particular conclusion is being supported or challenged by a specific claim. As a general rule, inferences, as well as strengthening and weakening conditions, should only be considered when directly featured in questions. See the discussion in Chapter 5 of Logic keywords for more on how to recognize arguments in passages.

Now, let's apply these skills of rhetorical and logical analysis to a sample passage. As you read the passage below, try to identify the elements of rhetoric and argument discussed in this chapter. Then compare your thoughts to the expert analysis that follows.

Sample Passage

Can we truly know anything with certainty? Since the dawn of the so-called "Early Modern" era in Western philosophy, this question has preoccupied both skeptics and their critics. Perhaps the most noteworthy challenger of the certainty-rejecting skeptics is René Descartes, who constructs in his seminal *Meditations* an elaborate argument that purports to ground all human knowledge on the indubitability of one's own existence. However, in his attempts, Descartes actually bolsters the case for skepticism. Indeed, later thinkers even cast doubt on the supposed surety of self-existence.

Dissatisfied with the dogmatism of his scholastic forebears, Descartes sought to clear away all the questionable but typically unquestioned "truths" handed down to him and his contemporaries, expecting that anything that remained after an onslaught of radical doubt would have to be known with certainty. Demolishing accepted opinions one by one would require volumes, so instead Descartes examines the basic categories of belief, rejecting any kind for which he can find plausible reasons for doubt.

The first *Meditation* begins the process by considering empirical knowledge, what is learned from experience by means of our senses. Descartes's most powerful argument relies upon the impossibility of distinguishing waking consciousness from sleeping. Who has not dreamt of "waking up" while still asleep? When I awaken, how do I know the "reality" around me is not just another layer of illusion, a dream-within-a-dream-within-a-dream? Thus, Descartes concludes that all knowledge that derives from sensation cannot be certain.

With the certainty of the *a posteriori* now eradicated, the *Meditations* turn to the *a priori*, knowledge that is independent of experience, such as mathematics and logic. This proves a more difficult task, so Descartes must introduce the possibility of a Great Deceiver, a malevolent being with godlike powers who deludes us at every turn. If I cannot prove that such an entity does not exist, then "how do I know that I am not deceived every time that I add two and three, or count the sides of a square, or judge of things yet simpler, if anything simpler can be imagined?" Readers of Orwell's *1984* might have an easier time of imagining this, recalling that Winston Smith under torture genuinely comes to believe that $2 + 2 = 5$.

If even arithmetic can be cast into doubt, then how could anything be known for sure? Descartes provides an answer in the second *Meditation*. Even if the Deceiver tricks me about everything else, he cannot delude me about my own existence: "Let him deceive me as much as he will, he can never cause me to be nothing so long as I think that I am something." Of course, this "I" that exists for certain does not include the physical body, which may just be an illusion, but is simply the thinking self or mind. Even so, Descartes builds on this proposition in the remainder of the *Meditations*, arguing first for the existence of a benevolent God who would not deceive us about anything perceived "clearly and distinctly," subsequently enabling him to claim certainty for all knowledge that results from clear perception and careful reasoning.

But is Descartes's foundation really so certain? Is it not possible to doubt the existence of one's own mind? In the *Genealogy of Morals*, Friedrich Nietzsche raises the possibility of an even more radical skepticism. Though Nietzsche uses the example of lightning, the same point might be clearer to English speakers with the statement "it is raining." Although "it" seems to suggest some agent independent of the action, the phrase simply means that raining is happening. Rather than saying with certainty that "I think," perhaps Descartes should have merely concluded that "thinking happens."

Passage Analysis

Paragraph 1

We shall see in this example that not all paragraphs are created equal! Here, paragraph 1 lays out three different arguments—that of the so-called "skeptics", that of Rene Descarte, and that of the author. Most of the rest of the passage is simply supporting evidence for the author's argument in particular. With such a rich first paragraph, let's make sure to spend some serious brain power teasing apart the beliefs of the skeptics, Descartes, and the author herself.

The passage opens with a question intended to introduce the reader to the author's central concern: "Can we truly know anything with certainty?" The goal of this passage, as is the case with many passages that open with a single question, will be to answer this question—but at this point it is too early to tell *how* the author will answer it. As might be suspected, this question could be answered *yes* or *no*, and the author presents in the second sentence a contrast between the skeptics (who would say, *no, we can be certain about nothing*) and their critics (who would say, *yes, we can know some things for sure*). At this point, the author then introduces the figure who will soon be the star of this passage, René Descartes, and his book, the *Meditations*. In this third sentence, Descartes is clearly identified as a critic ("challenger") of the skeptics, and the author spells out clearly (in case you were unsure) that the skeptics reject the possibility of certainty, while critics accept it.

This explicit identification of the sides of the debate is our first clue that the author is targeting a more generalized audience, though the relatively high difficulty of some of the language suggests the author expects this general audience to be well educated.

After reading this chapter, you probably paused when you saw the mention of Descartes's "elaborate argument" in the third sentence. We're not told yet precisely what that argument is, but we can get some sense of his conclusion (namely, that human knowledge can be certain) based on the fact that Descartes is described as a critic of skepticism. Remember that an argument is conclusion plus evidence, and all we're told about Descartes's evidence is that it has something to do with "the indubitability of one's own existence," whatever that means.

The next-to-last sentence is a crucial one. The use of the transition "However" is a clue; Contrast keywords like this one are discussed in Chapter 5. Despite the relatively neutral tone of this sentence, it reveals an important author opinion: that Descartes's argument doesn't actually do what Descartes wants it to do! Instead, according to this author, Descartes's argument actually strengthens the case of those skeptics he was supposedly criticizing! The final sentence of this paragraph acts as a first piece of evidence for the author's conclusion, suggesting there are even more reasons to doubt Descartes's argument, though they're not yet spelled out. Because the author is challenging Descartes's argument in these last two sentences, we can see that she is making what we have identified as a counterargument.

A well-constructed opening paragraph like this one can actually reveal a lot about the passage that follows, allowing you to set expectations that make the remainder of the passage more manageable to read. Not all CARS passages contain such a helpful introduction, but when they do you should try to make the most of them. This paragraph allowed us to get a sense of the topic (the certainty of knowledge), the author's tone (largely neutral, with some notable opinions), the intended audience (educated general public), and even a sense of the author's goal (answering the question about certainty of knowledge in some way that involves Descartes). We were also able to determine the conclusion in Descartes's argument (that some knowledge can be known with certainty), as well as the conclusion in the author's counterargument (that Descartes's argument actually backfires and supports the anti-certainty position, rather than the pro-certainty position he intended). As we'll see, with so much vital information already extracted, subsequent paragraphs can be analyzed more briefly.

Paragraph 2

Paragraph 2 presents us with some background on Descartes and his argument. The author suggests that Descartes was dissatisfied with the dogmatism that passed for knowledge because he would only be happy with what can be "known with certainty." The upshot of these two dense sentences is that Descartes uses a method that might seem counterintuitive: "radical doubt," which is more or less what "skepticism" is, the view that Descartes is supposedly against! When reading this, we might begin to get the sense of how the author is going to suggest that Descartes helps the anti-certainty position more than the pro-certainty one.

Paragraph 3

Paragraph 3 gives us some details about how Descartes uses his method of doubt, so this paragraph as a whole serves as evidence for Descartes's larger conclusion. Within the paragraph, though, there is a distinctive argument attributed to Descartes (this is a case of a piece of evidence serving as a kind of subconclusion with its own subevidence). The conclusion is that "empirical knowledge," which the author defines as knowledge that comes from sensory experience, can't be known with certainty. To support this, Descartes uses as evidence the claim that we can't really tell whether we're asleep or awake. This evidence, which might seem pretty controversial by itself, is in turn supported by its own subevidence, namely, the common experience of false awakenings in dreams. The author (channeling Descartes) poses this last evidence in the form of a rhetorical question: "Who has not dreamt of 'waking up' while still asleep?" This allows the author to be more persuasive than if she simply asserted that all people have had this experience, because a claim about all human beings will often seem controversial, while a rhetorical question will look like it has one obvious and simple answer.

We can sketch out the argument presented as follows (although it does not appear in this order in the paragraph):

We have all dreamt of "waking up" while still asleep.

↓

It is impossible to distinguish waking consciousness from dreaming.

↓

All knowledge that derives from sensory experience cannot be certain.

Paragraph 4

Paragraph 4 begins with some potentially frightening Latin terms—*a posteriori* and *a priori*. But the author helpfully defines the second one as experience-independent knowledge, and implies that the first term refers to the experience-dependent knowledge discussed in the previous paragraph. From that first sentence, we can expect to see something similar in paragraph 4 as we saw in paragraph 3: a subargument that concludes that experience-independent knowledge is uncertain. If you didn't follow all of the evidence presented in this paragraph, that's okay and not as important as recognizing the conclusion. But if a question required you to delve into it, you'd want to note that the central piece of evidence for paragraph 4's argument is "the possibility of a Great Deceiver," some kind of demon that tricks us whenever we try to do mental math. The author references a much later literary work (*1984*) to try to make this strange idea more plausible (in other words, to provide subevidence for the Great Deceiver subconclusion).

Paragraph 5

Paragraph 5 opens with a question that also helpfully rehashes the major point of paragraph 4 (that mathematics and other experience-independent knowledge are uncertain). Remembering paragraph 3 too, we've seen Descartes argue that both experience-dependent and experience-independent knowledge are uncertain. So what's left? The author (quoting Descartes) answers: self-knowledge, specifically the knowledge that I, a thinking being, exist. After all, how could some hypothetical Great Deceiver fool me, unless I existed in the first place? So again we have another argument, though presented quite briefly. The conclusion is that knowledge of self-existence is certain, and the evidence is that you can't be fooled about your own existence.

The remainder of the paragraph gives us more broad details about Descartes's big argument in favor of certain knowledge (as you can see, real-life arguments can contain many, many pieces of evidence; the passage here only points to a small fraction of the evidence Descartes uses in his book). The author tells us that he argues for the existence of God (without her telling us what evidence Descartes uses) and then asserts that this subconclusion is used to provide certainty for a lot more

MCAT EXPERTISE

On Test Day, you do not have to understand every component of an argument. If pressed for time, focus on identifying the conclusion and simply knowing the location of the supporting evidence.

knowledge (the references to "perception" and "reasoning" suggest both the categories of knowledge discussed in paragraphs 3 and 4). So, at the end of Descartes's argument, we see that there is reason to believe that some experience-dependent and some experience-independent knowledge can be known with certainty, despite what earlier steps in his argument seemed to suggest.

Paragraph 6

The questions at the beginning of the final paragraph are truly rhetorical questions (unlike many of the open-ended questions that appear in this passage, all of which the author tries to answer explicitly)—the author's use of "really" and "not" should be dead giveaways that she wants you to answer by saying *no, Descartes's foundation is not really so certain, and yes, it is possible to doubt the existence of your mind.* Unlike paragraphs 2–5, all of which presented Descartes's argument as he made it (with the author generally trying to make a good case for it, so it's clear that she's not being unfair to him), paragraph 6 at last tells us what the author thinks herself. And, in fact, the conclusion she reveals with these rhetorical questions is very much the one we saw at the end of paragraph 1: that Descartes's argument actually undermines the possibility of certain knowledge. To support this conclusion, the author presents a counterargument against the certainty of self-knowledge, which was the key piece of evidence at the very foundation of Descartes's argument. To support this counterargument, she draws on a point inspired by Nietzsche: just as there is no "it" that does the raining when you recognize "it is raining," there may be no "I" that does the thinking when you recognize "I am thinking." The author finishes the passage with the suggestion that Descartes could only really be sure about the existence of thinking, not about the existence of a self that thinks, a conclusion that follows from the evidence she just presented.

We can summarize the author's counterargument as follows:

There is no "it" that rains.

↓

There may be no "I" that thinks.

↓

Knowledge of even your own existence is uncertain.

↓

Descartes's argument fails to show that certain knowledge is possible.

Goal

Now that we've examined every paragraph and extracted their key rhetorical and argumentative features, we can reflect on the passage as a whole. As we expected, the author maintained a relatively neutral tone but was willing to be critical of Descartes, especially in the first and last paragraphs. The language was often high level, but the author continued to define abstract terms and break down complex arguments where she could, confirming an intended audience of educated non-experts. And now, we can more exactly formulate the author's goal and understand how she would answer the question that she opens the passage with. Her main purpose seems to be to make a counterargument against Descartes' argument, which was in favor of the possibility of certain knowledge. So she might answer that first question by saying, *no, knowledge cannot be certain, because Descartes' argument for certainty fails.* In short, we might summarize her goal as: to challenge Descartes' argument that knowledge can be certain.

Conclusion

As we saw in this chapter, rhetoric and logic can enhance our understanding of MCAT CARS passages. While CARS questions will seldom explicitly ask you to identify, say, the genre of a given passage, recognizing common rhetorical structures and features can aid your answering of other questions. Arguments, on the other hand, are often explicitly referenced in questions, so there's no debate that recognizing conclusions and their evidence is a vital Test Day skill.

Rhetoric and logic are not just ancient disciplines passed down from the Greeks, nor are they only useful when taking a standardized test like the MCAT. With the expansion of evidence-based medicine, you may find yourself answering clinical questions through a meticulous analysis of the research on the topic. For example, you may find yourself asking, *Which breast cancer screening guidelines should I follow? Do I opt for the United States Preventive Services Task Force (USPSTF), which tends to be more conservative with screening, or the American Cancer Society (ACS), which is more rigorous? Or what about the American Congress of Obstetricians and Gynecologists (ACOG), which is somewhere in between?* An effective physician would read each group's recommendations with an eye toward the identity of each group (the authors), their arguments, and each group's goals. In short, understanding logic and rhetoric is indispensable to success on the MCAT, in medical school, and beyond.

Concept and Strategy Summary

Key Components of Rhetoric

- **Rhetoric** is the art of effective communication through writing and speaking.
- **Rhetorical analysis** is the examination of a particular work for the sake of identifying its rhetorical elements (the components of rhetorical knowledge).
- The **author** is the individual or group writing the piece.
  - Authors who are experts in a topic and are writing for knowledgeable audiences may use **jargon**, which is vocabulary specific to a particular field.
  - Authors may use more opinionated words if they are passionate about the topic at hand. An author may use more neutral words if he or she is less invested or knowledgeable.
  - Tone is a reflection of how the author feels about the subject matter. Identifying the tone can be done by noticing the author's word choice and phrasing.
- The **audience** refers to the person or persons the author intended to read or hear the work and is closely related to the genre.
  - The **genre** is the category to which the written work belongs, such as a book, article, essay, letter, and so on.
- The **goal** is the reason why the author wrote the work.
  - The goal of many passages on the MCAT is to be persuasive, that is, to convince the reader to adopt new beliefs or to take action.
  - Other passages may have a goal of evoking an emotional response.
  - Authors may write with more than one goal in mind.

Key Components of Arguments

- At a minimum, arguments contain three parts: a **conclusion**, its **evidence**, and the one-way path of **support** between them.
- **Inferences** are unstated parts of arguments. They are claims that must be true given what else is said in the argument.
  - **Assumptions** are unstated pieces of evidence.
  - **Implications** are unstated conclusions.
- There are three main ways of **strengthening** an argument:
  - Provide a new piece of evidence that supports the conclusion.
  - Further support evidence that already exists to support the conclusion.
  - Challenge refutations against the conclusion.

- There are three main ways of **weakening** an argument:
  - Provide a new refutation that challenges the conclusion.
  - Further support (find new evidence for) an existing counterargument.
  - Directly challenge evidence for the conclusion.
- **Counterarguments**, also called refutations, objections, or challenges, are the opposite of evidence because they go against the conclusion.

Analyzing Passages with Rhetoric and Arguments

- Author, tone, and goal should be considered consistently while reading a passages; the intended audience should be considered only when demanded by a question.
  - Rhetorical clues can be used to anticipate what is coming up in a passage.
- Conclusions and counterarguments should always be explicitly identified because they are highest yield for questions, while evidence should be located and connected to particular conclusions. Inferences and strengthening and weakening should only be considered as prompted by questions.

CHAPTER 4

Previewing the Passage

CHAPTER 4

Previewing the Passage

In This Chapter

Introduction

LEARNING OBJECTIVES

After Chapter 4, you will be able to:

- Describe the timing in the CARS section
- Predict the difficulty of a CARS passage
- Determine the order in which to work the passages in a given CARS section

Great writers work with a plan, and so do great test takers. If you've ever had to make a plan to finish a project, you've already utilized this skill. Great doctors make a plan for the best way to attack a problem, be it large or small: whether in a trauma scenario or deciding on a long-term treatment plan for a patient, doing things in the proper order is a critical skill for any physician. Finding the proper order is equally important for the CARS section of the MCAT. In this chapter, we'll discuss how to take control of the section as a whole, passage by passage, so that you're driving the CARS section rather than letting it drive you.

This chapter begins by discussing the timing in the CARS section, common pitfalls in CARS, and how to overcome getting behind on time. Then we discuss how to assess a passage and techniques for knowing how difficult a passage will be for *you*. Finally, we discuss the order in which you can attack the passages in the CARS section to maximize your score.

4.1 Timing in the CARS Section

As mentioned in the introduction to this chapter, great test takers work with a plan! In particular, CARS requires that the test taker work against both a heavy cognitive challenge and a time challenge. This chapter will discuss the best ways to meet those challenges head on, starting with the time challenge.

The CARS section is the second section that you will encounter on Test Day. In the CARS section, you have 90 minutes to read nine passages and answer 53 questions. The number of questions per passage is variable, but the goal is generally to spend about one minute per question, just like in the science sections.

As you've probably already discovered, there are several important differences between passages from the three science sections and those found in CARS. The most apparent change is that CARS passages contain considerably more words, as well as having no images to break up the monotony of the text. Under the surface, CARS passages are much more variable, both in their range of topics and their diversity of writing styles. Moreover, unlike the science passages, which are nearly always impartial, CARS passages are often written by authors who take sides and express their opinions, although not always in a straightforward manner.

To account for these essential differences, the Kaplan Method for CARS Passages can be refined, and is shown in Figure 4.1. This chapter will focus on the very first step in this method. The remaining steps will be addressed in subsequent chapters of this book.

PREVIEW FOR DIFFICULTY

- Look for the big picture
- Assess the relative difficulty
- Decide to read *now* or *later*

CHOOSE YOUR APPROACH

- Highlighting: Best for high difficulty passages or if low on time
- Outlining: Gives a moderate understanding of the passage and allows for more time to work on the questions
- Interrogating: Gives a strong understanding of passage but allows for less time to work on the questions

READ AND **DISTILL** THE MEANING OF EACH PARAGRAPH

- Recognize keywords to identify the most important and testable content in each paragraph
- Use your approach from the choose step to extract your major takeaways from each paragraph:
  - Highlighting—highlight 1–3 key terms and phrases per paragraph you can use to quickly locate information later
  - Outlining—create a brief label for each paragraph that summarizes the main idea of that paragraph
  - Interrogating—thoroughly examine each major idea presented in the paragraph and determine why and how the author is using the information to build an argument
- Identify the reason the passage was written before moving into the questions

Figure 4.1 The Kaplan Method for CARS Passages

Note: The Kaplan Method for CARS Passages—as well as the Kaplan Method for CARS Questions, CARS Question Types, and Wrong Answer Pathologies—are included as tear-out sheets in the back of this book.

Passage Timing

The CARS section is conveniently timed to take 90 minutes to complete nine passages and their associated questions, which means that, on average, each CARS passage should take 10 minutes to complete. This is not necessarily the case in practice, as some passages will take more than 10 minutes and some will take considerably less. That is why it must be emphasized that 10 minutes per passage is merely an average. The best test takers will organize their approach to the CARS section to deliberately take advantage of those time differences, prioritizing passages with care. Most careful test takers opt to complete passages that will take less time up front and leave the more involved passages for the end of the section, though some may vary this approach slightly based on their individual strengths.

Your time spent within each passage will not be divided evenly, either. As you'll see in Chapter 6, there are three strategies to choose from that you can deploy to Distill information from the passage. The three methods— Highlighting, Outlining, and Interrogating—take different amounts of time to complete and produce distinct outputs. Nevertheless, each method has been designed to minimize the amount of time and effort you must expend overall in approaching the passage and questions, while maximizing the number of questions you can answer correctly. Based upon which passage method you choose, you'll divide your time a bit differently between distilling passage information and answering questions. Distilling information from the passage using these methods will be discussed in Chapter 6 of *MCAT Critical Analysis and Reasoning Skills Review.*

Timing Strategy

So why should you care about timing? Why not just take your time, start from the beginning of the section, go at your own pace, and just focus on getting the questions you encounter correct? Well, consider a common pitfall that occurs before many test takers master timing: you come across a hard passage early and you find yourself spending 15 or 20 minutes on it, but by the time you're done, you're not sure you got many questions right or even understood what you read. Now you have to scramble to finish the section on time, which adds to your stress, detracts from your focus, and leaves you susceptible to mistakes you otherwise wouldn't make. It's far better to avoid such unnecessary stresses by exercising a little foresight.

Getting a high MCAT score is a numbers game: your sole objective is to get as many questions correct as possible in the time allotted. If you had to choose between an easy passage where you'll probably get all six questions correct and a difficult passage where you're more likely to get four out of seven, your best bet is to tackle the easy passage early and get those points before running out of time. Moreover, having those quick points in the bag from tackling easy passages first gives you more confidence and extra time as you work on the harder passages that remain. The MCAT scoring model specifically rewards this type of strategic thinking.

MCAT EXPERTISE

Your time on Test Day is valuable, so make sure you're using it as efficiently as you can. It is not "against the rules" to change your plan if you realize midway that it isn't working.

4.2 Previewing CARS Passages

As we discussed in the previous section, using your time effectively means prioritizing easy and quick passages over difficult and time-consuming ones. But how do you know in the first place what's hard and what's easy? And how can you make this assessment quickly enough that it doesn't create its own timing problems? This is what the Preview step of the Kaplan Method for CARS Passages is all about! When you Preview a passage, you'll be looking out for a few specific factors that you can assess quickly, enabling you to make a swift, sound judgment about the passage's difficulty. Let's consider each of these factors in turn.

Passage Topic

The first indication of the difficulty of a passage on the MCAT will be the topic. As shown in Table 1.1 in Chapter 1, passages on the CARS section come in two basic types: social sciences and humanities. Many students find that identifying the broad category for a given passage is a good first filter to determine roughly when in the 90 minute section they should approach the passage.

Social sciences passages are more likely to rely upon empirical research and data, even if they don't share the heavy focus on controlled experimentation seen in the natural science sections of the MCAT. Also, social sciences passages are more likely to focus on cause and effect relationships and are often (but not always) more logically rigorous than their humanities counterparts. In contrast, humanities passages tend to rely on different types of evidence, such as quotations from famous writers and speakers, thought experiments, or interpretations of historical events. While both types contain jargon, the jargon in the humanities can seem more daunting to many pre-med students, with references to obscure thinkers, terms borrowed from foreign languages, and boatloads of "ism"s with similar-sounding names.

In addition to the broad humanities vs. social sciences distinction, the specific discipline (from among those listed in Table 1.1) of a passage can also be useful information. As you practice with more CARS passages, you might discover some disciplines come more easily for you, perhaps because of your educational background or interests, while other disciplines are more of an ordeal. For example, if you find yourself always struggling with philosophy passages but breezing through literature passages, then your default should be to assume a philosophy passage is hard and a literature passage is easy, at least as a first guess. We recommend compiling and refining your own list of strong and weak topics as you do more CARS practice.

MCAT EXPERTISE

Use caution when approaching a CARS passage on a topic with which you are familiar. Relying upon outside information that may be true, but is not in the passage, is a surefire way to pick an Out of Scope wrong answer.

Sentence Structure

A second important influence on passage difficulty is sentence structure. Some sentences are short. Others are very long and often seem needlessly wordy with the author repeating himself over and over seemingly for no reason at all! Most sentences are somewhere in between these two extremes. Unlike passage topic, which affects difficulty in a way that depends mostly on your personal interests and strengths, sentence structure tends to have a consistent influence: passages with more complex sentence structure are harder for just about everyone.

While the distance between periods (the average sentence length) is one good measure of sentence complexity, rare punctuation marks can also be helpful indicators. A passage strewn with semicolons, dashes, parentheses, or simply an excessive number of commas may suggest an author who is prone to go off on tangents, making the main points harder to follow. Remember on Test Day that you're trying to Preview passages quickly: it should only take you a few seconds to get a gut feeling of complexity based on sentence length and amount of punctuation. Any more effort than that is overthinking it.

Writing Style

While most CARS passages are standard academic prose, some authors (perhaps most notably those from earlier in history) adopt distinctive writing styles that can make the passage easier or harder to read. The exact effect usually depends on what you personally like to read, but a particularly archaic writing style can be a challenge even for a history buff. Style is not as easy to gauge as sentence structure, but you should nevertheless limit yourself to a gut-level check; taking just a few seconds when assessing writing style.

MCAT EXPERTISE

Just because a style is one you don't enjoy reading doesn't mean that you won't be successful reading it on Test Day. Don't avoid passages with simpler structures just because you find their topics or styles boring.

Length

The length of the passage is perhaps the most obvious clue that you should use to identify difficulty: longer passages take more time to read. MCAT CARS passages are typically 500–600 words in length, though apparent length can be distorted based on paragraphing: a passage with a greater number of shorter paragraphs will seem longer than one with the same word count but with fewer, longer paragraphs. Paragraph length, as well, has its own impact on difficulty; it's easier to get lost when in the middle of a long paragraph that never seems to end. Both overall passage length and average paragraph length are quick and easy to assess, but be careful not to let them be the only factors shaping your judgment of difficulty.

Questions

The last noteworthy influence on passage difficulty is the number of questions associated with the passage. Remember that questions appear only one at a time in the Test Day interface. However, you can easily determine the number of questions associated with a given passage because each passage will list the range of associated question numbers, for example, "Question 32–39." You may be tempted to think more questions necessarily means a harder, more time consuming passage. But remember: passages with more questions are also worth more points!

MCAT EXPERTISE

At the top of every passage on Test Day there will be a label that tells you the range of questions associated with that passage. Using that detail makes it easy to flip between passages using the test's navigation tools. It's a good idea to make a note of the question that starts each passage on your scratch paper to hop easily between passages in the order that suits you.

4.3 Determining Passage Order

Whether you've worked in a hospital, volunteered in a clinic, worked as a medic in the military, or just watched episodes of any medical television shows, you have probably heard the word *triage*. The idea behind triaging patients is that the sickest patients get to "skip the line" so that the physician can address the most urgent needs first, minimizing further damage and fatalities among patients in the waiting room.

The same idea can be applied to the MCAT and the CARS section. But instead of triaging to save lives, you'll be triaging to gain points, prioritizing passages to let you correctly answer as many questions as possible in the time you have. We will examine a few different ways to approach this triaging—and we recommend you experiment with each approach until you find the one that works best for you.

MCAT EXPERTISE

There will be stressful points during your exam; but try not to let these moments of stress "get in your head" and ruin the section as a whole. If a passage was more difficult than you expected, take a moment after you finish it to close your eyes, breathe deeply, and refocus before attacking the next passage.

Section Order Strategy

One proven technique is to order the entire section by difficulty at the start of the section. Every passage on the MCAT will have a label at the top indicating the passage number and range of questions (e.g., *Passage 1 [Questions 1–7]*). This label can help you to navigate the section efficiently by jumping between the first question of each passage using the Navigator tool, discussed in Chapter 2 of *Kaplan MCAT CARS Review*. You'll need to be able to navigate between passages quickly to use the section order strategy effectively.

To use this strategy, you'll need to find a place in your noteboard booklet to list a few pieces of information, as can be seen in Figure 4.2. For each passage, write down the passage number, the number of its first question, and your triaging decision. For your triaging decision, you have a few options for how much information you want to include.You will certainly want to record your perception of the passage's difficulty. Most students prefer to record difficulty as either Easy, Medium, or Hard. However, some students prefer to simplify the decision by recording only Easy or Hard. After recording the passage's difficulty, some students also prefer to execute the Choose step by deciding whether to Highlight, Outline, or Interrogate the passage. Because the Choose step often depends heavily on passage difficulty, pairing Choose with Preview is one potential way to increase your efficiency. However, other students prefer to save the Choose decision until they have committed to a passage. Regardless of how much or how little information you prefer to record with your triaging decision, give yourself just a few seconds to decide and jot down your decision. Then use the Navigator tool to jump to the next passage (see Chapter 2), and take a few seconds to record similar information for that passage. Triaging the section in this way should take around two minutes and not more than three.

MCAT EXPERTISE

In addition to your triaging decision, you can also write down a one-word clue about the passage's topic if you'd like a way to help keep the passages distinct in your mind. This can make it easier, for example, to decide which of those Hard passages to tackle first.

Once you've made your decisions for every passage (and have recorded something on your noteboard booklet that looks like one of the examples in Figure 4.2), then move in order of increasing difficulty: Easy passages first, then Medium (if any), and finally Hard. This approach lets you attack all of the easiest (and fastest) passages first and maximize the number of questions you answer. Further, starting with the easiest passages and slowly increasing difficulty can help you to manage stress within the section, and ensure you reach the hardest passages still full of energy. You can scratch out each passage as you complete it to have a visual representation of how much is left to do in the section.

MCAT EXPERTISE

When reviewing a practice CARS section that you've completed, don't just consider how many points you got, but take some time to evaluate your triaging decisions. Did a passage you thought would be easy turn out to be hard for reasons you didn't originally anticipate? Adjust future triaging decisions in light of your analysis.

The section order strategy is a time investment. You'll need to spend a couple minutes upfront, but the payoff will come when you move more quickly between passages as you complete them. The hard decisions primarily come at the start of the section, and then you can just focus on getting things done.

1	2	3	4	5	6	7	8	9
Q1	Q7	Q13	Q19	Q25	Q31	Q36	Q43	Q49
H	M	E	M	E	E	H	M	E
Econ	Lit	Psych	Ethics	Music	Art	Phil	Hist	Theat

1	2	3	4	5	6	7	8	9
Q1	Q7	Q13	Q19	Q25	Q31	Q36	Q43	Q49
H	H	E	H	E	E	H	E	E

1	7	13	19	25	31	36	43	49
H-H	M-1	E-0	H-0	E-1	E-0	H-H	M-1	E-1

E	M	H
Q13	Q7	Q1
Q25	Q19	Q36
Q31	Q43	
Q49		

Figure 4.2 Section Order Strategy Scratch Work

From the top: example scratch work with 3 levels of difficulty, passage type, passage and question number; example scratchwork with 2 levels of difficulty, passage and question number; example scratch work with 3 levels of difficulty, question number, and Choose step approach; and example scratchwork with 3 levels of difficulty and question number.

Now vs. Later Strategy

An alternative to the section order strategy is the Now vs. Later strategy. This strategy amounts to triaging as you work through the section, instead of doing all of your triaging upfront. If you use the Now vs. Later strategy, you'll still perform the Preview step (perhaps combined with the Choose step) when you first encounter a passage. You'll still assign a difficulty rating to the passage. And you'll still record the passage number, difficulty, first question number, and other important details in your scratch work. The difference is, with the Now vs. Later strategy, you'll immediately complete any passage you decide is Easy before moving on to the next passage. Those you triage as Hard (or Medium, if you use that label too) you will save for Later. See Figure 4.3 for examples of how your noteboard booklet should look with this strategy.

MCAT EXPERTISE

While our default recommendation is to work in order of increasing difficulty, some students find they prefer to structure the section like a hill, starting and ending with Easy passages and working the Hard passages in the middle. But regardless of which approach you adopt, you should always start with an Easy passage: data indicates that starting with a Hard passage first can negatively affect your performance across the entire section.

For students who are particularly concerned about timing, an alternate version of Now vs. Later involves working both the Easy and Medium difficulty passages "now", and saving only the Hard passages for a second pass. In this version, your scratch work will be minimal, because you will only need to make note of those Hard passages that you're saving for the end. While this variant of the strategy is a great fit for many students, be warned that it may not be the best fit for everyone; some students might find that they become less efficient and more stressed when they force themselves to work through passages of moderate difficulty too early in the section.

In short, experiment with each version of the triaging strategy. You'll quickly find the version that best suits your own personality. If you like having a detailed plan for the section clearly laid out at the start, you might prefer doing all your triaging up front with the Section Order version. On the other hand, if you suffer from decision fatigue trying to assign difficulty ratings to each and every passage, you might prefer to triage as you go with the Now vs. Later version of the strategy. The key is optimizing the triaging strategy to work for you.

Medium Passages

Q7 – LIT
Q19 – ETHICS
Q43 – HIST

Hard Passages

Q1 – ECON
Q36 – PHIL

M

Q7 – 1
Q19 – 0
Q43 – 1

H

Q1 – H
Q36 – H

H

Q1
Q36

Figure 4.3 Now vs. Later Strategy Scratch Work
From the top: example scratch work with 2 levels of difficulty, question number, and topic; example scratch work with 2 levels of difficulty, question number and Choose method; and example scratch work with 1 level of difficulty and question number.

Applied Examples

Now let's practice some of these triaging strategies with some applied examples. The following pages feature two test-like CARS passages. We've already set in bold important lines from each passage (ones that should stand out when completing the Preview step) to help aid your decisions. Remember that your goal is only to evaluate the difficulty of the passage for you; don't bother reading the whole thing!

Passage 1

The **world of contemporary art** is characterized by a growing number of **artists experiencing an entrepreneurial venture.** Especially in the context of performing arts, this has been lived both as a necessary and a voluntary **solution to the severe shortage of funds affecting the world of public institutions.** Thus a new actor emerges, represented by the artist-entrepreneur who lives a hybridization of roles and competences. But who are the artists-entrepreneurs? And how do they live the possible tensions emerging from the encounter of worlds that have been reputed as radically different for so long?

Artists-entrepreneurs' activity can be identified as a particular kind of **"cultural entrepreneurship."** This practice has been traditionally investigated adopting **two main perspectives**, based on different meanings of culture.

As a first meaning, *culture* refers to the sociological frame of reference identifying a set of habits, customs, traditions, and beliefs which constitute a shared way of life in a specific historical and political context. As a second meaning, culture identifies a complex set of processes, products, and actors involved in the design, production, and distribution of cultural and artistic goods and services.

Descending from the **first sociological perspective**, cultural entrepreneurship represents "the skill of certain entrepreneurs to use culture as a toolkit for constructing resonant identities and motivating resource-holding audiences to allocate their resources." Therefore, cultural entrepreneurship is instrumentally identified in the process of storytelling that gives shape and legitimates new ventures. Not referring to a specific industry, the adjective *cultural* is used to identify the process of legitimization that entrepreneurs sustain, giving shape to the story of their personal and professional life.

In a second perspective, cultural entrepreneurship identifies a set of processes through which a growing number of artists and cultural professionals assume an entrepreneurial role. Cultural entrepreneurship thus identifies the activity of conceiving, producing and marketing "cultural goods and services, generating economic, cultural and social opportunities for creators while adding cultural value for consumers." The artists-entrepreneurs combine their artistic attitudes with a deep sense of business, economically sustaining the cultural enterprise in coherence with their cultural vision.

Upon investigation, what emerges with more emphasis is the tough and complex relationship between the artistic and the entrepreneurial dimensions faced by the artists-entrepreneurs. The interdependence between the two spheres emerges as a matter of fact, emphasizing the dual nature of the cultural enterprise. But duality means adopting an integral perspective that results from an intense dialogue between the two languages.

Adapted from Clacagno, M., Balzarin, L. 2016. The Artist-Entrepreneur Acting as a Gatekeeper in the Realm of Art. Venezia Arti, 25. http://doi.org/10.14277/2385-2720/VA-25-16-3

As discussed earlier, the Preview step should take just a few seconds and should really be a "gut-check" decision. As we Preview this passage, we can see that it's a humanities passage, specifically one about art. Sentence structure is moderately complex, without tons of unusual punctuation and with an average sentence length of about two lines. The writing style is a bit unusual and the author uses some humanities jargon, but there seems to be a recognizable structure to some of the paragraphs (e.g., paragraphs 4 and 5 each deal with one of a pair of "perspectives"). The passage isn't extensively long, nor are its paragraphs, so the length is manageable. This is probably a Medium passage for most students, but students who struggle with art passages would likely triage this as Hard.

Now try the Preview step with a second passage, also featuring selective use of bold to help focus your assessment.

Passage 2

Bauls are a religious group living primarily in the rural areas of West Bengal (India) and Bangladesh. They come from a variety of backgrounds, including Hindu and Muslim, and span the castes, though most are low caste. Although their roots extend more deeply, Bauls have been around at least a century. In practice and belief, they have been influenced by local traditions of the more orthodox Gauriya (Bengali) Vaishnavism, the Tantric-influenced Buddhist and Hindu Sahajiyas, and Sufism. **However, unlike many other religious groups, Bauls intentionally reverse a number of orthodox practices.** They argue that the Divine is within all humans and thus people should respect and worship humans instead of going to the temple, mosque, or church to worship something that cannot be seen. Bauls sing and compose songs that critique societal divisions and allude to their philosophy and practice. They spread their messages door to door, on trains, and in performances at large public venues.

I sought the lived experiences and perspectives of Baul women.

In [the] first 18 months of work, I carefully framed how I presented my research to my interlocutors in order to avoid the usual pitfalls of my predecessors, such as having to exchange money or prestige for information, taking initiation, or losing integrity in my research methods in other ways.

[My interaction with the Baul] brings to the forefront the postcolonial predicament concerning the relationship between ethnographers and "informants," in which the ethnographer collects "data," which is then analyzed and presented for academic scrutiny.

I have argued elsewhere that as they traverse a Baul path, those who take the teachings seriously gradually adopt cognitive and spiritual models that shift their own understanding of the world around them. Thus, they learn (or aim to learn) to recognize the ways in which society creates divisions that lead to discrimination, and to recognize the divine in all human beings, regardless of caste, gender, or religion. By listening to their explanations about the micro- and macrocosmos, by hearing their songs, and by traveling with them, they expected that I, too, would experience these shifts in understanding. The cultural specifics of their experiences may not carry home, but elements of hierarchy, discrimination, and the inherent value of all human beings are as real and important in my American communities as they are in their Bengali ones. For Bauls, issues of hierarchy and domination are religiously meaningful, and knowledge of these realities constitutes important aspects of their religious experiences.

In these and other ways, the Baul women I worked with expected me to participate in their world, particularly when they knew I shared their views. Although moments of connection have shifted my own perspectives, [a specific Baul woman's] critiques ring loudest. In demanding that I (and the many Bengalis and foreigners who enjoy her singing) support her, she refuses to let me be a complacent ethnographer. She demands dignity, and she is right to do so. Several years later, her proclamations are forcing me to rethink my ethnographic work and to reveal, as a tentative step, that I have been moved, that I have taken their words and lifeworlds seriously.

Maybe language is inadequate to explain religious experience. Bauls would certainly say so, since they insist that one can only trust and know what is personally experienced. Perhaps, then, I should acknowledge moments of connection, for instance, when [the Baul] and I discussed our views of the world, its beauty and faults, and our struggle to make sense of suffering and find ways to improve the lives of those discriminated against. Maybe it's enough for me and my interlocutors to share meaningful experiences and conversations, to be open to being inspired and transformed in the field and also back home, and to be willing to put some of those ideas into action.

Knight, L. I. (2016). "I Will Not Keep Her Book in My Home": Representing Religious Meaning among Bauls. Asianetwork Exchange: A Journal for Asian Studies in the Liberal Arts, 23(1), 30–46. DOI: http://doi.org/10.16995/ane.159

This is a social sciences passage on studies of diverse cultures, or perhaps anthropology. The topic specifically concerns religion (which is a humanities topic) but references to social sciences jargon like "ethnographers" and "informants" indicate that the passage is not exactly a humanities passage. Sentence structure is again moderately complex, without tons of unusual punctuation and with an average sentence length of about 2 or 3 lines. However, unlike the art passage, the writing style is more conversational, being more of a narrative account of the author's experiences with this culture. Regarding length: both the paragraphs (with a few exceptions) and the passage itself are longer than we saw in Passage 1, but the conversational writing style might make this an Easy passage for many students. If you struggle with studies of diverse cultures, though, you might label this one Medium or even Hard.

Previewing and triaging are skills that require practice—to establish mastery, you'll need to repeat this process over and over, any time you work on a CARS passage. Break the habit (if you have it) of doing the passages in order. Try variations of both the section order strategy and the now vs. later strategy, and reflect on their effectiveness until you find the approach that works best for you. On Test Day, the higher score you're able to earn will be worth the effort!

Conclusion

Managing your timing, Previewing effectively, and triaging efficiently are all vital skills for Test Day that can help you "punch above your weight," and get more points on the CARS section. Practice these skills, along with those covered in the remaining chapters of this book, and you'll be well on your way to Test Day success!

The next three chapters continue the discussion of the Kaplan Method for CARS Passages with a more thorough examination of the Read and Distill steps.

Concept and Strategy Summary

Timing in the CARS Section

- You have, on average, 10 minutes per passage on the CARS Section of the MCAT.
- The AAMC rewards students who think strategically and approach the passages out of order.

Previewing CARS Passages

- The Preview step is a quick determination of the difficulty of the passage; it should be a gut-level check that takes only a few seconds.
- Several factors influence a passage's difficulty:
    - Passage Topic
    - Sentence Structure
    - Writing Style
    - Length
    - Questions

Determining Passage Order

- There is no single correct way to order a section. The most successful students will experiment with the different strategies and choose the option that is most effective for them.
- Practice is essential to learn how to triage effectively and efficiently.
- There are two basic triaging strategies:
    - Section Order Strategy: Go through the entire section, determining the difficulty of each passage. Then, work the passages from easiest to hardest or in the order most optimal for you.
    - Now vs. Later Strategy: Go through the section, stopping and working any Easy passages you encounter. Write down the difficulty of other passages and return to them later, working the Hard passages last.

CHAPTER 5

Reading the Passage

CHAPTER 5

Reading the Passage

In This Chapter

Introduction

LEARNING OBJECTIVES

After Chapter 5, you will be able to:

- Apply keyword strategies within a passage to locate important themes and elements
- Identify the relationship of a sentence to its surrounding context using Relation keywords
- Connect evidence and conclusion within a passage by identifying Logic keywords
- Use Author keywords to associate author tone and opinion with text

One of the biggest mistakes you can make as a student is to think that learning how to read is a one-time occurrence, like a switch that, once flipped, fully illuminates the darkened recesses of illiteracy and ignorance. It is always possible to improve your ability to read, both by refining your current approach and by broadening your comfort with a range of texts in a variety of settings. Would you read a novel for pleasure the same way you would read a textbook for homework? Once you recognize that there are many ways to analyze the written word, you can learn to customize your reading approach to fit your purpose, whether it be relaxing with a piece of fiction or reaching for that higher score on the MCAT!

KEY CONCEPT

Keywords are words and phrases commonly employed in CARS passages that serve as valuable clues for answering the accompanying questions. They fall into three broad categories: Relation, Logic, and Author.

In Chapter 3, we explored the nature of CARS passages by discussing the key components of rhetoric and argument. In Chapter 5, we'll explore strategies that will leverage keywords as a mechanism for better understanding passage structure and will build toward your ability to complete the Read and Distill step of the Kaplan Method for CARS passages. First, we will discuss the different modes of reading (reading for content, reading for purpose, and reading for reasoning) and their utility when analyzing a passage. We will then discuss keywords and their ability to clarify complex passages as part of your Read and Distill steps.

5.1 Reading Strategically with Keywords

KEY CONCEPT

Any CARS passage can be understood in three different ways, which we call the modes of reading. Each mode answers at least one vital question:

- Content—*What does the text say?*
- Purpose—*What does a particular sentence do? Why does the author write?*
- Reasoning—*How are claims supported? How are claims challenged?*

When it comes to reading dense academic prose—and this describes just about every passage in the CARS section—there are at least three ways to approach the text; that is, there are three distinct levels on which it might be appreciated. The first of these modes, **content**, is what you're most likely accustomed to looking for, whether reading for work or pleasure. On Test Day, you will want to pay attention to informational content, but you'll also want to broaden your approach to encompass the two other modes of reading: **purpose** and **reasoning**. Building on the concepts discussed in Chapter 3: Rhetoric and Arguments, reading for purpose involves identifying rhetorical components, while reading for reasoning involves identifying components of arguments.

Reading for Content

Before jumping into new ways to read CARS passages, let's briefly discuss what you are likely already familiar with, reading for content. Reading for content is focused on extracting the information from the text, discovering precisely *what* is being said. It is important to note that you will never be expected to have preexisting familiarity with the content of a passage in the CARS section; all information necessary to answer the questions is contained within the passage itself. This situation is in stark contrast to passages in the three science sections, in which you are expected to integrate outside knowledge with any new information provided in the passage.

When reading for the content, the question you should be trying to answer is *What is the author saying?* With that said, reading for content has its limitations for CARS passages, as it is unlikely that you will be a member of the passage's intended audience. Meaning, it's likely that you will not have the content background to fully understand the concepts presented in some of the harder CARS passages on Test Day. When this happens, don't panic! Instead, turn your attention toward rhetoric and arguments by reading for purpose and reasoning.

Reading for Purpose

Although there are many ways to talk about "purpose," we are specifically answering the question, *why did the author do that?* This general question can be applied to various portions of the passage and consequently reveal different aspects of the passage and author, such as:

- Why did the author use a particular word or phrase?
- Why did the author include this sentence? What does this sentence do?

- Why did the author include this paragraph? What does this paragraph do?
- Why did the author write this passage?

As you can see, we can get very specific when reading for purpose (possibly evaluating every word!), and although reading for purpose provides great insight, exploring each of these questions at every opportunity requires substantial amounts of time. Thus, it's vital to build efficient habits to ensure that reading for purpose will serve you on Test Day. As a general rule, determining the purpose of specific terms and sentences should be done when those components are major parts of the argument being built. Keywords will help you to establish when and where identifying the purpose of a portion of the passage is something you should spend time on. In contrast, determining the purpose of paragraphs and the passage as a whole should be done consistently, and is the focus of each of the Distill options in the Kaplan Method for CARS passages.

BRIDGE

The amount of time you spend reading for purpose when focused on a CARS passage will depend on the Distill method you've selected for that passage as part of your Choose step. If you're Distilling by Highlighting, you will likely do very little reading for purpose, and largely rely on keywords to identify components to highlight. If you are Distilling by Outlining, you will determine the purpose of each paragraph and write it down as part of your outline. If you are Distilling by Interrogating, you will focus on the purpose of separate chunks throughout the passage on a more granular level than once per paragraph. You will always consider the purpose of the passage overall, no matter the method you have chosen. The Distill methods, and their application in the Kaplan CARS Passage Strategy, are discussed in Chapters 6 and 7.

Purpose of a Particular Word or Phrase

Each CARS passage contains hundreds of words, so it is not worth your time to consider the author's intentions with every single word. Instead, you should only focus a particular term when the questions require it. Definition-in-Context questions (covered in Chapter 9) require you to focus on particular words and phrases, but they ask you for the meanings of terms (their content), not their purposes. In contrast, Function questions (covered in Chapter 10) focus exclusively on the author's intentions and will sometimes point to a brief phrase or even a single word. When assessing the purpose of a term, move outward, first thinking about how the specific sentence that contains the term functions in the larger paragraph, and then about how that paragraph functions in the passage as whole. If the term is consistent with these larger purposes, you need only look for the answer choice that best reflects the main themes of the passage. However, if something unexpected is introduced with the term, ask yourself why the author seems to going in a different direction. Pay special attention to language that conveys feeling, whether in the term itself or in the surrounding context. If nothing else, awareness of emotionally-laden language can help you to eliminate answer choices that contain the wrong sentiments.

Purpose of a Particular Sentence

Some question types, such as Function questions, will sometimes ask about specific sentences instead. Again, if you know the larger purpose of the paragraph that contains the sentence, as well as how that paragraph functions to accomplish the author's larger purpose in the passage, then such questions will be more straightforward. When you're confused about the author's larger purposes, though, you can still use context to infer the author's intentions. Consider the following example:

> The notion of realism in literature is based largely on the implicit belief that writers can accurately transform common objects or ideas from life into words on a page while maintaining an accurate representation of the object or idea. If an author writes a novel which seems believable, meaning that a reader can imagine events in the novel actually happening, then that book is often considered a "realistic" work of literature.

This is a dense paragraph, consisting of two convoluted sentences. But even if you had difficulty unpacking the meaning, you can still look at the general structure of each sentence to gather information about the author's intentions. The first sentence begins with the phrase, *The notion of realism in literature is based largely on . . .*, with a long string of words following. While it may be difficult to unpack the last two-thirds of the sentence, the first third tells us that the author is explaining the basis of literary realism or, in other words, giving us the definition of that concept. Thus, the first sentence functions as an introduction and definition of a particular concept.

The second sentence, in contrast, is structured as a big if-then statement. You may have noticed the author's use of quotation marks with "*realistic*" in the then portion of the statement. This sentence is effectively saying, "if a book's author does X, then that book is realistic." In other words, this sentence is again aiming to explain what it means to have realism in literature. Since the author already gave a definition in the first sentence, the second sentence functions as an elaboration upon, or perhaps a clarification of, that definition. While the second sentence doesn't include any "transition words" (a term you may have learned about in high school or college English), the author could easily have started with a phrase like "to clarify" or "in other words" without changing the meaning or purpose of the sentence.

Often authors will include **Relation keywords**, covered later in this chapter, to emphasize these relationships between sentences. You may have learned about these in high school English or first year composition courses as "transition words" because they facilitate the movement (or transition) between sentences. In the previous example, phrases like *To clarify* or *For example* could have been used to start the second sentence. Determining the purpose of a sentence assists with reading strategically, and it also helps provide a clear prediction for **Function** questions (covered in Chapter 10).

Purpose of a Paragraph

Reading for purpose is particularly high yield at the level of paragraphs. While Function questions commonly ask you about the purpose of specific paragraphs, you will generally find it worthwhile to investigate each paragraph's purpose as you read, irrespective of the questions. Knowledge of how each paragraph fits into the larger whole can better enable you to eliminate answer choices for a wide variety of question types, which means more points. In Chapter 6, we will explore Kaplan's three distillation approaches to CARS passages: Highlighting, Outlining, and Interrogating. Each of these approaches includes examination of paragraph purposes, but with varying levels of specificity.

When determining a paragraph's purpose, try to get a sense of it from the opening sentence. While CARS authors are less likely than many writers to open with a topic sentence, they are occasionally straightforward, and when they aren't, you can still often pick up a few clues from that first sentence. Watch out for keywords (discussed later in this chapter) that suggest a particular transition from the previous paragraph (if there is one). Look out for changes in direction in the sentences that follow, paying particular attention again to the concluding sentence, which may contain its own transition to a following paragraph. If the paragraph goes as expected, then your initial guess at its purpose is likely accurate. But if anything surprises you, take a moment to figure out why the author would include that unexpected element. In the end, find a good answer to one of these questions: *Why did the author include this paragraph? What does this paragraph do for the passage?*

Purpose of the Passage

Chapter 3 on Rhetoric and Arguments first introduced the concept of a passage's overall purpose or goal. Indeed, a major focus of rhetoric is the study of author intentions, of what authors hope to accomplish with their writing. Regardless of which approach you choose when reading a CARS passage, it will always be worth your time to seek out the passage's goal. Make a guess about this goal as early in the passage as you can, modifying this guess if you encounter anything you didn't expect. When you finish the passage, be sure to take a moment and reflect on what you read. Ask yourself: *Did the passage accomplish what I expected it to? If not, what was the author actually trying to do?* Your answers to these questions will enable you to articulate the passage's goal in your own words, which in turn will translate to points on Main Idea questions and other CARS question types.

Reading for Reasoning

The final mode of reading is reading for reasoning, which focuses on the structure of arguments. Chapter 3 introduced the topic of arguments, so return there if you need a more thorough review of the topic. For present purposes, recall that an argument, at its most basic level, consists of evidence (one or more supporting statements) and a conclusion (a supported statement). When you read for reasoning, you're on the lookout for these support relationships, as well as for refutations (also discussed in Chapter 3) that challenge particular claims.

When you read for reasoning, you should ask yourself questions like the following: *What is the author trying to convince the audience to believe? Does the author give any reasons to believe this sentence is true? Does this new claim make a previous one more plausible, or does it perhaps challenge an earlier one? What evidence or refutations does the author provide for this opinion? Which claims in the passage are given the most support?* Logic keywords, discussed later this chapter, can assist you in identifying the structure of arguments in a passage.

In addition to looking out for conclusions, evidence, and refutations within paragraphs, you should also consider whether the author uses a larger argument to structure the entire passage. For example, an author might introduce a controversial claim in the first paragraph, and then provide a few different types of evidence for that claim in each of the paragraphs that follows. Alternatively, an author may introduce a claim in one paragraph, support it in the next with some evidence, and then offer additional support (sub-evidence) for that evidence in a third paragraph. At other times, an author will present a claim, perhaps even give a little evidence for it, and then go on to refute the claim in subsequent paragraphs. Being mindful of these and similar argument structures can help you better understand a wide variety of passages.

Reading and Keywords

Keywords are words and short phrases that reveal a passage's structure and its author's intentions. Keywords are the connective tissue of a text, holding ideas together in a particular way that allows the passage to do what its author wants. Keywords do not refer to a passage's major ideas (its content), but are instead rather common words that many readers pay little attention to: conjunctions and other transition words, prepositions, adjectives, and adverbs. Paying attention to these often-overlooked words can give you an advantage on Test Day by enhancing your understanding of even the most difficult passages.

BRIDGE

The way the traffic light system should be applied to different types of keywords is discussed throughout the rest of Chapter 5. Generally speaking, Continuation keywords will be green lights, Contrast and Author keywords will be yellow lights, and Logic keywords will be red lights. However, there are exceptions to these rules, so review the following material carefully.

The following sections of this chapter will introduce Relation, Logic, and Author keywords, as well as advice on using them to read strategically. At the surface level, keywords can tell you what is important in a passage, letting you know where in a passage you should be focusing your attention. Towards this end, one useful approach is to assign colors to different categories of keywords: green, yellow, and red. In this traffic light system, *green* keywords indicate that you should keep moving (expecting to see more of the same), *yellow* keywords indicate that you should slow down and look out for author opinions or shifts in the direction of the passage, and *red* keywords suggest "*Stop!* There is something important to be found."

5.2 Relation Keywords

When tackling a CARS passage, it is essential to recognize how what you're reading now fits into the text as a whole. **Relation keywords** are words and phrases that reveal specific kinds of relationships between ideas. While there are many ways in which ideas can be related, the vast majority of Relation keywords will fall into one of two subcategories: Continuation or Contrast.

Continuation

Continuation keywords indicate that the coming material is an extension of a previous statement and include *and, also, moreover, furthermore*, and so on. In addition, phrases that indicate examples (*such as, for instance*, and *take the case of*) fall into this category, as do demonstrative pronouns (*this, that, these*, and *those*). Generally speaking, when a sentence or clause begins with a **Continuation keyword**, it will be continuing in the same vein as what came before. In other words, it won't be saying anything particularly new. As a consequence, if you understood the preceding material, you can generally read a clause that follows a Continuation keyword quickly, briskly moving through the text until a new keyword signals something different. On the other hand, if you struggled to understand a particular sentence, but see a Continuation keyword at the beginning of the next sentence, keep moving forward—the author is likely to continue with the same idea but may rephrase it in easier words. Thus, in our traffic light analogy, Continuation keywords would be *green lights*, signaling you to keep going!

Even though they are not technically *keywords*, certain punctuation marks can also indicate that a similar idea is coming up. Most notably, colons (:) and semicolons (;) are commonly employed to function as the verbal equivalent of the equals sign (=). The use of dashes—such as the ones surrounding this clause—and parentheses (the marks enclosing this phrase) also tend to indicate elaboration upon the same general theme. Finally, keep an eye out for quotation marks (""): while quotes can serve a variety of functions, one of the most common is to use another person's voice to restate the point the author just made.

Contrast

On the other hand, **Contrast keywords** will usually merit additional attention when you encounter them, as they tend to suggest more interesting (and therefore testable) relationships than Continuity keywords. Common Contrast keywords such as *but*, *yet*, *however*, *although*, and *otherwise* signal a change in the direction of the text. Like Continuation keywords, Contrast keywords serve as transitions between sentences, but they can also indicate deeper conceptual relations—the contrast between two solutions to a political problem, a point of disagreement between various critics of a literary work, or a rapid change in opinion from one time period to another, to give just a few examples. The connections between ideas are among the most commonly tested aspects of CARS passages, so strive to understand these relationships with as much specificity as the given clues allow. To this end, view Contrast keywords as *yellow lights* in our traffic light analogy. When you come across a Contrast keyword, be prepared to slow down and read carefully to ensure you follow the author's transition of thought.

While punctuation symbols often indicate continuation, this is not always true. If a punctuation symbol is accompanied by another type of keyword, the actual word or phrase generally takes precedence in determining the relationship (for example, a semicolon followed by *yet* usually suggests a point of contrast rather than continuation). This points out the greater importance of anticipating while you read; a great score in CARS depends on your ability to be a critical reader rather than a passive reader. When we passively read for pleasure, we tend to glide over text and often only understand the superficial message of the written work. In critical reading, we are continuously questioning the text and setting expectations for where the author will go with an idea. Even if they are not met, it is still worthwhile to set these expectations. In fact, when an author takes a starkly different route in a part of the passage than we expect, the test makers are more likely to ask about the sudden contrast from the text. That which is rhetorically unusual in the passage and frustrates expectations becomes excellent material for MCAT questions.

MCAT EXPERTISE

When you encounter a Contrast keyword, slow down your reading or (if needed) even stop! These keywords signify a change in the author's focus or a direct contrast between two things; either way, we need to know how the trajectory of the passage is changing to keep a step ahead of the author.

More Complex Relationships

While the keywords considered under this heading might broadly count as Contrast keywords, they designate special types of difference and are especially ripe for CARS questions. While by no means exhaustive of the types of relations you might see on Test Day, **Oppositions**, **Sequences**, and **Comparisons** are three of the most common.

Oppositions

Words and phrases like *not*, *never*, *on the contrary*, and *as opposed to* indicate not merely a contrast but an outright **Opposition** or conflict between ideas.

Many authors of passages used in CARS like to create **dichotomies**, which are divisions of entities into two categories. These categories are considered **mutually exclusive**, meaning that they don't overlap. The use of *either ... or*, *on one hand ... on the other hand*, and similar parallel phrase constructions are good indications of this phenomenon. Often, but not always, these dichotomies will be depicted as **exhaustive**, meaning that everything falls into one of the two categories. For example, an author writing about human behavior might claim that *all actions are either free choices or involuntary reflexes*, leaving no space for shades of gray. Because these dichotomies are frequently tested on the MCAT, treat Opposition keywords as *yellow lights*. Slow down to determine why the author is presenting this opposition and how it affects their overall argument.

Sequences

Some Relation keywords suggest a series of events advancing in time: *initially*, *first*, *second*, *third*, *next*, *subsequently*, *before*, *after*, *last*, and *finally*. These words will usually be spaced relatively evenly throughout the passage, or at least throughout a paragraph or two, so note how they organize the text into chunks. **Sequences** are something of a hybrid between Continuation and Contrast, with each word suggesting not only a connection to a larger process but also a departure from the other steps in the series. Generally, you're better off taking your time with these, at least until you have a good idea of how the sequence will unfold. However, you will often find Sequence keywords require less of a "slow down" than other yellow light keywords.

Some sequences can set up a clear contrast between time periods. *Historically*, *traditionally*, *used to*, *originally*, and—when used in comparison to a later time—*initially* and *before* can be used as time-based Contrast keywords when contrasted with words like *now*, *currently*, *modern*, *later*, and *after*. Such a setup often implies that new information was learned or discovered in the intervening time: *the traditional interpretation* vs. *a more modern understanding* or *historically, we thought* vs. *but now, we know*.

Comparisons

Sometimes, authors will evaluate ideas and rank them relative to other ideas. More often than not, authors will consider only two concepts at a time, contrasting them through the use of **Comparison keywords** like *more*, *less*, *better*, and *worse*. That said, occasionally authors will compare three or more items, or offer vague judgments of superiority (or inferiority) of one item over all others, reflected by superlatives such as *most*, *least*, *best*, and *worst*. When revealing attitudes, Comparison keywords function more like Author keywords, further explained later in the chapter.

In our traffic light analogy, Comparison keywords require more attention than most Contrast keywords and should be viewed as *yellow lights*. You should slow down and notice what entities are being compared and the author's opinion on them. A common wrong answer trap in CARS is to suggest that the author makes a comparison between two entities from the passage that were not actually directly compared. By taking special note of the comparisons the author does make, you can avoid such traps.

Table 5.1 lists examples of Relation keywords in each category. Note that some words can fit into more than one category; for example, *not* reveals a contrast, but it can also indicate a direct opposition.

Continuation	Contrast	Opposition
and	but	not/never/none
also	yet	either . . . or
moreover	however	as opposed to
furthermore	although	on the contrary
like	(even) though	versus (*vs.*)
same/similar	rather (than)	on the one hand . . . on the other hand
that is	in contrast	otherwise
in other words	on the other hand	**Sequence**
for example	otherwise	before/after
take the case of	nevertheless	earlier/later
for instance	whereas	previous/next
including	while	initially/subsequently/finally
such as	different	first/second/third/last
in addition	unlike	historically/traditionally/used to
plus	notwithstanding	now/currently/modern
at the same time	another	**Comparison**
as well as	instead	better/best
equally	still	worse/worst
this/that/these/those	despite	less/least
: [colon]	alternatively	more/most
; [semicolon]	unless	*–er*/*–est*
— [dash]	not	primarily
() [parentheses]	conversely	especially
" " [quotes]	contrarily	above all

Table 5.1 Common Relation Keywords

MCAT EXPERTISE

Simply memorizing lists of keywords is not sufficient to extract all of the information from a passage. A Contrast keyword can serve slightly different functions depending on the context in which the word is found. Recognize that MCAT CARS success is not about memorizing the fact that *but*, *yet*, and *however* are Contrast keywords, but is instead about understanding how Contrast keywords reveal changes in direction in a passage.

5.3 Logic Keywords

Reading for reasoning is perhaps the most difficult of the reading modes because the one-way support relationship between a conclusion and its evidence is among the most complex you'll encounter on Test Day. **Logic keywords** are words and phrases that signal these support (and refutation) relationships. Unfortunately, they tend to be relatively rare, occurring less frequently than either Relation or Author keywords in most passages. Notwithstanding these difficulties, Logic keywords are a powerful tool: once you gain proficiency in recognizing them and understanding what they entail, you'll find the many CARS questions on reasoning much less daunting.

Evidence and Conclusion

In Chapter 3: Rhetoric and Arguments, we defined an argument as a relationship between two claims, a conclusion and its evidence. To review, a **conclusion** is a claim that the author (or whomever the author is speaking for) is trying to convince the audience to believe, while pieces of **evidence** are the reasons that are given for believing it. Determine whether the conclusion is one that the author would endorse or whether it is intended to represent some other viewpoint by paying attention to nearby Author keywords (discussed later) and other clues.

Typical examples of **Evidence keywords** are *because, since, if, why, the reason is, for example, on account of, due to, as a result of, is justified by*, and *after all*. There is a bit less variety in **Conclusion keywords**, which also tend to occur less frequently than Evidence keywords. The most important conclusion words to know are *therefore, thus, then, so, consequently, leading to, resulting in, argue*, and *conclude*.

In our traffic light analogy, both Conclusion and Evidence keywords should be viewed as *red lights*. When you encounter Evidence keywords, stop and ask yourself, *What is this supporting? Where is the conclusion?* When you encounter Conclusion keywords, take note of whose conclusion it is (the author's or some other view's) and be on the lookout for its evidence (if any is provided).

Refutation

Refutation keywords will not always be included in the presentation of an argument, but they are effectively the opposite of evidence—countervailing reasons for rejecting a conclusion. They include words such as *despite, notwithstanding, challenge, object, counter, critique, conflict*, and *problem*. Given the opinionated nature of Refutation keywords, you should consider *why* the author included the refutation. Was it is to attack an idea that the author has a negative view on? Or perhaps the author intentionally put forward a weak refutation only to later refute the refutation, thus supporting the original contention. No matter the purpose of the refutation, it is important that you notice it and consider its purpose—refutations are relatively rare, so if a passage features one or more of them, expect to see questions on them. For this reason, Refutation keywords should be viewed as *red lights*, as with the other Logic keywords.

MCAT EXPERTISE

Having trouble figuring out what part of the argument a Logic keyword is indicating? Try this simple substitution test: If you can replace the word or phrase with *because* or *because of*, then whatever follows is a piece of evidence. If, instead, *therefore* would preserve the meaning, the subsequent claim is a conclusion.

MCAT EXPERTISE

Sometimes Evidence and Conclusion keywords are used to signal cause-and-effect relationships, rather than arguments. This most commonly occurs in scientific and historical explanations of phenomena. Fortunately, while causation differs from argumentation in important ways, they share a consistent relationship to the Logic keyword subcategories. When used for causation, Evidence keywords will always indicate a cause and Conclusion keywords will always indicate an effect.

Table 5.2 lists examples of Logic keywords in each category.

Evidence	Conclusion	Refutation
because (of)	therefore	despite
since	thus	notwithstanding
if	then	challenge
for example	so	undermined by
why	consequently	object/objection
the reason is	leading to	counter (argument)
as a result of	resulting in	critique/criticize
due to	argue	conflict
as evident in	conclude	doubt
justified by	imply	problem
assuming	infer	weakness
after all	suggest	called into question by

Table 5.2 Common Logic Keywords

5.4 Author Keywords

Author keywords can be among the most subtle clues that you'll encounter on Test Day, but they are crucial for answering the many questions you'll face that ask about the author's attitudes, whether directly or indirectly. Authors of passages used in CARS rarely say *I believe* or *it seems to me* (and if you do find this language, it's more often in the humanities than the social sciences). Instead, they are more likely to hint at their opinions by selecting verbs, nouns, adjectives, and adverbs that carry a particular emotional valence—a connotation of either approval or disapproval. Moreover, authors will use characteristic words and short phrases to make their claims more extreme (indicating emphasis and strengthening ideas), as well as others that moderate their claims (qualifying or limiting what they are saying and weakening ideas).

Given the importance of establishing the author's tone and opinions, it may seem appropriate to make them red lights in our traffic light analogy. However, because Author keywords are so common, they are typically best viewed as *yellow lights*: slow down and take notice, particularly noting the context of the keyword so you know *who* feels a certain way and about *what topic*. In passages with multiple arguments or viewpoints, you may have to treat some Author keywords as red lights, simply because their context is complicated enough to require you to stop and clarify which perspective holds what opinion.

Positive vs. Negative

Understanding the author's attitude becomes a much simpler matter if we employ a metaphor taken from the sciences. Just as an atom or molecule might possess an electrostatic charge, so too can a word or phrase contain a kind of emotional charge that may be positive or negative. Because most of the language in CARS passages tends to be "uncharged" or neutral, you'll predominantly pay attention to the exceptions—those cases in which terms have clear positive or negative connotations.

Positive keywords include nouns such as *masterpiece*, *genius*, and *triumph*; verbs such as *excel*, *succeed*, and *know*; adjectives such as *compelling*, *impressive*, and *elegant*; and adverbs such as *correctly*, *reasonably*, and *fortunately*. Among **Negative keywords** would be nouns such as *disaster*, *farce*, and *limitation*; verbs such as *miss*, *fail*, and *confuse*; adjectives such as *problematic*, *so-called*, and *deceptive*; and adverbs such as *questionably*, *merely*, and *purportedly*.

MCAT EXPERTISE

Most CARS passages contain strong, but not necessarily extreme, opinions. For example, a typical CARS passage might provide several refutations to challenge a particular theory, but its author likely wouldn't claim that the theory is "utter nonsense." Passages with neutral authors, such as might be found in a typical textbook, are not usually selected for the CARS section because they feature less testable material.

Keep in mind that, just as there is a difference between a cation with a +1 charge and one with a +3 charge, so too is there a difference between a moderately positive opinion and an extremely positive one. For instance, an author probably approves more strongly of a novelist described as *a masterful artist* than one portrayed merely as *a quality writer*. Consequently, it may be helpful to think of the author's attitude as varying along a spectrum or continuum, with extremely positive opinions on the one end and extremely negative on the other, as in Figure 5.1. Note that most authors' attitudes in CARS fall in a comfortable middle ground between being too extreme and being too moderate, as implied by the relative widths of the sections in the diagram.

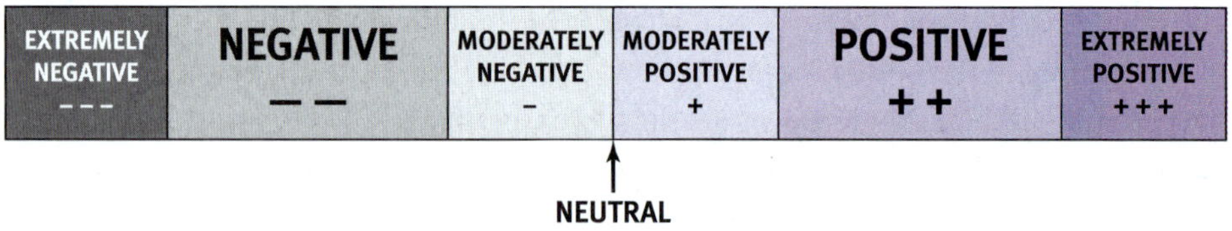

Figure 5.1 The Spectrum of Author Attitudes (Linear)

Note that in addition to positive, negative, or neutral, an author can also be **ambivalent**. Ambivalence literally means *feeling both ways*, and it is as different from **impartiality**—having no strong opinion one way or another—as the set of 1 and −1 is from the number 0. Continuing the analogy with an electrostatic charge, an ambivalent attitude is like an amino acid in its *zwitterionic* form, with both a positively charged and a negatively charged end, and an impartial attitude is like an uncharged, unpolarized atom. Describing the net zero charge of a zwitterion as merely "neutral" would mean neglecting its distinctive properties. The MCAT won't let you get away with such oversimplifications!

KEY CONCEPT

While they are both neutral overall, these two attitudes are very different:

- Ambivalent = having both a positive and negative opinion
- Impartial = having neither a positive nor negative opinion

Extreme

Placing a particular idea on the author-attitude spectrum above becomes easier by paying attention to **Extreme keywords**, a type of Author keyword that you can imagine as enhancing the charge of what the author is saying, forcing the author into one or the other extreme. These words and short phrases are functionally equivalent to exclamation points (!), offering insight into what the author feels passionately about and regards as important.

Examples of Extreme keywords include *indeed*, *very*, *really*, *quite*, *primarily*, *especially*, *obviously*, *foremost*, *always*, *in fact*, *above all*, and *it is clear that*. Note that words that indicate necessity, like *need* and *must*, also serve as Extreme keywords, as do words that indicate value judgments like *should* and *ought*—these tend to be rare in CARS passages, so they deserve special consideration when they do appear.

Moderating

Authors will sometimes modify the strength of their claims in the other direction by using qualifying language, also known as *hedging*. **Moderating keywords** are those words that set limits on claims in order to make them easier to support (because a stronger statement is always more difficult to prove than a weaker one). For example, it would be an extreme claim to say that *human beings are motivated only by greed*. Though some might agree with this formulation, the bulk of MCAT authors would sooner water it down by saying something like *in many aspects of life, humans are predominantly motivated by greed*, or even further limit it to a subset of human beings, such as *investment bankers are often motivated by greed*. Such modifications transform a controversial claim into one that is much more plausible.

Among the most important Moderating keywords are those that use the language of possibility, such as *can*, *could*, *may*, and *might*. Claims about what is possible are always weaker than claims about what is definitely true. Other Moderating keywords include limits on time or place, whether stated specifically or in vague phrases such as *now*, *here*, *at times*, *in some cases*, and *in this instance*. Still others will impose general constraints on meaning; examples include *in this sense*, *according to this interpretation*, and *in a manner of speaking*.

Accounting for Opposition

One final consideration when working with Author keywords is contradiction or opposition, a special type of Contrast keyword mentioned above. It can be particularly tricky to figure out the authors' attitudes in CARS when they use double negatives (or worse!). The key is to remember that the opposite of an extreme statement will typically be a moderate statement of the reverse charge. In the diagram below, which is just a slight rearrangement of the author attitude spectrum presented in Figure 5.2, a word of opposition will typically serve to flip the author's view 180 degrees.

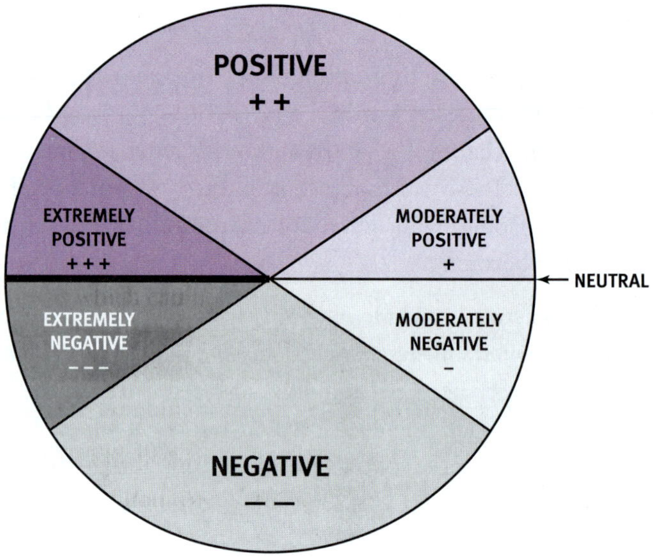

Figure 5.2 The Spectrum of Author Attitudes (Circular)

So, if an author were to claim that a particular event is *impossible*, this statement would fall somewhere in the extremely negative range. On the other hand, suggesting the situation was *not impossible* would be making a moderately positive claim, effectively saying that it was *possible*. Be aware, however, that some words retain their strength even when accompanied with *not* or some other word of opposition—for instance, *must* is extremely positive in charge, while *must not* is extremely negative in charge.

Table 5.3 lists examples of Author keywords in each category.

Positive	Negative	Extreme	Moderating
masterpiece	disaster	must	can/could
genius	farce	need/necessary	may/might
triumph	limitation	always	possibly
excel	miss	every	probably
succeed	fail	any	sometimes
know	confuse	only	on occasion
compelling	problematic	should/ought	often
impressive	so-called	indeed	tends to
elegant	deceptive	very	here
correctly	questionably	especially	now
reasonably	merely	obviously	in this case
fortunately	purportedly	above all	in some sense

Table 5.3 Common Author Keywords

Conclusion

We began this chapter with a discussion of the modes of reading and finished with a discussion of keywords, but these two topics are closely interrelated. Reading for purpose is aided by close attention to Relation and Author keywords, while reading for reasoning is made simpler when authors use Logic keywords. As you continue your CARS practice, try the reading strategies discussed in this chapter (the modes of reading, the keyword categories, and the traffic light system) until you have a solid grasp of them. In conjunction with the other Kaplan strategies, they'll help you get that higher score in CARS. The next chapter focuses on another CARS strategy: distilling passage information.

Concept and Strategy Summary

Reading Strategically with Keywords

- **Keywords** are words and short phrases that reveal passage structure and author intentions.
- When we read for **content**, we ask, *What does the text say?*
  - You are never expected to have prior familiarity with the content of a passage in CARS.
  - When reading for content is a struggle, reading for purpose and for reasoning can help you make sense of the passage.
- When we read for **purpose**, we ask, *Why does the author include this?* and *Why does the author write?*
  - Reading for purpose is guided by **Relation keywords** and **Author keywords**.
- When we read for reasoning, we ask, *What is the conclusion? How are claims supported?* and *How are claims challenged?*
  - Reading for reasoning is guided by **Logic keywords**.
  - Differentiating between conclusions, evidence, and refutations can help you make sense of the author's larger argument in the passage.
- Keywords serve as signs that indicate where you should focus your attention. As an analogy, you can consider different categories of keywords to be different colors of a traffic light (green, yellow, and red).

Relation Keywords

- **Relation keywords** are words and phrases that reveal specific kinds of relationships between ideas.
- **Continuation keywords** indicate that the following material continues in the same vein as the preceding material (*green light*).
- **Contrast keywords** signal a change in the trajectory of the passage (*yellow light*).
- **Opposition keywords** are particularly strong Contrast keywords that create a dichotomy, or divisions of entities into two categories (*yellow light*).
  - **Mutually exclusive** categories do not overlap.
  - When dichotomies are **exhaustive**, all relevant entities fit into one or the other category.
- **Sequence keywords** suggest a series of events advancing in time. They may also be used to set up a contrast between two time periods (*yellow light*).
- **Comparison keywords** rank ideas relative to each other (*yellow light*).

Logic Keywords

- **Logic keywords** indicate relationships of support and refutation (all types are *red lights*).
- **Conclusion keywords** signal what the author is trying to convince the audience to believe.
- **Evidence keywords** indicate reasons why the audience should believe the author's claim.
- **Refutation keywords** signal reasons for rejecting a conclusion.

Author Keywords

- **Author keywords** indicate the author's feelings or opinions about a topic (*yellow light*).
- **Positive** and **Negative keywords** indicate whether an author likes, agrees with, or supports a topic, or dislikes, disagrees with, or opposes a topic. An absence of either type suggests an author is neutral with respect to a topic.
  - An author with an **ambivalent** attitude has both positive and negative opinions on a topic.
  - An author with an **impartial** attitude has neither positive nor negative opinions on a topic.
- **Extreme keywords** enhance the charge of what the author is saying.
- **Moderating keywords** permit the author to qualify a claim, or hedge.
- The opposite of an extreme statement tends to be a moderate statement with the opposite charge.

CHAPTER 6

Distilling Passage Information

CHAPTER 6

Distilling Passage Information

In This Chapter

Introduction

LEARNING OBJECTIVES

After Chapter 6, you will be able to:

- Differentiate between the *Highlighting*, *Outlining*, and *Interrogating* approaches to CARS passages
- Identify the types of passages best suited for each method
- Apply each of the three methods to sample MCAT passages
- Step beyond the text to analyze the underlying themes of the passage

Not all passages on the MCAT are created equal. As we discussed in Chapter 4 of the *MCAT CARS Review*, you will use things like passage topic, sentence structure, writing style, and passage length to triage each passage on the test and decide which ones you want to save for later. However, once you do decide to attack a passage, you must have a battle plan. Just like you wouldn't treat a five year old with a cold the same way you would treat an eighty-five year old with chronic heart failure and asthma, you are not necessarily going to approach a long, complex humanities passage the same way you might approach a more concise, clear social sciences passage. Remember, one of our main tasks when working through a CARS passage is to truly understand the big picture of the passage. So, it is essential to read actively and methodically so as not to miss key details or insights. During this chapter, we'll explore different ways to attack passages so that, come Test Day, you will be equipped with both the ability to gauge the best way to attack a passage and to actually use your chosen method to get through the passage.

This chapter will begin by introducing three different approaches to attacking MCAT CARS passages and discussing their respective pros and cons. We will then carry out a more in-depth discussion of each approach, and we will end by seeing how each approach can be applied to a sample MCAT passage.

6.1 Choosing Your Approach

Interviewing a new patient can feel like an uphill battle against an onslaught of information, and so can reading through a CARS passage. Having a solid plan for how to Distill and organize the high-yield information from a CARS passage allows you to gather the context you need without getting overwhelmed by extraneous material. In this section, we will introduce the three different methods for distilling a passage (each of these will be covered in depth later in this chapter as well). Note that you should plan to practice all three of these methods throughout your prep so that you can easily choose the best method for each passage on Test Day.

Triaging the Passage

Remember, we're not reading CARS passages for pleasure, the way we might read a novel or a magazine. In CARS, certain parts of the passage will inevitably contain more testable information than others. So, given the tight time constraints, you need to develop strategies to navigate through passages effectively and to identify the most important pieces of information as you read.

In Chapter 4, we discussed how to determine the order in which you will work the passages. Once you have decided to commit to and read a passage, use the conclusions you made while triaging to influence the approach you use to work through that passage. For example, you will probably not approach a long, complex philosophy passage the same way you would approach a more straightforward literature passage. While a long, complex passage is often best met with a quick reading focused on identifying big picture concepts, a straightforward passage is often better handled with a more detailed read and with a greater focus on identifying the rationale behind the author's arguments. One of the best things about having multiple strategies in your arsenal is that you can pick and choose the best approach for each passage rather than sticking to a "one size fits all" approach for every passage in the section; so don't be afraid to change it up as you go. As was referenced in Chapter 4, you can either wait and Choose your approach when you are actually working the passage, or you can preemptively Choose as you triage each passage. In either case, you will want to make an active decision about how you are going to Read and Distill the passage.

BRIDGE

Chapter 4 focused heavily on Previewing the passage, the process by which you will quickly gain information about the passage and determine difficulty. This allows you to triage each passage (decide whether to work it now or later), but that same information gathered in the Preview step is also used for Choosing your passage approach. Refer to Chapter 4 for more information on the Preview step.

Keywords

A key strategy for identifying where to read thoroughly, as opposed to skimming quickly, is noting the use of keywords and punctuation. For example, since Continuation keywords indicate that more of the same is coming, if you understand what has already been stated, then save yourself time by reading quickly through that portion. Conversely, Contrast keywords indicate that either a change, a shift in meaning, or a new piece of information is up ahead, essentially telling you to slow down and to ensure that you don't miss something important. Furthermore, Conclusion keywords indicate that the major point of an argument is ahead, so it is worth stopping and ensuring that you completely understand the point before moving forward in the passage.

Approaching the Passage

The ultimate goal of the Distill step of the CARS passage method is to help you identify what is most critical in each paragraph of the passage. You want to have some concept of the main idea and overall argument of each paragraph. This will save you time on questions, because you will already have a solid idea of what is contained in each paragraph and how that information fits together. As a result, you will be able to find specific details or supporting evidence in the passage more easily. We can compare this process to the scientific process of distilling a liquid: you are removing the impurities and extraneous information from the passage. And you are left with a purified product consisting of only the information relevant to answering the questions associated with the passage. This step can be done using one of three approaches. We'll describe each approach in broad strokes over the next couple pages. Then we'll drill down into each approach and examine how and when to implement that approach, the pros and cons of that approach, and worked examples.

Highlighting the Passage

For some students, and especially in time crunch situations, a thorough read of the passage that revolves around taking notes or questioning the purpose of every sentence is going to be too much to accomplish in the limited time allotted for each passage. If you are in a time crunch situation, if you are unable to completely understand the content of a paragraph and you know you won't understand even with more time, or if the passage is simply too long to accomplish with the other methods, then Highlighting can be a great choice. Focus on highlighting a phrase that points to the main idea or purpose of each paragraph, so that when you return to the passage while working through a question, you can easily identify the high yield information. Highlighting doesn't allow you to create a true paraphrase of the main idea, so you'll need to rely on finding the most helpful phrase or word already located in the paragraph. Note that if you're using the Highlighting method, you should not be highlighting random words that you don't know, or highlighting details. Instead, highlight words that will allow you to quickly relocate where main ideas were found throughout the passage. Finally, when you reach the end of the passage, take a moment to consider why the author wrote the passage. Some examples include: to argue a point, to discuss a topic, to contrast two different ideas. From a timing perspective, the Highlighting approach is the quickest of the three methods we'll discuss. Highlighting allows you to work through the passage in 2–3 minutes. Furthermore, the highlighting you did should allow you to find main themes and search the passage more efficiently. However, you should be aware that the Highlighting method is going to require that you spend more time on the questions, as compared to the other two methods. This time will be spent returning to the passage for further research. With the Highlighting approach in particular, it is imperative that you work all passage questions immediately following the passage, with no intention of coming back to any questions after looking at other passages. Your highlighting will make the most sense immediately after you've done it, and if you look at other materials in the interim, it's unlikely that you'll be able to return to the passage and decipher your highlighting.

Outlining the Passage

When composing a paper for your high school or college English classes, you likely were asked to first create an outline. In this outline, you recorded the purpose of each paragraph you intended to write, and then you used that outline as a template for the full paper. The Outlining strategy for CARS passages is essentially the reverse of that: you will take a full passage and reverse engineer the outline that captures each paragraph's purpose. Like with the Highlighting approach, the goal of Outlining is to identify the main idea and the purpose of each paragraph. However, the Outlining approach goes a step further, asking you to use your own words to summarize the main points and to record these main points on your scratch paper. To Outline, simply jot down a summary of the purpose of each paragraph as you read. You may find a paragraph's purpose toward the beginning, middle, or end of the paragraph. But in any case, read all the way to the end of each paragraph, and only then pause to record your outline of that paragraph—this is, after all, the Outlining method, not the "note-taking" method! Finally, as with the Highlighting method, when you reach the end of the passage, take a moment to consider why the author wrote the passage. Compared to the fast, lightweight Highlighting method, the Outlining method adds two large pieces of value: first, Outlining allows you to use your own words to clearly and concisely summarize main ideas and themes, meaning you are not relying on the sometimes obtuse language used in the passage. Second, your Outline creates a "table of contents" for the passage, which should make it easier to go back and find information necessary to answer questions. Perhaps surprisingly, you can often find important information more rapidly and more precisely using your outline than you could using highlights. From a timing perspective, the Outlining approach will take you 3–4 minutes per passage.

Interrogating the Passage

The Interrogation method is based on this truth: if you take the time to fully understand why an author has written a passage, and if you reframe the passage content in a context that is relevant to your own life, then you will be extremely competent in answering questions about that passage content. In the Interrogation method, you will approach the passage by questioning why portions of the text were included and how they connect to the rest of the author's argument. Interrogating is active reading. To interrogate the passage is to continually ask "why" and "how" questions as you read. This strategy amounts to a deeper read even than Outlining. When Outlining, you are attempting to capture the main purpose at the paragraph level. But when Interrogating, your goal is to capture the purpose of each individual conclusion, piece of evidence, example, counterexample, and so on. And, as with the other methods, when you reach the end of a passage while Interrogating, you'll stop to consider why the author wrote the passage overall, in light of everything you have read. The Interrogation method sounds time consuming, and it is. But there are two major ways you'll gain time back when you Interrogate a passage. First, unlike the Outlining method, you won't write anything down while Interrogating. You won't need to. The goal of Interrogating is to generate enough comfort with the author's argument that you rarely need to refer back to the passage at all. You just won't need the reference points that Highlighting or Outlining provide. Which leads to the second way that Interrogating saves time. When you have developed the skill of Interrogating, you'll be able to answer most questions confidently and immediately, usually without referring back to the passage at all. Expect that Interrogating a passage will take 4–6 minutes per passage, or peharps longer for the toughest/longest passages you could see on Test Day.

6.2 Highlighting the Passage

Each of these three approaches comes with its own strengths and weaknesses. It is up to you to determine which approach will best fit each passage. The first of these approaches is the Highlighting method, which should enable you to work through the passage in as little as 2–3 minutes.

What Is Highlighting?

The goal of the Highlighting approach is to make it easier for you to locate main ideas and high yield information as you return to the passage to answer questions. As you read the passage, keep an eye out for main themes of each paragraph, as these themes are likely to be the focus of multiple questions. Rather than highlighting as you go, aim to read a few sentences, process the information from them, and then highlight the key phrase or phrases, if there are any. It is important to note that highlighting is for your eyes, not for your brain. You need to fight the impulse to highlight full sentences, long phrases, or every other word, and instead use your highlighting tool sparingly. By highlighting less, your focus will be drawn to keywords and main ideas, making it easier to track down correct answers. Do your best to be stringent with your highlighting, and always remember the mantra "Highlighting everything is like highlighting nothing." One more benefit to targeted and thoughtful highlighting is that it forces you to actively analyze what passage information is worth highlighting, giving you a deeper understanding of the passage as a result.

Finally, although the Highlighting method requires the least amount of time to perform on Test Day, Highlighting does require a good deal of practice beforehand in order to learn how to highlight well.

How to Highlight as a Method

The words and phrases discussed below will be used in passages to depict major changes or unique aspects, which are highly likely to be tested. Remember, in order to easily locate important ideas when you return to the passage, only highlight the first few words of a phrase of interest.

As you work through the passage, the following are MUST highlights:

- **A term that captures the essence of the sentences around it.** For example, *magnanimous* to describe the actions of a victor after an important battle.
- **A detail that departs from the central idea of the paragraph.** For example, a complete change in subject.
- **A word that captures or indicates a position shift.** For example, the introduction of a counterargument.
- **A word that indicates the start of a new major perspective.** For example, the introduction of a new school of thought.

In contrast with the MUST highlights, the following types of words have the potential to be tested, but it is up to you to determine the likelihood and highlight or not highlight accordingly.

The following are MAYBE highlights:

- **Specific, new terms that are defined within the passage.** For example, the word *troika* to mean a group of three people working together.
- **Names and Dates.** Often, names and dates are simply included as supplemental details. Names and dates should only be highlighted if they are a key part of an argument in the passage.

Finally, the following are things you absolutely do NOT want to highlight in the passage. While these suggestions may depart from what you might typically find yourself highlighting as you work through a textbook or academic paper, remember that the goal of Highlighting on the MCAT is to understand the main idea and to make the passage more searchable once you get to the questions.

The following are do NOT highlights:

- **Background information in the first paragraph** if it seems likely that this information will be discussed in greater detail through the rest of the passage. This situation occurs most commonly when the first paragraph is short, serving to merely introduce the topic at hand without providing significant analysis.
- **Anything in the first or last line of a paragraph.** By nature of its location, information in the first or last line of a paragraph is already easy to find, should you need to return to it.
- **Words that have the same meaning or represent the same concept as a previously highlighted word.**
- **Full sentences** or more than one highlight per sentence.

Note that while keywords like "because," "therefore," "argue," etc. indicate that a transition to new information is imminent, the words that come directly after such transition words tend to be most important to the passage. So when you identify these transition keywords in the text, consider highlighting the words that follow in order to capture the unique, highly testable passage details.

There may be instances where you are unable to identify main ideas as you read, or you simply may not understand what you are reading. In these instances, the best plan of attack is to highlight words that will help you locate information quickly when questions ask about it.

As you become a more effective highlighter, you will find yourself able to skim portions of the text and feel confident that your highlighting will allow you to find relevant information without having to reread large parts of the passage.

BRIDGE

Remember, you can still highlight in the other two approaches, when applicable, but the rules you use to determine what should be highlighted will differ from those listed above. With the Highlighting approach, your highlights are meant to draw your eye to main ideas and major shifts. In the other two approaches, highlights can be used to help you isolate specific ideas or details that support the information in your outline or analysis.

When to Choose Highlighting

Advantages

Though the Highlighting method can technically be used for any passage, there are certain passages where Highlighting is the preferred approach. For example, this method is specifically geared toward passages that would otherwise take a large amount of time. Highlighting is a good choice for passages that are very long. It is often the best strategy to use if a passage appears to be very dense, which indicates that an in depth analysis would take far too long to carry out. And because highlighting does not require paraphrasing, it is the perfect choice for passages where you might not completely understand what you are reading.

Highlighting also effectively counters passages with strange formatting. Some passages feature long paragraphs lacking natural breaks or transitions. Because highlights draw your eye back to the information that is worth re-reading, Highlighting can help you navigate such passages. At the other end of the strange-formatting spectrum, Highlighting is also surprisingly effective on passages that feature a large number of unusually short paragraphs! Such a passage would take a very long time to Outline, due to the number of trips to your scratch paper after each paragraph. But the Highlighting method lets you cruise smoothly through such passages.

In summary, Highlighting allows you to utilize the computer-based testing tools to quickly parse long, dense, or strangely formatted passages. This method does not require in-depth comprehension, or that you spend time making notes on your scratch paper.

Disadvantages

However, this approach is not without potential pitfalls. If you do opt to use the Highlighting method on a passage, be aware that Highlighting's rapid pace and lack of paraphrasing can lead to limited understanding of the text. So, for passages that require in-depth analysis (more common among humanities passages), a different approach might be more suitable.

Additionally, because Highlighting relies on the exact wording from the text, this method can make Faulty Use of Detail answer choices seem exceedingly tempting. Therefore, when you use this method, take the time as you work through the questions to investigate the passage and rephrase passage text, before matching to an answer choice. This method requires you to frequently refer back to the text when answering questions, so choosing to Highlight a passage does not mean you're totally off the hook for in-depth passage analysis. You will just do the majority of your in-depth passage analysis as part of your investigation of the questions.

Highlighting Method Example

The passage below has long paragraphs with deep discussions of abstract ideas; as such, the Highlighting method is ideal for this passage. Highlighting will help you to locate major high-yield concepts and to keep track of multiple perspectives. Your aim is to work through the passage as quickly as possible while still capturing some idea of the author's purpose. That way, you will have time to perform the deep analysis of important sections of the passage that some questions require. Remember that highlighting goes hand in hand with active reading. Do not get lazy and just highlight keywords. Instead, read a couple of sentences and then challenge yourself to highlight the word or phrase that accurately describes the main idea of what you just read.

Notice that not every detail is highlighted in the passage below. Rather, only the high yield, likely-to-be-tested ideas are targeted.

HIGHLIGHTING METHOD

Passage	Expert Thinking
Alienation is a fundamental concept in present-day thought about the human being and his/her spatial and social environment. Alienation depends on transformation; without change-for-the-better, alienation cannot exist, and therefore the concept of transformation—and not mere change—must be understood before alienation can be understood. The concepts of transformation and alienation have only been current since the nineteenth century. The theories that form the original concept of alienation are those of Hegel, Marx, and Brecht.	What did the first paragraph reveal about the passage? What are the key terms that we need to know? *Alienation is the theme of this paragraph and possibly this entire passage. However, since it is the first word of the passage, there is no need to highlight it. I need to highlight more important details, like "transformation," which according to the paragraph is critical for "alienation."* *More importantly, the end of the paragraph reveals that the passage will discuss three names. This is a cue that I should be prepared to compare and contrast the three ideas. No need to highlight the names, as the future paragraphs will soon talk about them.*
To Georg Wilhelm Friedrich Hegel, a German philosopher who lived through the French Revolution, alienation was a value-adding process of self-creativity and self-discovery, which is the origin of Hegel's idea of *entfremdung* (alienation and estrangement) as an action to become other than oneself, to enter into what is other than the spirit or to become an alien to oneself. Hegel saw alienation as an experience with positive or value-adding, long-term consequences.	What can you highlight that can portray the most information about Hegel's definition of alienation? *While I wanted to highlight* "entfremdung" *as that is a new term, I know that would not benefit me as much because that term is isolated. The better highlight is the definition: "become other than oneself." I also highlighted "positive" because I know it is important to note the author's perspective.*
To be invited into a desired social or cultural space, which is not the individual's local culture, requires extreme discipline in courtesy, education, and achievement. Hegel calls the self-sacrifice and strenuous effort it requires alienation. Hegel's alienation is the extra effort needed to raise an individual to another and desired level of culture—like the newly acquired social mobility of the social groupings he observed after the French Revolution. Only a few individuals attain alienation consciousness, because appreciating a foreign culture seems remote and unreachable, and even humiliating for the average individual. However, alienation consciousness bridges the gap(s) between cultures. Hegel's alienation is a conscious and intrinsic choice of an individual with a positive outcome for the individual and his/her society and culture over the long term. It is alienation from fellow human beings, which is different from Marx's iconic response to Hegel's theory of alienation.	This paragraph is complicated. How can the details of this paragraph be related to the previous paragraph? *This paragraph shows one application of Hegel's alienation principle. For Hegel, alienation is necessary to move from one culture to another. The phrase "bridges the gap(s)" best captures this main idea, which is why I chose to highlight it. Though other relevant phrases exist, they are too long to concisely capture the same idea. I need the highlights to be efficient so I don't highlight everything and I can quickly locate the information later on.*

HIGHLIGHTING METHOD	
Passage	**Expert Thinking**
To Karl Heinrich Marx, philosopher, economist, and sociologist, alienation was a forced separation of things—not other people—that naturally belong together, or antagonism between those who should be in harmony, with damaging long-term consequences. Marx's well-known illustration describes the production-line worker who is alienated—and deprived—from the products of his labour: profit, satisfaction, feedback, and creativity. Marx attributes alienation to the capitalistic economic system where the worker is a puppet in the hand of the capitalists. The individual workers do not have any choice in what to produce and how to produce and are therefore alienated from the product they produce. Marx's alienation is a negative experience of perpetual deficit.	What is Marx's definition of *alienation*? *I highlighted phrases and words that parallel the highlights in the Hegel paragraph. Unlike Hegel, Marx does not view alienation as positive, and sees it as "forced separation," not a way to evolve as Hegel had suggested.* *Since I am unfamiliar with Marx's philosophy, I also highlighted the detail "alienated from the product," to identify a practical example of Marx's alienation, similar to my previous paragraph's highlights.* *While this may be a lot of highlighting in such a short paragraph, I know these highlights will be impactful because it will allow me to quickly compare the three theories directly.*
Hegel's and Marx's alienation refers to unintentional and gradual processes, whereas Brecht's alienation refers to a deliberate and immediate process. Whether alienation is the result of a gradual or an immediate and deliberate process, the experience is the same: an experience of dissonance and a desire to move toward the old or a new space of comfort. To the playwright, Eugen Berthold Friedrich Brecht, alienation is the distancing and estrangement effect or *Verfremdungseffekt*, when the audience was hindered from simply identifying itself with the actions and utterances of the performers when techniques designed to distance the audience are used or when using innovative theatrical techniques to de-familiarize the familiar in order to provoke a social-critical audience.	Here we are introduced the the last opinion holder. Highlight his definition of alienation as compared to the previous two definitions. *To follow the pattern of the previous highlights, I quickly found the key characteristic of Brecht's definition of alienation. "Deliberate and immediate" is a characteristic distinct from the previous two philosophers' definitions. The rest of the paragraph elaborates on this idea, so I highlighted the term "dissonance," which summarizes it well.*

HIGHLIGHTING METHOD	
Passage	**Expert Thinking**
The Brechtian approach identifies acceptance or reception as the opposite of rejection or alienation. To alienate is, therefore, the deliberate choice of an artist on the reaction of the audience: to receive or to not-receive. He describes it as playing in such a way that the audience was hindered from simply identifying itself with the characters in the play. Acceptance or rejection of their actions and utterances was meant to take place on a conscious plane, instead of in the audience's subconscious. The artist challenges the audience-member therefore to venture into an unfamiliar space and to experience alienation in order to encounter the unknown. Brecht's alienation might be an instantaneous distancing experience, which might have a slow-developing positive or negative long-term effect. To Hegel, Marx, and Brecht, alienation is the process whereby people become foreign to the world they are living in.	What new idea is introduced here? *This paragraph talked about abstract and complicated topics. To stay focused, I stuck with concrete details. I first highlighted "reception as the opposite" to capture Brecht's view of the relationship between acceptance and alienation. The paragraph seems to suggest that the playwright wants the audience to experience alienation, and I highlighted "venture into a unfamiliar space" to capture that idea.* *The last sentence of the passage provides a nice summary of what alienation is to the three authors. I did not need to highlight it because it is at the end and therefore easy to find.*

Adapted from Smit, E., & Nel, V. (2016). Alienation, reception and participative spatial planning on marginalised campuses during transformational processes. *Cogent Arts & Humanities*, 3(1), 1154715. https://doi.org/10.1080/23311983.2016.1154715

Choosing what to highlight and what not to highlight is a learned skill that improves with time and experience. The way to master this skill is to practice frequently and to constantly ask yourself, *Was this actually worth highlighting?* Over time, this habit will fine-tune your ability to gauge what is likely to be tested and what can be written off as "fluff." And this skill will carry over even when you choose a different method for Distilling a passage.

6.3 Outlining the Passage

As we saw in the last section, Highlighting is going to be an approach that is optimal for particularly long or dense passages, or for when you are low on time. Naturally, the Outlining method is the best choice in different situations. The Outlining approach ensures that you have identified the main idea of each paragraph, and of the passage as a whole, before you move on to the questions. This approach is useful because the MCAT requires that you draw conclusions that often reference a bigger picture. Using the Outlining method, you should take about 3–4 minutes to work through a passage.

What Is Outlining?

The goal of the Outlining approach is to extract the most important information from the passage and to summarize that information on your scratch paper. Doing so enables you to better understand the relationships among the many ideas in a passage, as well as how those ideas interlock to form the bigger picture of the passage. As a result, before you even get to the questions, you will already have done some of the work required to answer them!

A second benefit of the Outlining approach is that you can use your outline to identify the specific part of the passage you should target as indicated by a given question. You can think of the outline as a map for the questions. Your outline helps guide you to the information you need.

A good rule of thumb to use when determining whether to use this approach on a particular passage is to ask yourself, *Would I likely benefit from a table of contents on this passage?*

How to Build an Outline

As you read, focus on determining the main idea of each paragraph. Once you reach the end of the paragraph, paraphrase the main idea into your own words and write your paraphrase on your scratch paper. Optionally, you can add key features within the paragraph—major details or examples, key shifts in scope or tone, and so on. Just keep these additional notes short, if you add any at all. Also, keep in mind that your outline is for your eyes only. As long as you can understand it, shorthand and symbols are fair game. And feel free to throw grammar and spelling to the wind!

As you Outline, you do want to mentally note the existence of details, evidence, and examples, but don't spend too much time trying to understand these features of the passage. Use the red/yellow/green light keyword system (Chapter 5) to guide you quickly through the passage. Should a question ask about one of the passage's details or examples, a quick glance at your passage outline will tell you which paragraph to return to for further research. As you read and outline, also be on the lookout for words or phrases that seem more testable than others. Spotting highly testable passage content is certainly a learned skill. But as you complete more passages, you'll have an easier time determining what material is likely to return in the questions and what can be categorized as "fluff."

MCAT EXPERTISE

Since you won't be writing down minor details, evidence, and examples on your scratch paper, these details, evidence, and examples could be a great place to take advantage of the highlighting tool, to make it easier to find them when you return to the passage.

When to Choose Outlining

Advantages

The Outlining approach forces you to identify main ideas in your own words, which helps you predict answers quickly. As you develop the ability to make your outlines concise yet specific, you will find that you can answer many questions using your outline alone, and this will save you time. Additionally, rephrasing the key points while you construct your outline will make you less likely to select Faulty Use of Detail answer choices.

Pausing each paragraph to construct your outline may sound time consuming, but building an outline actually saves time and lowers stress. Outlining helps you move through the passage quickly, enabling you to skim details and examples, rather than spending time deciphering parts of the passage that may not even be referenced in the questions. The Outlining strategy asks you to return to the passage as needed, so you don't have to get stressed if you don't understand every detail in every paragraph, provided you write something in your outline (even just a name or theory) that indicates where to go if a question references it.

Outlining is especially useful for highly detailed or list-driven passages that require you to return to the passage frequently to identify details in order to answer questions. It can also help you keep different points of view organized, especially in those social science passages that describe, compare, and contrast multiple groups' opinions. Organizing details and opinions in this way is particularly effective in passages with a straightforward structure, where one main theme is covered per paragraph. In short, Outlining is usually quite effective for the typical CARS passage.

Finally, if you consider yourself to be a visual person, then having a physical map of the passage can help you organize what you read and keep you from getting lost in the minutia.

Disadvantages

There are certain instances where Outlining may not be the best strategy to use. For example, some authors do not make their main points obvious. Time spent trying to paraphrase a rambling author is often time wasted, which can lead to stress. Likewise, if a rambling author causes you yourself to ramble in your own outline, that's more wasted time! For these reasons, passages with wordy, dense paragraphs often lead to outlines that are wordy and dense. And getting lost in your own outline is certainly even worse than getting lost in the passage! At the other extreme, Outlining a passage with unusually short paragraphs means making many trips to your scratch paper, just to paraphase a small amount of information per paragraph. As we saw in the Highlighting section, for passages with strange formatting—very large paragraphs or unusually short paragraphs—both are better attacked using Highlighting.

Additionally, though some questions can be answered directly from the outline, the Outlining method still requires you to refer back to the text when answering many question types. So, you will need to be very careful to limit the time you spend outlining in order to allow enough time to Investigate those questions that require additional investigation.

Building an outline, if done carelessly, can eat up a significant amount of time. And spending too much time creating one's outline or trying to understand the passage at too deep of a level are common pitfalls of the Outlining strategy. In order to reliably identify when Outlining is the best strategy, you must consistently practice all three methods during your prep.

Outlining Method Example

The passage in this section is an excellent fit for the Outlining method because the passage features a simple structure with a clear argument. Each paragraph is organized around a single goal, and we can identify each paragraph's goal from just the first few sentences. A number of paragraphs also contain Conclusion keywords, helping us identify these paragraphs' main ideas. In addition, the author of the passage avoids using uncommon words, which helps us to understand the author's argument. All of these structural clues allow us to read with a focus on the big ideas, which helps us to develop an efficient outline that captures these big ideas.

OUTLINING METHOD

Passage	Expert Thinking
The world of contemporary art is characterized by a growing number of artists experiencing an entrepreneurial venture. Especially in the context of performing arts, this has been lived both as a necessary and a voluntary solution to the severe shortage of funds affecting the world of public institutions. Thus, a new actor emerges, represented by the artist–entrepreneur who lives a hybridization of roles and competences. But who are these artists–entrepreneurs? And how do they live the possible tensions emerging from the encounter of worlds that have been reputed as radically different for so long?	With a lot of passages, the first paragraph is where we want to start looking for the author's main idea. What is being introduced to us here? *The author introduces the idea of artists-entrepreneurs.* Does the author give us a direction of where we will be going with this main idea? If so, how? *This comes in the form of two rhetorical questions at the end of the paragraph. We are asked about who these artists-entrepreneurs are and how they manage to fulfill such radically different roles at the same time.* P1: art. - entrep -> who and how?
Artists–entrepreneurs' activity can be identified as a particular kind of "cultural entrepreneurship." This practice has been traditionally investigated adopting two main perspectives, based on different meanings of culture.	Paragraphs of this length will typically serve as either a quick transition or a continuation of previous paragraph. What is the author accomplishing here? *The author introduces a new term: "cultural entrepreneurship." The author also mentions that this idea is typically investigated using two main perspectives that depend on the definition of* culture. P2: Cult. Entrep.: -> 2 persp., based on diff defns of culture
As a first meaning, *culture* refers to the sociological frame of reference identifying a set of habits, customs, traditions, and beliefs, which constitute a shared way of life in a specific historical and political context. As a second meaning, *culture* identifies a complex set of processes, products and actors involved in the design, production and distribution of cultural and artistic goods and services.	This paragraph has a pretty clear structure, as indicated by keywords. What is the author doing in this paragraph? *The author provides two definitions of* culture. *The first is that it refers to a sociological phenomenon, and the second refers to a process in design and production of goods.* P3: 2 defns of culture
Descending from the first sociological perspective, cultural entrepreneurship represents "the skill of certain entrepreneurs to use culture as a toolkit for constructing resonant identities and motivating resource-holding audiences to allocate their resources." Therefore, cultural entrepreneurship is instrumentally identified in the process of storytelling that gives shape and legitimizes new ventures. Not referring to a specific industry, the adjective *cultural* is used to identify the process of legitimization that entrepreneurs sustain, giving shape to the story of their personal and professional life.	Using the first sentence of this paragraph as a clue, what is the author introducing to us here? *The author discusses how cultural entrepreneurship can be viewed using first perspective and how it becomes a version of storytelling.* P4: 1st perspec. on cult. entrep. -> storytelling

OUTLINING METHOD

Passage	Expert Thinking
In a second perspective, cultural entrepreneurship identifies a set of processes through which a growing number of artists and cultural professionals assume an entrepreneurial role. Cultural entrepreneurship thus identifies the activity of conceiving, producing and marketing "cultural goods and services, generating economic, cultural, and social opportunities for creators while adding cultural value for consumers." The artists–entrepreneurs combine their artistic attitudes with a deep sense of business, economically sustaining the cultural enterprise in coherence with their cultural vision.	Continuing from the last paragraph and using the first sentence again as a guide, what new information is provided here? *The author now looks at cultural entrepreneurship through a second perspective and how this converts "Cultural goods and services" into economic opportunity for the creator and cultural value for consumer.* How can we outline this without making our statement too long? P5: 2nd perspec. on cult. entrep. -> business
Upon investigation, what emerges with more emphasis is the tough and complex relationship between the artistic and the entrepreneurial dimensions faced by the artists–entrepreneurs. The interdependence between the two spheres emerges as a matter of fact, emphasizing the dual nature of the cultural enterprise. But duality means adopting an integral perspective that results from an intense dialogue between the two languages.	As we come to the last paragraph, we are typically expecting the author to provide some sort of conclusion to the main idea of the passage. What is the author's conclusion? *The author states that the actual result is a combination of two perspectives, a dual nature of cultural entrepreneurship.* Is there any additional information provided by the author? *Yes! The author finishes with the contrast that adopting this dual nature also means adopting a new perspective that encompasses both main views.* P6: Author: result -> comb. of 2 views, but -> new perspec.

Adapted from Clacagno, M., Balzarin, L. 2016. The Artist-Entrepreneur Acting as a Gatekeeper in the Realm of Art. *Venezia Arti*, 25. http://doi.org/10.14277/2385-2720/VA-25-16-3

Effectively outlining passages on the MCAT is not a skill you master overnight. Learning to move through the passage and summarize each paragraph is challenging, so you'll need to practice this method. Many passages in this book will have a sample passage outline so that you can compare your work to that of an expert. Another good drill to improve your outlining skills is to review and re-outline each passage after you have completed and reviewed the questions. In this drill, when you re-outline the passage, you are doing so knowing what ends up actually being tested. Each time you practice this exercise, you will naturally identify important information that you should have included in your original outline, which trains you to spot that type of information the next time you outline a passage.

6.4 Interrogating the Passage

Now that the first two methods have been discussed in detail, you see that different passages call for different approaches, and that each approach comes with its own pros and cons. The final method is the Interrogating approach. This approach asks you to identify whether or not each new piece of information is important and, if so, why that piece of information was included by the author. Many questions specifically ask you to analyze the relationships presented in the passage. So, if you use the Interrogating method, then you will have done most of the work needed to answer the questions before you even reach them, eliminating much of the need to return to the passage. Interrogating a passage should take you about 4–6 minutes.

What Is Interrogating?

The goal of Interrogating, also referred to as **Elaborative Interrogation**, is to understand not only what each important piece of information in the passage means, but also how that piece of information fits in with the rest of the passage. This method is valuable not only for CARS, but as a method validated by learning science for strengthening working memory with any content presented in a similar, text-heavy format. In other words, not only is Interrogating valueable in CARS on the MCAT, but you might find that it is a useful way to approach any content that you must read and understand, including textbooks, science content, and even MCAT science passages. Interrogating is a step beyond the other two methods. The other two methods focused more on *what* the author said rather than *why* the author said it. By contrast, while Interrogating a passage, you will constantly ask yourself, *Why is this information here?* and *How does this information impact what came before?*

How to Interrogate a Passage

The biggest thing to remember as you Interrogate a passage is that you have the okay to spend a lot of time working through the passage, provided that the majority of that time is spent thinking about the passage, rather than just reading it. Remember: Interrogation will put you in a position where you are preemptively generating predictions for many questions that will be associated with the passage, so it's okay to take some serious time to Interrogate!

There are two steps involved in the Interrogation approach. In the first step, you will identify a "chunk" of information as you read. Ask yourself, *Can the sentence I am currently reading be grouped together with the sentence that I just read?* Use your Continuation keywords to identify chunks of the current paragraph that essentially say the same thing or build on the same concept. When you hit a yellow or red light keyword, it's time to assess whether or not you've just reached the end of a chunk. Be careful though! Sometimes these keywords are false alarms. That is, an MCAT author will sometimes use these keywords inappropriately by throwing in such a keyword even when the author doesn't actually intend to transition to a new idea. Even more dangerous: sometimes an author will transition to a new topic with no warning at all! Traps like these are why you must consistently be asking yourself *Can I group the current sentence together with the previous one?* These dangers aside however, a yellow or red light keyword usually means you have reached the end of a chunk.

BRIDGE

Keywords, and the keyword stoplight system, are described in Chapter 5 of *MCAT CARS Review*

Once you have identified a chunk, the second step when Interrogating is to ask yourself what role that chunk plays in the passage. The reasoning behind this step is two-fold. First, asking and answering questions as you go makes you an active reader, which makes it easier for you to remember the information in the passage. Second, forming an answer to a *Why?* question helps you contextualize and mentally group the information in the passage, rather than thinking of each sentence as unrelated to the ones around it.

MCAT EXPERTISE

Most of the time in CARS, the focus is on avoiding using outside information to answer questions. However, in the Interrogating method, asking about how information in a passage relates to other things you know about the topic or about the author can be a helpful way to contextualize and remember what's going on in the passage. The key to bringing in outside thoughts is to only use them to better understand what's happening in the passage generally, and to help you remember key themes. You do not want to use that outside information to directly answer questions, as that would be Out of Scope!

Listed below are some examples of the types of questions you should ask yourself as you Interrogate:

- Why is this piece of information being used in this way in the argument?
- Why is this piece of information important?
- How does this relate to other information the author has shared?
- How does this chunk differ from the previous chunk?
- Why did the author choose to reference this example?
- How does this support or refute a claim made previously?
- How can I relate this content to other related content, or even to the outside world?

When to Choose Interrogating

Advantages

Once you master this approach, you will no longer need to return back to the passage for most questions. As a result, even though this strategy consistently requires the most amount of time to work through the passage, you will be able to answer questions more rapidly since you will have already performed the critical thinking required by higher-level CARS questions. Questions that may have seemed daunting without this method will be very approachable since, technically speaking, once you reach the end of the passage, you have already identified the main idea, processed the supporting information, and evaluated new ideas as they arose. The end result of this active analysis is that you've effectively already predicted the answers to most questions.

BRIDGE

When you use the Outlining approach, at the end of each paragraph, you should be asking yourself *What?* questions—*What was the point of this paragraph?*, *What did the author say?*, etc. The Interrogation approach takes that a step further, making you ask *How?* and *Why?* and demanding that you think about the purpose of each chunk and the connections between them as you read.

By asking and answering questions as you Interrogate the passage, you also build a far stronger memory of the passage. To borrow a familiar example from the sciences, the difference between Interrogating and Outlining is similar to the difference between knowing *why* a nucleophile attacks an electrophile, rather than only memorizing *that* it occurs. When you understand *why* phenomena occur, you are much more likely to remember them. And so it goes with CARS. When you understand *why* the author felt she needed a certain example, you are much more likely to remember that example. As you improve your elaborative interrogation skills, you'll find that you can easily recall major ideas, destructive counterarguments, important examples, and often even small, random details... at least for a long enough time to answer the passage's questions!

Unlike the other two methods, Interrogating does not require you to spend time making use of the computer-based testing tools or writing on your scratch paper. So, if you are uncomfortable with either resource, Interrogation can be a solid alternative.

Finally, this approach is most ideal for tackling shorter passages, and passages that you triage as Easy or Medium difficulty. In particular, social sciences or humanities passages with a large number of inferences or logic-based arguments, especially on topics like law and philosophy, tend to be ideal for this approach since they contain a large number of relationships and subpoints.

Disadvantages

Some students pick up the Interrogating technique rapidly, or find that it is already similar to their natural style of reading. For most, however, learning to use this approach can be frustrating, especially at first. Undoubtedly, even for experienced test takers, this method takes the longest out of the three Distill approaches, and it also requires the most substantial amount of analysis and critical thinking. So for those first learning this approach, there are many potential pitfalls early on. You may find that the passage is harder to understand than you expected, and now you're trapped in an over-long interrogation. Or, you might get to the questions and realize that you spent too much time focusing on parts of the passage that are not tested.

It should come as no surprise, then, that becoming a good interrogator is not a quick journey. But remember, the more you practice, the more natural it will feel to constantly be asking yourself *How?* and *Why?* And as you practice more with the method, you will develop a better sense for timing and for what is testable. Even an expert interrogator will likely not use Interrogation for every passage they see. So be sure to practice the Choose step and select the Interrogation approach only when it is your best option.

Avoid using this approach on passages that seem incredibly complex or long, and on passages that appear to be long lists of details or information. In particular, for an incomprehensible passage, the Interrogation method will likely take more time than you can afford.

Interrogating Method Example

A quick glance at this passage reveals that it is a social science passage. The use of personal pronouns reveals that the author is personally invested, making it likely we will see many opinions and arguments in this passage. That observation alone makes the Interrogating method a great choice.

INTERROGATING METHOD	
Passage	**Expert Thinking**
Bauls are a religious group living primarily in the rural areas of West Bengal (India) and Bangladesh. They come from a variety of backgrounds, including Hindu and Muslim, and span the castes, though most are low caste. Although their roots extend more deeply, Bauls have been around at least a century. In practice and belief, they have been influenced by local traditions of the more orthodox Gauriya (Bengali) Vaishnavism, the Tantric-influenced Buddhist and Hindu Sahajiyas, and Sufism. However, unlike many other religious groups, Bauls intentionally reverse a number of orthodox practices. They argue that the Divine is within all humans and thus people should respect and worship humans instead of going to the temple, mosque, or church to worship something that cannot be seen. Bauls sing and compose songs that critique societal divisions and allude to their philosophy and practice. They spread their messages door to door, on trains, and in performances at large public venues.	S1–8. Why has the author decided to write about and focus on the Bauls? *The author is focusing on the unique aspects of the Baul religious group, and we can infer that the author finds the Bauls to be particularly interesting because of those unique traits. Specifically, this group is unique because they show influence of other religious and ethnic groups, but they have opted to reverse a number of traditional practices. We can infer that the author is going to continue to talk about what makes this group unique, and that describing the Baul religious group will be the purpose of this passage.*
I sought the lived experiences and perspectives of Baul women. In [the] first 18 months of work, I carefully framed how I presented my research to my interlocutors in order to avoid the usual pitfalls of my predecessors, such as having to exchange money or prestige for information, taking initiation, or losing integrity in my research methods in other ways. [My interactions with the Baul] bring to the forefront the postcolonial predicament concerning the relationship between ethnographers and "informants," in which the ethnographer collects "data," which is then analyzed and presented for academic scrutiny.	S1. Why is the author using "I" and "my"? *This author is describing their personal experience with people of the Baul religion. Within this context, we can infer that the passage is likely going to contain more tone and opinion than the average MCAT passage, and can stay vigilant looking for those Tone keywords, as there are likely to be questions about tone.* S2. Why does the author talk about "avoiding the usual pitfalls"? *Implicit in this statement is that falling into pitfalls might skew the "lived experiences and perspectives" that the author is seeking in the first sentence. The author is trying to convey a value judgment about previous work in this field, and set their work apart as having more validity.* S3. Why does the author bring up the "postcolonial predicament," and how does it relate to their argument? *The author is continuing to establish a contrast between themselves and researchers previously integrated with this group. While it is unclear exactly how the author has overcome this predicament, it appears that they carefully considered the problem and they feel that their first 18 months (from the beginning of this paragraph) helped them avoid the issues experienced by others.*

INTERROGATING METHOD	
Passage	**Expert Thinking**
I have argued elsewhere that as they traverse a Baul path, those who take the teachings seriously gradually adopt cognitive and spiritual models that shift their own understanding of the world around them. Thus, they learn (or aim to learn) to recognize the ways in which society creates divisions that lead to discrimination, and to recognize the divine in all human beings, regardless of caste, gender, or religion. By listening to their explanations about the micro- and macrocosmos, by hearing their songs, and by traveling with them, they expected that I too would experience these shifts in understanding. The cultural specifics of their experiences may not carry home, but elements of hierarchy, discrimination, and the inherent value of all human beings are as real and important in my American communities as they are in their Bengali ones. For Bauls, issues of hierarchy and domination are religiously meaningful, and knowledge of these realities constitutes important aspects of their religious experiences.	S1–5. How did the Baul expect the author to respond to their religion? How did the author actually respond? *The Baul believe that those who learn about their religion will shift their understanding. The author states that this happens for those who take the teachings seriously and "traverse a Baul path." As a researcher, the author is not in the position of someone who is looking for spiritual guidance, but rather is looking to collect data. The researcher does seem to have identified values and features in the community that they recognize parallels for in their own home country. We don't yet know exactly how the author responded, but it seems like at least some of the Baul messaging was effective.*
In these and other ways, the Baul women I worked with expected me to participate in their world, particularly when they knew I shared their views. Although moments of connection have shifted my own perspectives, [a specific Baul woman's] critiques ring loudest. In demanding that I (and the many Bengalis and foreigners who enjoy her singing) support her, she refuses to let me be a complacent ethnographer. She demands dignity, and she is right to do so. Several years later, her proclamations are forcing me to rethink my ethnographic work and to reveal, as a tentative step, that I have been moved, that I have taken their words and lifeworlds seriously.	S1–4. Why does the author mention the specific Baul woman's critique? *The woman's critique is a call to action to the author to become an active participant in the religion, rather than simply an observer. This particular example is presented as a piece of a larger argument, that the author of the article feels they must actively participate, rather than observe, and so the demands that the author do just that resonated strongly. This serves in contrast to paragraph 2, where the author describes the research and role of the ethnographer. The author's position is now becoming more clear: they have in fact changed and responded to the information they gathered when researching the Baul, which is what the Baul expected in paragraph 3.*

INTERROGATING METHOD	
Passage	**Expert Thinking**
Maybe language is inadequate to explain religious experience. Bauls would certainly say so, since they insist that one can only trust and know what is personally experienced. Perhaps, then, I should acknowledge moments of connection. For instance, when [the Baul] and I discussed our views of the world, its beauty and faults, and our struggle to make sense of suffering and find ways to improve the lives of those discriminated against. Maybe it's enough for me and my interlocutors to share meaningful experiences and conversations, to be open to being inspired and transformed in the field and also back home, and to be willing to put some of those ideas into action.	S1–4. Why does the author conclude by writing about explaining religious experiences? *The final paragraph holds that language isn't enough to explain experience, but connecting to others and sharing experiences is. This directly supports the position the author has been building in previous paragraphs, as they argued against a detached, traditional approach to ethnographic research and validated their own choice to take a more active role in the community. The author talks about their research as more of a personal experience than as something to generate data, and concludes that data collection via traditional methods is not sufficient for understanding the beliefs of some religious groups.*

Adapted from Knight, L. I. (2016). "I Will Not Keep Her Book in My Home": Representing Religious Meaning among Bauls. Asianetwork Exchange: A *Journal for Asian Studies in the Liberal Arts*, 23(1), 30–46. DOI: http://doi.org/10.16995/ane.159

Now that you've seen all three methods, remember that there is no "one size fits all" when it comes to MCAT passages. As you Preview and Choose, think about which method will allow you to work through the passage quickly and accurately. But don't be afraid of choosing the wrong method as you start out. All things are hard before they are easy. And as you practice choosing a Distill method for each passage, you'll find you often learn as much or more from your wrong choices as you do from your right choices! The key is to make many mistakes now, while it's safe to make then, so that you have mastered this skill by Test Day. And finally, know that what someone else may consider the best approach for a passage may be different from your judgement. Ultimately, the best choice for you is the one that earns you the most points you can get. So, on each individual passage, choose the method that works best for you, based on your history of successes (and mistakes!) practicing all three methods.

Conclusion

Each of the three Distill methods discussed in this chapter provides you with a way to effectively navigate each passage while reading strategically. Each strategy has its own pros and cons, and each is particularly well suited for certain situations. But let's make one final connection between the three. Perhaps as you read the description of each method, you observed that each method trains you in its own way to become a better CARS test-taker overall. Highlighting trains you to rapidly identify testable information, even when you're not Highlighting. Outlining trains you to quickly summarize the main point of each paragraph, even when you're not Outlining. And Interrogating trains you to analyze the author's arguments and examples, even when you're not Interrogating. So it is imperative that you do not utilize a "one size fits all" approach. Learning the three different methods will expand your playbook on Test Day, and it will make you a much stronger test taker along the way. The following chapter will help you further develop your ability to choose the right method for each passage and show you how each strategy can ultimately be leveraged to help you effectively answer the questions.

Concept and Strategy Summary

Choosing Your Approach

- As you triage, use things like passage length, topic complexity, and sentence structure to choose your Distill approach.
- As you work through the passage, aim to identify main ideas and key points using either the **Highlighting**, **Outlining**, or **Interrogating** approach.
- Use Contrast and Logic keywords to help you identify regions of the passage that contain high-yield, testable information and note them down accordingly, based on your approach.

Highlighting the Passage

- Highlight the main ideas in each paragraph so they are easier to find as you work through questions.
- Use your highlight tool sparingly. Avoid highlighting full sentences or wordy phrases.
- Keep an eye out for keywords that indicate structural shifts or the author's opinion and highlight a word or phrase that sums up the ideas that follow.
- Use this method on:
  - dense or complex passages that you will want to work through quickly (2–3 minutes)
  - passages that you have triaged until the end of the test
  - passages without natural transition points
- Avoid using this method on passages that require in-depth analysis (usually humanities passages).

Outlining the Passage

- Determine the main idea of each paragraph, rephrase it, and note it down on your scratch paper.
- Avoid getting caught up in details, evidence, and examples as you read.
- The act of rephrasing main ideas helps you better internalize them and read more quickly (3–4 minutes)
- Use this method for:
  - passages for which having a table of contents would be highly useful
  - highly detailed or list-driven passages
  - passages where multiple opinions are expressed and discussed
  - passages that you do not completely understand (simply jot down a relevant name or theme on your scratch paper and move on)

- Though the Outlining method is intended to help you when you come across a passage you do not fully understand, avoid using this method if you find yourself getting stressed over forming a good passage label or that you are spending too much time doing so.

Interrogating the Passage

- Identify a "chunk" of information within the text that essentially says the same thing or builds on the same idea.
- Then ask yourself questions about the role that the chunk plays in the passage. Example questions include:
  - How does this impact what I have read so far?
  - Why was this detail included here?
- Answering questions as you go makes you a more active reader and helps you give context to the information in the passage.
- Use this method for:
  - shorter passages
  - passages you triage to be of either Easy or Medium difficulty
  - passages with a large quantity of inferences or logic-based arguments, like those in the social sciences
- Since this method takes 4–6 minutes, avoid using it on long or complex passages, and on passages that are composed of lists of details or information.

Worked Example

Use the Worked Example below, in tandem with the subsequent practice passages, to internalize and apply the strategies described in this chapter. The Worked Example matches the specifications and style of a typical MCAT *Critical Analysis and Reasoning Skills* (CARS) passage. Note that this is the first Worked Example in the book, and for this chapter alone, all three methods have been demonstrated on the same passage to better display how each method could be applied.

HIGHLIGHTING METHOD	
Passage	**Expert Thinking**
One of the more well-known female writers to adopt a pen name was George Sand, born Aurore Dupin, who became one of the most prolific and admired French authors during the nineteenth century. The true identity of George Sand did not remain a secret for long, for the author used this name in her everyday life, and close friends commonly referred to her as "George."	From the first paragraph, what can you tell about what the rest of the passage will be about? What should you highlight to capture that? *This paragraph is just a short introductory paragraph to the subject and doesn't necessarily have any term that jumps out at me. However, I highlighted "not remain a secret" to capture the essence of having an obvious pseudonym.*
Most portraits of the author as an adult are simply George Sand and make no reference to her given name. Her son, too, adopted this new last name, even though his association with his famous author–mother did not bring him any obvious benefits. Given the name "George Sand" is radically different from Aurore Dupin's birth name, many readers have wondered how the author formulated her masculine pen name.	What is the purpose of this paragraph? What should we highlight to show that? *This paragraph mentions several details in the beginning, but I did not see it as particularly high yield. All that serves as an introduction to the idea of how deeply ingrained the name is associated with the female author. I highlighted "wondered" because the question of why she chose this name to be associated so intimately with her life is an idea the rest of the passage will likely explore.*
At least two possible answers spring to mind. The first, as indicated in Curtis Cate's biography *George Sand*, is that the pseudonym arose from a collaboration with her first lover, Jules Sandeau, with whom she co-authored several articles as well as a full-length novel entitled *Rose et Blanche*. On the advice of their publisher, the authors signed this latter work under the name "J. Sand." Once Aurore's writing began to overshadow that of Jules, she decided to sign her solo works as "Georges Sand," which eventually became simply "George Sand." Because her own literary output was a great success, she quickly became known by this name and began to use her pen name on a daily basis.	Here, the different theories are introduced. How should we highlight to efficiently capture these theories? *The first theory revolves around collaboration, so I chose to highlight that term. In addition, I highlighted "overshadow" because it is another detail of this theory not captured by the previous "collaboration." I know that there are other details, but I need to be careful to not over-highlight the paragraph, and only choose the terms that can capture the most about the details around them. Also, I know while words like "first" are critical indicators for important ideas, highlighting that word won't actually provide me with useful information.*
By continuing to use the name initially assigned to collaborative writings with her lover, perhaps Aurore hoped to maintain their connection. Perhaps she fondly remembered their time together and wished to have a permanent reminder. Or perhaps she simply realized that it would be much more expedient to continue to write under a name which was already familiar to her audience thanks to their joint works.	What did you notice about the shift in the author's voice? What should you highlight in this paragraph? *Tone is important to note in a CARS passage. In this paragraph, the author starts to discuss the hidden motives for why Dupin decided to keep the masculine name. Since the previous paragraph did not address this, I need to capture the ideas behind this. I highlighted "connection" and "familiar" as they each respectively capture one motive and parallel previous paragraphs' highlights.*

HIGHLIGHTING METHOD	
Passage	**Expert Thinking**
Given that George Sand began writing under this masculine name around the same time as she began to roam around Paris in pants and a jacket—typically male clothing—it is not hard to understand why she chose a masculine pseudonym because, like her clothes, this male identity gave her more freedom of expression, both literally and figuratively. And once she became known as a successful author under this name, there was no reason to change. Writing under a false name allowed her to distance parts of her character—her roles as wife, mother, and lover—from the creative and literary parts that formed her role as an author. Using a male name set her apart and added to her persona as an unusual and fascinating woman. In the end, the reason why she chose this particular pen name is not nearly as important as the vast quantity of writing—articles, letters, novels, and plays—that forms her legacy to the field of French literature.	What is this paragraph about? *Although the transition is quite subtle, I noticed that this is the other theory that the author mentioned previously. This theory revolves around how the male identity was actually beneficial to Dupin's career. "More freedom" captures this notion. I did not highlight "male identity" because that would not give me more information than what the previous paragraphs have already established.* *I also highlighted "roles" to capture the idea that Dupin had two roles that she kept separate. This is a good detail, as it embodies a new idea that goes further than simply having more freedom.* *Furthermore, "legacy" is worth highlighting as it reveals the author's opinion of Dupin. The MCAT loves to test the author's opinions!*
The name could have a more symbolic meaning as well which would give more deserved credit to the author herself. Taking each letter of "SAND" as an allusion to names, places, or people from Aurore's life, this name can be seen as a representation of Aurore's childhood and early married life. Even if George created the name, however, she was well aware of the similarity to her lover's name, and was equally aware that many readers would make this connection. As an intelligent and perceptive woman, she recognized that such an association with a male author would help to validate her early writing career before she had succeeded in establishing her own reputation as a talented and publishable author.	In this last paragraph of the passage, what was the author's message? *This paragraph seems to talk about yet another theory, although not as developed. I highlighted "representation" to capture the essence of the symbolic meaning theory of her pen name. I also chose to highlight "validate" because it captures the practical aspect of having a masculine pen name, similar to what is mentioned in the previous paragraph.*

OUTLINING METHOD

Passage	Expert Thinking
One of the more well-known female writers to adopt a pen name was George Sand, born Aurore Dupin, who became one of the most prolific and admired French authors during the nineteenth century. The true identity of George Sand did not remain a secret for long, for the author used this name in her everyday life, and close friends commonly referred to her as "George."	The first paragraph will often give us a hint toward the author's main idea. What seems to be the author's focus? *The author introduces a writer with the pen name "George Sand" who was actually Aurore Dupin.* Are there any keywords to tell us how author feels about the Aurore Dupin? *Yes! "... one of the most prolific and admired French authors..."* How can we express all this information in a concise format? P1: George Sand = Aurore Dupin → +ve tone
Most portraits of the author as an adult are simply George Sand and make no reference to her given name. Her son, too, adopted this new last name, even though his association with his famous author-mother did not bring him any obvious benefits. Given the name "George Sand" is radically different from Aurore Dupin's birth name, many readers have wondered how the author formulated her masculine pen name.	While this is another short paragraph, make sure not to glaze over it. How does the author further narrow the main point of the passage? *The author is wondering how the pen name "George Sand" came to be, as well as why it was more popular than writer's real name.* What would be the best way to briefly express this? Note we can abbreviate names by their initials! P2: G.S. > A.D. → How was G.S. created?
At least two possible answers spring to mind. The first, as indicated in Curtis Cate's biography *George Sand*, is that the pseudonym arose from a collaboration with her first lover, Jules Sandeau, with whom she co-authored several articles as well as a full-length novel entitled *Rose et Blanche*. On the advice of their publisher, the authors signed this latter work under the name "J. Sand." Once Aurore's writing began to overshadow that of Jules, she decided to sign her solo works as "Georges Sand," which eventually became simply "George Sand." Because her own literary output was a great success, she quickly became known by this name and began to use her pen name on a daily basis.	Following the logic of previous paragraphs, what are we expecting to find in this one? *The answer to the question of how G.S. name came to be. The writer originally used the name "Georges Sand" in connection to her first lover, which later evolved into "George Sand".* While there is a lot of detail in this paragraph, we still want to come up with a short statement to outline it. P3: G.S. → from first love

OUTLINING METHOD

Passage	Expert Thinking
By continuing to use the name initially assigned to collaborative writings with her lover, perhaps Aurore hoped to maintain their connection. Perhaps she fondly remembered their time together and wished to have a permanent reminder. Or perhaps she simply realized that it would be much more expedient to continue to write under a name which was already familiar to her audience thanks to their joint works.	Based on the first line of the paragraph, what can we expect to see in the rest of it? *Since there is a Continuation keyword present, we are expecting further development of the idea in the previous paragraph and why the writer kept the name G.S.* P4: G.S. kept name → familiar
Given that George Sand began writing under this masculine name around the same time as she began to roam around Paris in pants and a jacket—typically male clothing—it is not hard to understand why she chose a masculine pseudonym because, like her clothes, this male identity gave her more freedom of expression, both literally and figuratively. And once she became known as a successful author under this name, there was no reason to change. Writing under a false name allowed her to distance parts of her character—her roles as wife, mother, and lover—from the creative and literary parts that formed her role as an author. Using a male name set her apart and added to her persona as an unusual and fascinating woman. In the end, the reason why she chose this particular pen name is not nearly as important as the vast quantity of writing—articles, letters, novels, plays—that forms her legacy to the field of French literature.	This is another long paragraph, but focus on what new idea is being introduced. *The author provides further justification why the writer chose a more masculine pen name. There are multiple reasons stated and we do not want to try and write them all down in the outline. Just specifying where we can find them is enough.* P5: Further reasons to keep G.S.
The name could have a more symbolic meaning as well which would give more deserved credit to the author herself. Taking each letter of "SAND" as an allusion to names, places, or people from Aurore's life, this name can be seen as a representation of Aurore's childhood and early married life. Even if George created the name, however, she was well aware of the similarity to her lover's name, and was equally aware that many readers would make this connection. As an intelligent and perceptive woman, she recognized that such an association with a male author would help to validate her early writing career before she had succeeded in establishing her own reputation as a talented and publishable author.	Thinking back to paragraph 3, we are still waiting for a second possibility for where the name George Sand came from. And this is exactly what this last paragraph gives us. What is that reason? *The writer perhaps picked the name as homage to people and places from her life.* Is there any additional information the author provides us with? *Yes! We are also told that even if this was the reason for picking the pen name, G.S. knew that readers would still think of connection to writer's first lover.* How can we best outline this paragraph and still keep it short? P6: 2nd reason for G.S. → connec. to prev. reason

INTERROGATING METHOD	
Passage	**Expert Thinking**
One of the more well-known female writers to adopt a pen name was George Sand, born Aurore Dupin, who became one of the most prolific and admired French authors during the nineteenth century. The true identity of George Sand did not remain a secret for long, for the author used this name in her everyday life, and close friends commonly referred to her as "George."	S1–2. Why is the author discussing the different names of the writer George Sand? *The writer, Aurore Dupin, is female, but used a male pen name (George Sand) in the 19th century. I know it was common for women writers to use male pen names at this time because it was not appropriate for women to be writers at this time.* S2. Why is it important to note that the true identity wasn't a secret? *If her true identity was no secret, then she must not have been using it to mask her gender. I expect the author will go on to explain why she used this pen name.*
Most portraits of the author as an adult are simply George Sand and make no reference to her given name. Her son, too, adopted this new last name, even though his association with his famous author–mother did not bring him any obvious benefits. Given the name "George Sand" is radically different from Aurore Dupin's birth name, many readers have wondered how the author formulated her masculine pen name.	S1–2. Why does the author use the example of Sand's son taking her pen name? *The passage says that the son gained no benefit from taking her name and that the author is almost never referred to by her birth name, so it seems that this pen name was more of an identity to the writer than an attempt to disguise her gender. This makes sense based on the last paragraph.* S3. Why does the author mention that many readers wonder about the origin of Sand's pen name? *The last two paragraphs discuss how this pen name was part of the writer's identity and was clearly important to her. I expect that in the next few paragraphs, the author will discuss possible reasons for how Sand came up with her pen name that might explain why the name was so important to her.*
At least two possible answers spring to mind. The first, as indicated in Curtis Cate's biography *George Sand*, is that the pseudonym arose from a collaboration with her first lover, Jules Sandeau, with whom she co-authored several articles as well as a full-length novel entitled *Rose et Blanche*. On the advice of their publisher, the authors signed this latter work under the name "J. Sand." Once Aurore's writing began to overshadow that of Jules, she decided to sign her solo works as "Georges Sand," which eventually became simply "George Sand." Because her own literary output was a great success, she quickly became known by this name and began to use her pen name on a daily basis.	S1. Why is it important to pay attention to the fact that there are two possible answers? *Two answers means that we can anticipate the majority of the rest of the passage will be structured around those answers to the question posed in the last paragraph.* S2–5. Why does using Jules Sandeau's name justify the pen name? *Jules Sandeau is Aurore's former lover, with whom she coauthored some works. The pen name could be a derivative of her former lover's name.*

INTERROGATING METHOD	
Passage	**Expert Thinking**
By continuing to use the name initially assigned to collaborative writings with her lover, perhaps Aurore hoped to maintain their connection. Perhaps she fondly remembered their time together and wished to have a permanent reminder. Or perhaps she simply realized that it would be much more expedient to continue to write under a name which was already familiar to her audience thanks to their joint works.	S1-3. How do the claims in this paragraph link to the previous paragraph? Why is the author not sure about the reasons? *There are three possible reasons (each signaled by a "perhaps") for why Aurore took on that specific name. It's clear that Aurore did not state the reasons for the pen name, so the author is using historical and contextual clues to come up with reasons.*
Given that George Sand began writing under this masculine name around the same time as she began to roam around Paris in pants and a jacket—typically male clothing—it is not hard to understand why she chose a masculine pseudonym because, like her clothes, this male identity gave her more freedom of expression, both literally and figuratively. And once she became known as a successful author under this name, there was no reason to change. Writing under a false name allowed her to distance parts of her character—her roles as wife, mother, and lover—from the creative and literary parts that formed her role as an author. Using a male name set her apart and added to her persona as an unusual and fascinating woman. In the end, the reason why she chose this particular pen name is not nearly as important as the vast quantity of writing—articles, letters, novels, plays—that forms her legacy to the field of French literature.	S1–5. Why would having a male identity allow for more freedom? *After Aurore adopted the pen name, she also wore male clothing, further enhancing the male identity. Given that this is the 19th century, females are typically known to have less freedom of expression and movement. Additionally, the male name allowed her to separate her creative life from her other roles.*
The name could have a more symbolic meaning as well which would give more deserved credit to the author herself. Taking each letter of "SAND" as an allusion to names, places, or people from Aurore's life, this name can be seen as a representation of Aurore's childhood and early married life. Even if George created the name, however, she was well aware of the similarity to her lover's name, and was equally aware that many readers would make this connection. As an intelligent and perceptive woman, she recognized that such an association with a male author would help to validate her early writing career before she had succeeded in establishing her own reputation as a talented and publishable author.	S1–4. How does this information structurally fit with the earlier paragraphs? Why does the author spend very little time on this reason? *This is the second of two reasons for how Aurore came up with George Sands as a pen name (P3). The author postulates that SAND can be an acronym, but doesn't discuss what that acronym can be. It's clear that the author believes the first reason (association with her former lover's name) has a lot more support, as indicated by the last sentence of this paragraph.*

Question	Analysis
1. According to the passage, the following were all possible reasons for George Sand to create a pseudonym EXCEPT:	The phrase *according to the passage* indicates that this is a Detail question; the word *EXCEPT* makes it a Scattered Detail question, and we are looking for what isn't mentioned. We will need to reference the passage and use process of elimination to eliminate all of the answer choices that were mentioned in the text.
A. she began publishing collaborative works with Jules Sandeau.	We can eliminate **(A)** because it was mentioned in paragraph 3 where the author discussed Cate's theory.
B. her new name reflected important parts of her life.	**(B)** was described in the final paragraph where the author mentioned that each of the letters in SAND could represent allusions to parts of Dupin's life.
C. she was not able to publish any works under her own given name.	**(C)** is the correct answer because the author never mentioned that Dupin was unable to publish under her given name. While the author states that her pseudonym may have helped validate her writing and provided greater freedom of expression, no direct evidence is given that Dupin could not publish any works under her name.
D. the works published under her pen name sold well.	**(D)** was mentioned in paragraph 3 where Dupin had published previously with her lover under the name of "J. Sand," adopted the name "George Sand," and published successfully.

Question	Analysis
2. With which of the following statements would the author most likely agree?	Because this question is asking us what the author would most likely agree with, we are looking at an Inference question. We will want to make a general framework for what we expect the correct answer to look like based on what the author has discussed. The pen name "George Sand" had multiple possible origins—a connection to an old lover or a more symbolic meaning—but regardless of the origin, the pen name gave Dupin more freedom of expression as discussed in paragraph 5.
A. Aurore Dupin should have written works under her own name once the secret of her pseudonym was revealed.	**(A)** is an Opposite. The author stated that *once she became known as a successful author under this name, there was no reason to change*, so the author would not think she should have changed her name.
B. By writing under a pseudonym, George Sand created for herself a new identity that allowed her to transcend the limitations of society.	**(B)** closely matches the prediction, highlighting the freedom Dupin's pseudonym gave her. This is the correct answer.
C. George Sand owed her early success to her partner, Jules Sandeau.	**(C)** is a Distortion that misconstrues Dupin and Sandeau's literary output. While Dupin began her writing career by collaborating with Sandeau, the author does not imply that her early success was due to him.
D. The choice of a masculine pseudonym was restrictive for George Sand and forced her to live as a man throughout her life.	**(D)** is both a Distortion and an Opposite. Although the pseudonym "George Sand" certainly was masculine, it did not force Dupin to live as a man; she wore pants and a jacket of her own free will. Furthermore, the masculine pseudonym was freeing, not restrictive.

Question	Analysis
3. The author mentions Curtis Cate in order to:	This is asking us why the author mentioned a particular example, so it is a Function question. We will need to refer back to the passage where Curtis Cate is discussed. The author refers to Cate in paragraph 3. In this paragraph, the author mentions that there are at least two possible reasons why Dupin chose the name George Sand and introduces Cate's theory as one possible reason. The author does not say anything negative about Cate's theory, so we expect the tone of the correct answer to be neutral.
A. refute his claims about the reason for Aurore Dupin's choice of a male pseudonym.	**(A)** is a Distortion because the author never refutes Cate's claims. While the author does present another possible reason, the author does not imply that Cate's theory is invalid.
B. provide support for a plausible explanation of the creation of Aurore Dupin's pseudonym.	**(B)** matches the prediction, both in terms of content (*creation of Aurore Dupin's pseudonym*) and tone (*provide support*). This is the correct answer.
C. advocate the reason for Aurore Dupin's pseudonym as presented in this particular biography.	**(C)** is also a Distortion because the author does not suggest that Cate's theory is the best; the author just mentions it as one of many possible reasons. *Advocate* is too strong a word to describe the author's writing.
D. show that biographers do not always write accurately about their subjects.	**(D)** is Out of Scope because the author never makes any claims about the accuracy of biographers on their subjects.

Question	Analysis
4. According to the passage, there was widespread use of the pseudonym "George Sand." Based on the points the author brings up, which of the following is NOT proof of this widespread use?	The question stem contains the phrase *according to the passage*, which would usually indicate a Detail question. However, the question is asking us to identify evidence (related to logic), so this is actually a Function question. The word *NOT* means this is a Scattered Function question. We will need to use the passage to find the pieces of evidence and eliminate them. The author discusses the widespread use of the pseudonym "George Sand" mainly in paragraphs 1 and 2.
A. Members of her family used part of her pseudonym for themselves.	**(A)** can be eliminated because the author stated that Dupin's son adopted the last name Sand in paragraph 2.
B. Aurore Dupin's lovers and close friends called her "George."	**(B)** can also be eliminated because the author mentioned that many close friends referred to her as "George" in the last sentence of paragraph 1.
C. Portraitists and the general public knew her predominantly by her pen name.	**(C)** was also mentioned in paragraph 2, where the author stated that portraits of the author went by the name "George Sand" and didn't mention the author's given name.
D. Early book reviews of her works never referred to her given name.	By process of elimination, the answer must be **(D)**. The author did not mention early reviews of her books, so **(D)** was not used as evidence.

Question	Analysis
5. Regardless of the facts set forth in the passage, the author implies that the second possible reason for George Sand's pen name is:	The word *implies* in the question stem indicates that this is another Inference question—specifically, an Implication question. The first theory, Curtis Cate's, predominates in paragraphs 3 and 4, whereas the second theory, SAND as an acronym, is presented in paragraph 6. The author introduced this second possible theory as one that *g[a]ve more deserved credit* to Dupin, suggesting that the author somewhat agrees with this theory.
A. appealing because it demonstrates Dupin's creativity and independence.	The word *appealing* in **(A)** matches the author's positive opinion of this theory. The use of an acronym (and a pseudonym at all) would be good evidence of *Dupin's creativity and independence*, which the author already praised in paragraph 5. This is the correct answer.
B. equally plausible as the first reason even though it has no relevance to the Dupin's family.	**(B)** is an Opposite wrong answer choice, as the second reason states that the name originated from *names, places, or people* in Dupin's life.
C. too sentimental for such a rational and innovative writer.	**(C)** is a Distortion because the author believes that Dupin, as an "intelligent and perceptive woman," would recognize the benefit of using the pseudonym—rather than shying away because it seemed *too sentimental.*
D. based on reading she did during her childhood and early married life.	**(D)** is Out of Scope. The author does not mention any references to books that Dupin read.

Question	Analysis
6. According to information put forth by the author within the confines of the passage, George Sand's male pen name and her choice of clothing are related because:	The phrase *according to* hints that this is a Detail question, so our first step is to locate the relevant information. The author mentioned the male clothing and masculine pen name in paragraph 3, saying that "it is not hard to understand why she chose a masculine pseudonym because, like her clothes, this male identity gave her more freedom of expression." In other words, she was able to overcome limitations by taking on a male identity.
A. both acknowledge her strong masculine side.	**(A)** is a Distortion because although Dupin wore more masculine clothing, the author never suggested Dupin herself had a *strong masculine side.*
B. both provide evidence of her androgyny.	**(B)** is also a Distortion as there is nothing to suggest that she was androgynous (having equal characteristics of both masculinity and femininity).
C. both freed her from stereotypical female constraints.	**(C)** matches closely with the prediction and is the correct answer; Dupin's adoption of a male identity (both pseudonym and clothing) allowed her to overcome limitations based on her sex.
D. both permitted her to succeed in a patriarchal society.	**(D)** is a nuanced Distortion of the author's points. The adoption of a male identity helped Dupin overcome obstacles in her way, but did not itself lead to her success—her success is attributable to "establishing…her reputation as a talented and publishable author," as stated at the end of paragraph 5.

PRACTICE QUESTIONS

Passage 1 (Questions 1–5)

GO ONLINE

Because it impinges upon so much—from bilingual education, political correctness, and Afro-centered curricula to affirmative action and feminism—the current discussion on multiculturalism is essential to understanding Western academic culture today. Charles Taylor's account of the development of multiculturalism out of classical liberalism traces multiculturalism through changing conceptions of what he terms "the politics of recognition."

Deft as his historical account may be, any analysis of the motivations for multiculturalism solely in terms of "recognition" must remain fundamentally incomplete. In his analysis are two central demands for recognition underlying classical liberal thought: the demand for the equal recognition of human dignity, and the demand for the recognition of, and the respect for, all human beings as independent, self-defining individuals. Multiculturalism, according to Taylor, rejects both of these ideals and their political application in an official "difference-blind" law (which focuses on what is the same in us all). Instead, it embraces laws and public institutions that recognize and even foster particularity—that cater to the well-being of specific groups. These two modes of politics, then, both having come to be based on the notion of equal respect, come into conflict.

Taylor acknowledges that it can be viewed as a betrayal of the liberal ideal of equality when the multiculturalist calls for a recognition of difference rather than similarity, and seeks special treatment for certain groups—such as aboriginal hunting privileges or the "distinct society" of Québec. However, he plausibly argues that to recognize only sameness is to fail to recognize much that is necessary for real "recognition" because we are all cultured individuals with personal histories and community ties. Still, Taylor does not stray far from classical liberalism, insisting that multiculturalism be able to "offer adequate safeguards for fundamental rights."

The more extreme forms of multiculturalism, which Taylor disavows, commit the crucial error of reducing all ethical and normative standards to mere instruments of power because in doing so, any distinctly moral arguments for these positions become absurd. Though Taylor seems correct to reject this diminution, he's wrong to think that the "recognition" model alone can sufficiently account for the demands made by various minority groups for both the promotion of discrete cultural identities and the transformation of the dominant culture. What many in these groups desire is much more than mere recognition or approval: it is the power to more effectively and independently control their own destinies.

It's even become common to disdain the respect or solidarity professed by those in the dominant group in an attempt to consolidate separate cultural identities. How Taylor misses this fact is not clear because even his favorite example of Québec's distinct society presents a case in which the primary function of the demand for recognition is to acquire the power necessary for those within to maintain, promote, and even enforce their way of life. Taylor understands that the Québécois want more than to merely preserve their culture, or to have others appreciate it. They also want to create a dynamic, autonomous society in which future generations will participate as part of a common project. Unfortunately, he does not consider how this fact undercuts the notion of "recognition" as an adequate lens through which to view their project.

1. According to the passage, extreme multiculturalists make which of the following mistakes?
   - **A.** They wrongly disdain the solidarity professed by those in the dominant group.
   - **B.** They undercut their position by eliminating any moral arguments that they could make for their view.
   - **C.** They overestimate the distinctness of their position from classical liberalism.
   - **D.** They argue for their position on moral grounds, rather than pointing to the practical benefits of their views.

2. According to the argument posed by the author in the passage, multiculturalism may be seen as a betrayal of liberal ideals because:
   - **A.** classical liberalism is not concerned with the well-being of minority groups.
   - **B.** it abandons the demands for equality that characterize classical liberalism.
   - **C.** a failure to recognize what is different about individuals can be a failure to fully recognize individuals.
   - **D.** it is not capable of respecting diversity or offering adequate safeguards for basic rights.

3. The author's two references to the "distinct society" of Québec are primarily intended to:
   - **I.** give an example of a multiculturalist demand.
   - **II.** give an example for which Taylor's analysis is inadequate.
   - **III.** give an example of a group for which special treatment is sought.

   - **A.** I only
   - **B.** III only
   - **C.** I and II only
   - **D.** I, II, and III

4. Which of the following can most reasonably be inferred from the passage about the author's attitude toward the two classical liberal ideals of equality?
   - **A.** They are adequate for most contexts in which recognition is demanded.
   - **B.** They do not safeguard fundamental rights for individuals in aboriginal groups.
   - **C.** They reflect a disguised attempt by a privileged group to maintain its power over other groups.
   - **D.** They reflect an impoverished conception of the individual person.

5. Based on the information provided in the passage, it would be most reasonable to expect the author to agree that Charles Taylor's "politics of recognition" model is:
   - **A.** admirable, but flawed.
   - **B.** of limited use.
   - **C.** historically deft.
   - **D.** an unmitigated failure.

Passage 2 (Questions 6–11)

Peter Gay's book, *The Education of the Senses*, reexamines Victorian bourgeois attitudes about sensuality and sexuality in an attempt to discredit the pervasive and negative view of the Victorian bourgeois as repressed and repressive people whose outward prim public appearances often hypocritically masked inner lascivious thoughts and private behaviors. One of the most interesting facets of Gay's study is his discussion of the necessary, yet taboo, issue of birth control during the latter part of the nineteenth century.

Gay points out that the very process of giving birth was dangerous to both the newborn and the mother—that most women suffered greatly during the birth, that many children and new mothers died within five years of a birth, and that many women approached the child-bearing process with trepidation although they believed that producing offspring was a woman's ultimate fulfillment. Advice or assistance from the medical profession—whether licensed doctors or self-trained midwives—was sorely lacking and inconsistent, hardly capable of reassuring the expectant mother and father as to the woman's and the baby's safety.

In fact, the medical profession itself was largely responsible for promulgating myths and rumors about the dangers of attempting to limit family size through use of some forms of birth control, regardless of the fact that so many women and children died each year due to complications of pregnancy or birth. Some medical, religious, and social experts did acknowledge that the continual cycle of birth was not only detrimental to the health of the mother but also could take a toll on the quality of life of the family because multiple children increased the financial burden and responsibility of the father. Furthermore, women were often caught in this cycle of pregnancy and childbirth well into their late 40s, greatly increasing the health risks and mortality rate of mother and child alike. It seemed appropriate and even necessary, then, to make efforts to limit the number of offspring in order to benefit the family unit and thus the greater good of the larger society itself.

Nonetheless, open discussion of birth control methods, both natural and device-assisted, was rare, even between a doctor and his patient. Most information was passed along by word-of-mouth, which inevitably led to a great deal of unchecked misinformation that was, at times, deadly. Gay maintains that a primary motivation for this reticence was deeply ingrained in the Victorian bourgeois mindset that emphasized the value of family and traditional roles and thus encouraged women to be productive—in the very literal, procreative sense. Though concerned husbands certainly did take steps to assure that their wives and families were not jeopardized by an overabundance of offspring, a widespread effort to limit family size was not firmly rooted in society until the advent of a strong women's movement, which did not make many real and meaningful strides in changing public attitudes and behaviors until the early twentieth century. Thus, Victorian bourgeois women were obliged to fulfill their societal role as child bearers despite very real fears about the toll this could take on their health and on the well-being of their family.

6. Which of the following is NOT, according to the author, a reason that most Victorian women did not use any form of birth control?
   - **A.** The medical profession did not offer any methods of birth control, but only vague suggestions for limiting the number of pregnancies a woman had during her lifetime.
   - **B.** Although the risks to women were widely recognized, women were still held responsible for continuing the family line.
   - **C.** Accurate information about safe forms of birth control was difficult to ascertain, even from doctors or midwives.
   - **D.** Many medical experts were reluctant to discuss birth control with their patients because they themselves were not well informed about the options and consequences.

7. Which of the following, if true, would most seriously WEAKEN Gay's argument that deeply ingrained social attitudes were responsible for the lack of open discussion about birth control?
   - **A.** The majority of Victorian women were satisfied with their role as wives and mothers.
   - **B.** Limiting family size was inconsistent with bourgeois perceptions about social status.
   - **C.** Bourgeois Victorians believed the ideal family consisted of fewer than five children.
   - **D.** Many Victorian physicians actually knew that family planning would improve women's overall health.

8. The passage suggests that which of the following was/were commonly associated with childbirth during the Victorian era?
   - **I.** Death of the child
   - **II.** The need for additional family income
   - **III.** Appreciation of the mother's suffering

   - **A.** I only
   - **B.** I and II only
   - **C.** I and III only
   - **D.** II and III only

9. Which of the following general theories would be most consistent with Gay's arguments as presented in the passage?
   - **A.** Examination of sociohistorical context offers the modern historian little useful information about prevailing attitudes and behaviors of a certain time period.
   - **B.** Understanding the psychological basis for social actions can help to explain apparently contradictory behavioral patterns.
   - **C.** Social historians can best analyze past cultures by applying modern theories to the work of earlier critics.
   - **D.** Critics should not attempt to rework prior studies of social classes because they cannot properly understand the historical context of their predecessors.

10. The passage suggests that Peter Gay is LEAST likely to agree with which of the following statements?
    - **A.** Social critics should occasionally reexamine existing beliefs to make sure that they are appropriate.
    - **B.** The prevailing views of Victorian bourgeois society were based on accurate sociocultural perceptions.
    - **C.** The women's movement of the early twentieth century helped to bring the issue of birth control into the public consciousness.
    - **D.** Sometimes, common mind-sets must be altered before real social change can be accomplished.

11. Based on the passage, which of the following does the author consider was the most important factor contributing to the high danger associated with Victorian childbirth?
    - **A.** False and misleading information about the risks of childbirth
    - **B.** Unsanitary conditions in the birthing rooms
    - **C.** Successive cycles of pregnancy and childbirth
    - **D.** Oppressive social attitudes that forced women to procreate

PRACTICE ANSWERS

For this and all following sets of Practice Questions, examples of the text you should be Highlighting or writing down in your Outline will be included in the explanation for each passage.

Passage 1 (Questions 1–5)

Sample Highlighting

P1. "multiculturalism out of classical liberalism";
P2. "incomplete" and "rejects both"; **P3.** "fail to recognize";
P4. "wrong to think"; **P5.** "Québécois want more."

Sample Outlining

P1. Introduction to Taylor's analysis of multiculturalism; "politics of recognition"

P2. Taylor: classical liberalism's two demands for equal recognition; multiculturalism rejects these (recognizes differences instead)

P3. Multiculturalism seems anti-liberal by recognizing difference; Taylor: "recognition" also depends on differences, but safeguard rights

P4. Extreme multiculturalism too narrow; Auth: Taylor's "recognition" model insufficient to explain minority groups' goals

P5. Author's criticism of Taylor: groups, desires are more than just recognition, gives example (Québécois)

1. **B**

This is a Detail question looking for the mistakes that extreme multiculturalism makes. Extreme multiculturalism is discussed in paragraph 4. The author states that extreme multiculturalists "commit the crucial error of reducing all ... standards to mere instruments of power ... and in doing so make moral justifications absurd." We can make a prediction that the extreme form eliminates moral arguments. **(B)** best matches this prediction. **(A)** describes a mistake some minority groups make, as described at the beginning of the fifth paragraph, but doesn't refer to extreme multiculturalism. **(C)** could potentially describe Taylor, given the author's claim that *Taylor does not stray far from classical liberalism*. Still, this choice does not describe extreme multiculturalists. Finally, **(D)** is an Opposite—indeed, the author argues that moral arguments are lost in extreme multiculturalism.

2. **B**

This is another Detail question, so we will need to reference the passage. Paragraph 3 indicates that multiculturalism differs from liberalism by recognizing differences—so what does liberalism recognize? As noted in the first sentence of this paragraph, the liberal ideal that is betrayed is equality. **(B)** matches cleanly with this prediction. **(A)** Distorts the author's argument and makes it extreme. Although liberalism does not concentrate on minority groups to the same extent as multiculturalism, there is no evidence that suggests that it has no concern at all for minority groups. **(C)** is the justification Taylor gives for multiculturalism, not the reason why multiculturalism betrays liberal ideals. Finally, **(D)** is an Opposite, as Taylor insists that multiculturalism "offer[s] adequate safeguards for fundamental rights."

3. **D**

Question 3 is a Scattered Function question because we are asked to interpret the author's motives for including a particular example. The example of Québec is mentioned in paragraphs 3 and 5, so we will need to analyze the content in which it is mentioned. In paragraph 3, the author notes that the "distinct society" of Québec sought special treatment, so Statement III is true. We can eliminate **(A)** and **(C)** because they don't have III. The Québécois are mentioned again in paragraph 5 as an example of where Taylor's analysis is inadequate, matching what is said in Statement II. **(D)** is the only answer choice that has both II and III, and thus must be the correct answer. Statement I is also true because the Québécois are mentioned as an example of one society's *multiculturalist calls for recognition*.

4. **D**

This is an Inference question as it directly asks what can be inferred from the passage. Liberal ideals were discussed in paragraphs 2 and 3. The author states that Taylor acknowledged the deviation from liberal ideals and that Taylor's justification is *plausibl[e]*. According to Taylor, concentrating on sameness—the focus of liberalism—"fail[s] to recognize much that is necessary for real 'recognition.'" The liberal ideals fail to fully recognize the individual, which makes **(A)** incorrect and finds a match in **(D)**. **(B)** is a Distorted version of the concern some may have about multiculturalism's rejection of equality and does not describe the author's attitude toward the liberal ideals. Finally, **(C)** is Out of Scope because the topic of privileged groups maintaining power over others is never discussed. The dominant group was briefly mentioned in paragraph 5, but not in the context of maintaining power.

5. **B**

This is another Inference question in which we need to identify the author's feelings toward Taylor's model. We can think about the overall goal of the passage to make a preliminary prediction: because the author is critiquing Taylor's ideas, the author believes there are some flaws. More specifically, the author introduces Taylor's historical account as "deft," but claims that his "analysis…[is] fundamentally incomplete." **(B)** is the best match for this prediction. **(A)** and **(C)** are subtle wrong answers. For the former, the author certainly does believe that Taylor's ideas are flawed, and the language does not suggest admiration. For the latter, while the author describes Taylor's historical account as *deft*, he or she has a different opinion about the model itself—it is incomplete. Finally, **(D)** is a Distortion as it is extreme: while the author views Taylor's model as incomplete, he or she does not believe that it is an absolute failure.

Passage 2 (Questions 6–11)

Sample Highlighting

P1. "discredit"; **P2.** "trepidation"; **P3.** "responsible" and "efforts to limit"; **P4.** "unchecked misinformation" and "obliged."

Sample Outlining

P1. Introduces Gay's reexamination of Victorian bourgeois attitudes; will focus on birth control

P2. Describes dangers to women and their babies; Victorian women: "childbirth = ultimate fulfillment"

P3. Doctors exacerbated the misinformation that women received, but recognized some issues with constant cycle of birth

P4. Discussion of birth control remained hidden in Victorian society—women had to fulfill their roles (until women's movement)

6. **A**

This is a Scattered Detail question asking for what is *not* stated as a reason women did not use birth control. We will need to use the process of elimination to eliminate the answers discussed in the passage. Starting with **(A)**, the medical profession's attitude about birth control is mentioned in paragraph 3. Doctors did not provide even vague advice for limiting family size, but rather "promulgat[ed] myths and rumors about the dangers of attempting to limit family size." Thus, **(A)** is the correct answer. **(B)** is mentioned at the end of paragraph 4. **(C)** is mentioned in detail in paragraph 2 and briefly mentioned again in paragraph 4. **(D)** is discussed in paragraph 3 and again at the beginning of paragraph 4.

7. **C**

This is a Strengthen–Weaken (Beyond the Passage) question focusing on what would *weaken* the author's argument. The author argues that deeply ingrained social attitudes were responsible for the lack of open discussion of birth control. We will need to consider what the social attitudes were at the time, as the right answer will contradict them. The author stated that women "believed that producing offspring was a woman's ultimate fulfillment" and that "a widespread effort to limit family size was not firmly rooted in society." A prediction for the answer would likely be something that implies that women felt that they had a different role or wanted to limit the size of the family. **(C)** best matches this prediction. **(A)** is consistent with Gay, reflecting on the idea that Victorian women "believed that producing offspring was a woman's ultimate fulfillment," and does not significantly weaken Gay's argument. **(B)** would strengthen the author's argument, as it gives an additional reason why women would not aim to limit family size—and would likely not discuss birth control openly, by extension. Finally, **(D)** would also strengthen the author's argument because it removes one of the alternative explanations why birth control was not discussed—if doctors actually knew that limiting family size was beneficial, and yet still did not discuss it openly, then it increases the likelihood that prevailing social attitudes limited discussions of birth control.

8. **B**

Because the question asks what the passage *suggests*, this is an Inference question; the Roman numeral answer choices make it a Scattered Inference question. We will need to evaluate the Roman numerals one by one to arrive at the answer. Starting with Statement I, the process of childbirth and the consequences of it were mentioned in paragraph 2. The author stated that many "children ... died within five years," so Statement I is true, and we can eliminate **(D)**. For Statement II, the author mentioned in paragraph 3 that additional children "increased the financial burden and responsibility of the father." This suggests that the family would need a larger income to support a larger family, so Statement II is also true. When we look at the answer choices, there is only one answer that has both Statements I and II, so **(B)** is the correct answer. For reference, Statement III is false because while some husbands were concerned about their wives during childbirth, it was not widespread.

9. **B**

This is an Apply question in which we need to use our understanding of the author's argument to determine which statement would be most consistent with it. The author mentioned the problems associated with childbirth in the second paragraph and the lack of action by medical professionals in paragraph 3. The author then explained that the reason birth control was not discussed was due to the current social mindset in which women were expected to be productive. This argument structure is most consistent with **(B)**. **(A)** and **(D)** are Opposites; Gay uses sociohistorical context and reworks prior studies about Victorian bourgeois attitudes—so these statements would be inconsistent with Gay's argument in the passage. **(C)** is Out of Scope as Gay does not apply modern theories to earlier critics, but rather reevaluates information and research on the Victorian bourgeoisie themselves.

10. **B**

This is an Apply question (specifically of the Response subtype) asking us what Gay would be least likely to agree with. Gay's study was a reexamination of the stereotypical beliefs about Victorian bourgeois attitudes and behaviors with regard to sexuality. We can predict that if Gay found it necessary to reevaluate the information of the Victorian bourgeois, he must have found the existing studies and opinions to be inaccurate. **(B)** best matches this prediction. **(A)** is an Opposite because this is exactly what Gay did—so he would definitely agree with reexamining existing beliefs. **(C)** and **(D)** both reflect on the end of the passage, where the author states that widespread attempts to limit family size did not occur until a strong women's movement made meaningful strides in changing public attitudes and behaviors.

11. **A**

This is another Inference question asking what the author thought was the most important factor contributing to the danger of childbirth in the Victorian era. While the author did not explicitly state which factor they thought was the most important, we can look at how the argument was structured for insight. The predominant focus of the passage, which plays a role in paragraphs 2, 3, and 4, is the misinformation from the medical field. This matches closely with **(A)**. **(B)** is Out of Scope as sanitation was never discussed. **(C)** is indeed mentioned in the passage but merely as an example of one of the few things doctors recognized was problematic—it is too narrow and restricted of an answer when compared to the author's overall argument in the passage. Finally, **(D)** is Extreme; social attitudes certainly encouraged women to procreate, but the word *forced* is simply too strong.

CHAPTER 7

Applying the Kaplan Passage Strategy

CHAPTER 7

Applying the Kaplan Passage Strategy

In This Chapter

Introduction

LEARNING OBJECTIVES

After Chapter 7, you will be able to:

- Categorize MCAT passages by type and difficulty
- Choose the best Distill approach based on passage features
- Predict questions that are likely to be asked about a given passage

Now that we have examined how to Preview, Choose, and Read and Distill passages in the CARS section, it's time to put these strategies into action. All of the strategies will come together in this chapter. We'll start with a summary of the Kaplan Method for CARS Passages. Then we'll work through the Method with examples from the two categories of passages. The chapter concludes with a Worked Example and two practice passages.

7.1 The Kaplan Passage Method

The Kaplan Passage Method is a set of passage strategies that you will apply to every passage you encounter on Test Day. It begins with a scan of the passage and a determination of the order in which the passages will be attacked. As you dive into a passage you'll pick an approach from the three Distill options. Then, using keywords, you will read the passage and extract the most question-relevant information.

As discussed in previous chapters, there are a variety of passage topics you will encounter on the MCAT. In fact, the AAMC lists eleven different fields in the humanities and a dozen in the social sciences (as shown in Table 7.1) that most CARS passages can be classified into. Preview, Choose, and Read and Distill (PCRD for short) is designed to address the variety of passages you will see on Test Day: Previewing enables you to work the passages in an efficient manner, Choosing ensures you match your approach to the demands of the passage, and Reading and Distilling will focus your attention on the information most likely to show up in the questions.

MCAT EXPERTISE

According to the AAMC, 50 percent of the questions in the CARS section will come from the humanities, and 50 percent will come from social sciences.

Humanities	Social Sciences
Architecture	Anthropology
Art	Archaeology
Dance	Economics
Ethics	Education
Literature	Geography
Music	History
Philosophy	Linguistics
Popular Culture	Political Science
Religion	Population Health
Studies of Diverse Cultures*	Psychology
Theater	Sociology
	Studies of Diverse Cultures*

* Note: Studies of Diverse Cultures can be tested in both humanities and social sciences passages.

Table 7.1 Humanities and Social Sciences Disciplines in the CARS Section

The goal of determining the passage subtype is to help you set your expectations appropriately, just as you would in the science section. For example, you would expect a science passage on rate kinetics to discuss rate laws, rate-of-formation tables, and reaction mechanisms. You would likely also anticipate that the associated questions will ask you to calculate a rate law using the tables and to consider topics like rate-limiting steps and catalysis. Similarly, you can set expectations about the trajectory of a CARS passage and predict what topics are more likely to be tested than others. In the following two sections, we will work through three different CARS passages together, breaking down our approach as we go, in order to give you a more comprehensive understanding of how to apply the Kaplan CARS Passage Method.

KEY CONCEPT

Think about trends you notice in passages that discuss similar topics or fall into similar disciplines. Applying these trends to future passages will give you insight into both their structure and the questions you will likely be asked, allowing you to become more efficient in navigating and distilling the passage and helping you predict and answer questions more rapidly and accurately.

Steps of the Method

As stated previously, this method can be applied to every CARS passage you encounter, so it is important to make a habit of automatically executing it as you begin each passage. Figure 7.1 shows the steps of the method.

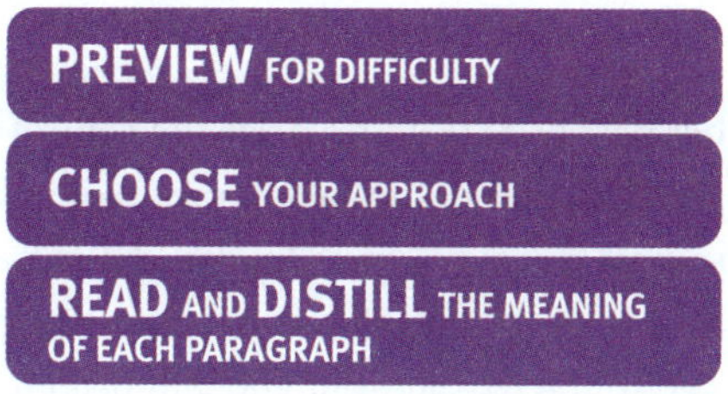

Figure 7.1 The Kaplan CARS Passage Method

Preview

The first step of the Kaplan Passage Method is to Preview the passage. As described in Chapter 4, you will look at the topic, sentence structure, writing style, length, and the first question to determine whether you want to attack the passage now or save it for later. Remember, there is a lot of variety among test takers in how they make those triage decisions and how they track those decisions on their noteboard. A topic that may be interesting to you may be kryptonite for the person sitting next to you, and as such, your preview is always going to be a very personal assessment.

Triaging and saving harder passages for the end provides two key benefits. First, it ensures that you are spending your time on the questions that you are more likely to answer correctly. It's unwise to spend excessive amounts of time on harder passages only to run out of time while working on an easier one. Second, it allows you to build confidence and a time buffer, both of which are key for when you do finally attack the harder passages.

Choose

Once you have decided to move forward with a passage, it's time to choose your approach to Distilling that passage. As discussed in Chapter 6, you can choose to Highlight, Outline, or Interrogate the passage. Each method has its own advantages and disadvantages. Rely on the analysis from your Preview step to help make this decision. In fact, many students will make their decision *as they preview* and note their choice on their noteboard. Should you choose to save a passage for later, you can either determine the approach you want to use before moving on to the next passage, or make that call when you return to the passage. Choosing the first option removes one of the steps you will have to perform when you return to the passage; however, you should remain open to changing your decision based on the time you have left.

Highlighting

The **Highlighting** approach is the quickest method for Distilling. When using this method, you should take about 2–3 minutes to work through a passage. As you read, plan to highlight important pieces of information, major shifts, and central ideas.

This approach tends to work best for passages that appear to be dense and confusing during your Preview step. You will typically use this strategy on passages you save until the end, because it enables you to move through the passage quickly. It is also a good option for passages that don't follow a clear structure, that is, those that lack distinctive transition points and breaks.

Outlining

The **Outlining** approach is best used to summarize the central ideas of each paragraph, while also noting the location of important pieces of information. You should be able to work through a passage in 3–4 minutes using this method. Your outline will be especially helpful in forming predictions for questions that test your understanding of the main ideas discussed in the passage, but you will need to return to the passage to answer questions asking about specific details. This is a useful approach for passages that appear to contain a large number of examples or details.

Interrogating

The **Interrogation** approach takes the most time but allows you to carry out a deep analysis of the passage as you read, enabling you to answer most questions with little rereading of the passage. However, at 4–6 minutes per passage, this is the most time consuming approach. In general, it is best used on passages you have triaged as *Now* and on shorter passages, to avoid timing becoming an issue. Passages that have complex central arguments or several distinct lines of argument are also well-suited for this approach because their questions will often ask about the logical relationships established in the passage.

Read and Distill

After deciding which approach to use, it is finally time to begin reading the passage. Note that the Read and Distill steps are grouped together because, functionally, they happen at the same time: you read with keywords, and use those keywords to guide your distillation of passage information. This is an active process of thinking and analyzing as you read. By staying engaged with the text, you help prevent a loss of focus and the need for rereading.

In Chapter 5, we discussed the importance of identifying and evaluating keywords in the passage. As you work your way through the passage, use keywords to help you identify notable shifts or key points in the text, along with author intentions and parts of arguments.

The goal of this step is to combine your knowledge of the passage types (and the questions commonly associated with them) with the most applicable passage approach in order to efficiently extract the major takeaways of each paragraph.

Highlighting

As you read, aim to highlight 1–3 key terms and phrases per paragraph that you think are likely to show up in the questions. Since the goal is to read quickly while still capturing the main points, your Distill step will primarily be limited to highlighting the main idea (or, in some cases, ideas) of each paragraph, and highlighting words that describe, support, or refute that idea. The keywords discussed in Chapter 5 will help guide you toward words worth highlighting, but remember that keywords typically indicate connections between ideas, rather than the content of the ideas themselves. So instead of highlighting keywords, focus on the concepts surrounding the keywords and note the progression of ideas through the passage.

Outlining

As you read, actively predict (and adjust your prediction as you encounter new ideas) the main idea of each paragraph. When you reach the end of a paragraph, take a moment to rephrase this main idea in your own words and write it on your noteboard. By identifying and rewording the main idea of each paragraph, you are already performing some of the analysis required to answer the questions. Thus, Outlining increases the accuracy of your predictions and saves you time on the questions. Because you are writing these main ideas on your noteboard with the Outlining method, you can use the highlighting tool to emphasize Relation, Logic, and Author keywords that you'd typically ignore when using the Highlighting method, making it easier to see important relationships and author intentions in the text. Remember that a major benefit of this method is that it allows you to read without having to memorize details. Once your paragraph label is on your noteboard, you can move on to the next paragraph without being concerned about forgetting the main idea of the previous one. It is important to remember that as you outline, you don't necessarily want to rephrase the entire paragraph. Rather, you want to select the information that answers questions like *What new information was added here?* and *What is the most testable information in this paragraph?* to include in your passage label.

Interrogating

As you read, actively examine the major idea presented in each paragraph to determine why and how the information was used by the author. Remember, this step goes beyond the Outlining method because instead of simply asking, *What information is new?*, you also want to understand the reasons for its inclusion. Use keywords to help you identify the information most likely to be tested. Also be sure to take advantage of the "Chunking" method. Don't only work through this method sentence by sentence. Rather, group similar ideas, often linked by Continuation keywords, into a chunk, before asking yourself questions about each chunk and why it was included.

This method has the most involved Distill step, as your goal is to do the majority of the work analyzing and understanding the passage before you even get to the questions. Once you've done this deep analysis, though, you won't have to spend a lot of time referring back to the passage.

In all three of the methods, the underlying goal is to help you identify the most important and testable content in each paragraph. As part of that, regardless of the method you choose to use, by the time you reach the end of the passage, you should be able to identify the reason the passage was written (the overall purpose or goal of the passage, as was discussed in the rhetoric portion of Chapter 3). If it is not clear, take a moment to consider it before jumping into the questions.

In the next section, you will get to see the Kaplan Passage Strategy applied to several examples. Use them as guides for understanding the advantages and disadvantages of each method.

7.2 Humanities Passages

Now that we have discussed the components of the Kaplan CARS Passage Method, let's apply it to a passage similar to the ones you will see on Test Day. As previously mentioned, the two broadest passage categories are humanities and social sciences. The next section will discuss the unique challenges of humanities passages and will demonstrate how the Kaplan Passage Method can be applied to these passage types.

Humanities Passages

Passages in the humanities tend to fall into two broad categories. The first category, which includes most of the passages from architecture, art, dance, literature, music, popular culture, and theater, can broadly be considered **arts passages**.

Many arts passages are characterized by the heavy use of quotations from other artists, writers, or critics, in addition to strong Author opinions and the use of descriptive language when discussing artistic examples. Often, each paragraph is structured with a claim, followed by evidentiary support. While not every Arts passage will contain these elements—literature passages, in particular, tend to have a lot of variation—the majority do, making them easy to identify and triage.

Most of the other passages in the humanities fall into the second category, which includes ethics, philosophy, religion, and studies of diverse cultures. These are broadly categorized as **philosophical passages**.

Philosophical passages tend to be very abstract, with an emphasis placed on logic. They focus on concepts and the relationships between them, and they often appeal to the reader's memory or imagination, drawing on common experiences or thought experiments (asking a question with no definite answer to help you delve deeper into the nature of things).

Keep in mind that there will be plenty of humanities passages that mix characteristics of arts and philosophical passages, as well as some passages that don't properly fit in either category. Nevertheless, this distinction is useful for setting expectations about the kind of support that a passage will use as well as the types of questions that will accompany it.

MCAT EXPERTISE

Passages with a lack of support or argumentation are likely to include *Reasoning Beyond the Text* questions, which bring in a new element of information and ask you to apply the information in the passage to a new scenario (Apply questions) or ask how the new information would impact the passage (Strengthen–Weaken [Beyond the Text] questions). These question types are discussed in Chapter 11.

Applying the Method to Humanities Passages

Passage 1

Now, let's take a look at how to apply the Kaplan CARS Passage Method to a humanities passage, keeping the unique challenges of this passage type in mind.

Preview

Preview the passage below (do not read through it completely; there will be an opportunity for that later in the chapter), then read the analysis that follows to assess what you are already noticing and what other factors you could have considered.

The Monkey Kid, written and directed by Xiao-Yen Wang, is probably one of the best Chinese feature films ever made but few Americans have seen it. Released in 1995 by the Beijing-San Francisco Film Group, the film "was an Official Selection at the 1995 Cannes International Film Festival and received the Grand Prize at the 1996 Aubervilliers International Children's Film Festival, awards for Best Film and Best Director at the 1995 Danube Film Festival, Best Foreign Film at the 1995 Fort Lauderdale International Film Festival, the Young Jury Award at the 1996 International Women's Film Festival at Créteil, and the Critic's Prize at the Cinestival 97 at Marseille."

The film is the autobiographical account of director Xiao-Yen Wang's childhood in China during the Cultural Revolution, the great political upheaval that gripped Chinese society and politics from 1966–1976. Ms. Wang is not just a casual observer of the times. She is an accomplished writer and director who graduated from the Beijing Film Academy in 1982. That same class of directors came to be called the famed Fifth Generation of Chinese filmmakers and included Chen Kaige (*Farewell My Concubine*, *The Emperor and the Assassin*) and Zhang Yimou (*Curse of the Golden Flower*, *Hero*, *House of Flying Daggers*, *Ju Dou*, *Raise the Red Lantern*, *Red Sorghum*, *The Story of Qiu Ju*, and *To Live*). Ms. Wang has also written and directed two other films, *The Blank Point* (1991) and *I'm Seducible* (2006).

At first glance, *The Monkey Kid* may seem like a simple slice-of-life story about one family during the Cultural Revolution, but it is much more than that. This feature film shows just how much Maoist ideology affected everyone in China, especially mothers, fathers, and children. Every family felt the turmoil in some way. What makes this film about the Cultural Revolution different from all others is that it offers a way to understand revolutionary society from a child's perspective.

The main character of the film is Shi-Wei. For much of the film, Shi-Wei and her sister, Shao-Qiong, are alone in the family's apartment. Her parents live apart in two different locations in rural China as part of the campaign to have intellectuals learn from the peasants. During the film, both the father and the mother return home for short periods of time. Otherwise, Shi-Wei and Shao-Qiong must take care of themselves. They do their homework, keep the apartment tidy, shop for vegetables from street vendors, and even cook the way their parents have taught them, without any adult supervision.

The film depicts Shi-Wei as a model child. She takes on a lot of the responsibility at home while her parents are away. She also does well in school and serves as a class leader. Yet even Shi-Wei is not perfect. In one of the first scenes of the film, Shi-Wei arrives at school late because she and her friends were playing in the snow. Other students would have been punished, but Shi-Wei counts on the fact that she is a "teacher's pet." Shi-Wei just whispers some excuse to the teacher and all is forgiven.

The Monkey Kid is not a new film, but educators should not be afraid to use it in the classroom. It is especially helpful for illustrating the many ways in which Mao's cult of personality dominated this era. It offers a view of the Cultural Revolution that is both compelling and entertaining. I have used the film with great success in both undergraduate and graduate classes on modern China. Students report that more than any other film about the Cultural Revolution, *The Monkey Kid* stays with them long after its showing in class because of the story line, the acting of the children, and the effective direction. The film always generates much discussion about class divisions, ideological education, and mass mobilization.

Adapted from Andrew, A.M., 2011. The Monkey Kid: A Personal Glimpse into the Cultural Revolution. ASIANetwork Exchange: A Journal for Asian Studies in the Liberal Arts, 18(2), pp.108–111. DOI: http://doi.org/10.16995/ane.190

A quick peek at the first sentence indicates that this is likely an arts passage, since it begins by discussing a movie. The second sentence is easily identifiable as a list of the movie's accolades. Glancing at the first sentence of the next two paragraphs further supports the idea that we will be spending the passage learning more about the movie and its origins. Looking down the passage, the sentence structure appears to be generally simple.

Since this passage appears relatively straightforward and does not contain long, complex sentences, advanced vocabulary, or highly abstract ideas, this is an excellent passage to do now, rather than later, and should rank as Easy or Medium for most readers.

Choose

None of the paragraphs are extremely long and many appear to focus on examples and specific details, so the Outlining method is ideal for this passage. Furthermore, the language and the structure are not overly complex, so an in-depth reading is not required to determine the author's goal.

Read and Distill

Given the simplicity of the wording and sentence structure, you should be able to create a well-structured outline while reading through the details more quickly. Expect to work quickly through the passage, labeling the key points and returning back to it as needed to revisit specific details and examples if asked about them in a question.

OUTLINING METHOD	
Passage	**Expert Thinking**
The Monkey Kid, written and directed by Xiao-Yen Wang, is probably one of the best Chinese feature films ever made but few Americans have seen it. Released in 1995 by the Beijing-San Francisco Film Group, the film "was an Official Selection at the 1995 Cannes International Film Festival and received the Grand Prize at the 1996 Aubervilliers International Children's Film Festival, awards for Best Film and Best Director at the 1995 Danube Film Festival, Best Foreign Film at the 1995 Fort Lauderdale International Film Festival, the Young Jury Award at the 1996 International Women's Film Festival at Créteil, and the Critic's Prize at the Cinestival 97 at Marseille."	Ideally, the first paragraph will let us know the general direction for the rest of the passage. What is being introduced to us here? *The author introduced a Chinese film* The Monkey Kid. Does the author have an opinion on this film? *Absolutely! The phrase "one of the best" clearly indicates the author's high opinion of the film.* Notice also that the bulk of the paragraph is dedicated to describing why the author considers this one of the best films. But we don't need to outline that because it's just evidence for the conclusion we already identified. P1: The Monkey Kid = one of the best films
The film is the autobiographical account of director Xiao-Yen Wang's childhood in China during the Cultural Revolution, the great political upheaval that gripped Chinese society and politics from 1966–1976. Ms. Wang is not just a casual observer of the times. She is an accomplished writer and director who graduated from the Beijing Film Academy in 1982. That same class of directors came to be called the famed Fifth Generation of Chinese filmmakers and included Chen Kaige (*Farewell My Concubine*, *The Emperor and the Assassin*) and Zhang Yimou (*Curse of the Golden Flower*, *Hero*, *House of Flying Daggers*, *Ju Dou*, *Raise the Red Lantern*, *Red Sorghum*, *The Story of Qiu Ju*, *To Live*). Ms. Wang has also written and directed two other films, The *Blank Point* (1991) and *I'm Seducible* (2006).	Based on the previous paragraph, we can expect further description of the film to come next. What information are we given? *The author states that the film is an autobiography of the director. We are also told further information about the director.* Does the author express an opinion about the director? *Yes! Words like "accomplished" show that the author feels positive.* This paragraph contains a lot of details that simply support the author's opinion and shouldn't be included in the outline. P2: Author: Director = accomplished
At first glance, *The Monkey Kid* may seem like a simple slice-of-life story about one family during the Cultural Revolution, but it is much more than that. This feature film shows just how much Maoist ideology affected everyone in China, especially mothers, fathers, and children. Every family felt the turmoil in some way. What makes this film about the Cultural Revolution different from all others is that it offers a way to understand revolutionary society from a child's perspective.	The first sentence provides us with a contrast. What is being contrasted? *The author states that the film may seem simple at first but shows how Moaist ideology affected families. In addition, it is done from a child's perspective.* P3: Film = Maoist ideol. from child eyes

OUTLINING METHOD

Passage	Expert Thinking
The main character of the film is Shi-Wei. For much of the film, Shi-Wei and her sister, Shao-Qiong, are alone in the family's apartment. Her parents live apart in two different locations in rural China as part of the campaign to have intellectuals learn from the peasants. During the film, both the father and the mother return home for short periods of time. Otherwise, Shi-Wei and Shao-Qiong must take care of themselves. They do their homework, keep the apartment tidy, shop for vegetables from street vendors, and even cook the way their parents have taught them, without any adult supervision.	The beginning of this paragraph quickly tells us that the author is about to provide more information on what the film is about. Look out for contrasts or anything else unexpected, but otherwise don't spend too much time on this paragraph. P4: Main char description
The film depicts Shi-Wei as a model child. She takes on a lot of the responsibility at home while her parents are away. She also does well in school and serves as a class leader. Yet, even Shi-Wei is not perfect. In one of the first scenes of the film, Shi-Wei arrives at school late because she and her friends were playing in the snow. Other students would have been punished, but Shi-Wei counts on the fact that she is a "teacher's pet." Shi-Wei just whispers some excuse to the teacher and all is forgiven.	At first glance, this paragraph may seem like a continuation of the previous one. However, we do see a contrast made in the middle of it. What has changed? *The author states that even though main character is depicted as a model child, she is not perfect.* P5: Main char = model child but not perfect
The Monkey Kid is not a new film, but educators should not be afraid to use it in the classroom. It is especially helpful for illustrating the many ways in which Mao's cult of personality dominated this era. It offers a view of the Cultural Revolution that is both compelling and entertaining. I have used the film with great success in both undergraduate and graduate classes on modern China. Students report that more than any other film about the Cultural Revolution, *The Monkey Kid* stays with them long after its showing in class because of the story line, the acting of the children, and the effective direction. The film always generates much discussion about class divisions, ideological education, and mass mobilization.	Since this is the last paragraph, we expect the author to come up a with a conclusion and perhaps a reason why this film was discussed. What is the author's conclusion? *The author is advocating for this film as a great educational tool on Maoist ideology. In addition, there is evidence provided to back up the conclusion. We do not need to outline the evidence but we can make note of it.* P6: Film = great educ tool on Maoist ideol

This passage has a lot of details about the movie and its accolades, but our outline allows us to stay focused on the big ideas while being confident that we can find those details if a question requires it. Overall, the author wrote this passage to argue that this excellent movie can be used to teach students about the Maoist era in China. Given all of the details, we should certainly expect to see some Detail questions, but the CARS section is usually more interested in your understanding of the author's reasons for writing. Thus, we should also expect to see Function questions about particular details, such as why the author describes the main characters in such depth.

Passage 2

Now, let's apply the Kaplan method to another humanities passage. Pay particular attention to how we might vary our approach to this passage from that of the previous passage.

Preview

Preview (don't read) the passage below, then read the following analysis to assess what you are already noticing and what other factors you could have considered.

> The poet lives by words as the musician by sounds. One or other of the threads with which both build, investigate, decipher and seek to express wonderment refers back to the order of the originary. In the beginning was what? The network of ancient cosmogonies and the imbricate lexical and semantic web of documents and sources that give shape to our concepts bear witness to the universal operative power of both sound and syllables. The word is the purest symbol of the manifestation of being, of the being that thinks himself and expresses himself or the being that is known of, and communicated by another. Expressing the person, the word is part of its dynamism. Even if by analogy, does the same claim not apply to music?
>
> Nothing authorises us to say that Guerra Junqueiro (1850–1923) knew music or was able to read music beyond the most elementary level—even though, from a certain point he lived in a house inhabited by music. Both his daughters played piano and the youngest, Julia, excelled at the violin, to the point that, for example, in a public concert in 1898, she played the allegro from Mozart's 3rd Quintet with Gilhermina Suggia on cello; and even if he openly, to write some poetry, sought inspiration in certain pieces by Beethoven, which reminded him of those "immortal melodic souls of great epic tales that have died. . . ."
>
> Guerra Junqueiro has been a unique case in Portuguese literary history. No other poet, whatever the period being analysed, attained such popularity. The effectiveness of the poetry-reader communication meant a mighty power of persuasion, able to seduce, charm, convince of its truth. This happened in a society marked by a high level of illiteracy in which music was the fundamental language and collective means of communicating feelings and meanings. Many, not knowing how to read, learned by ear and recited by rote extensive compositions of Guerra Junqueiro (such as "O Melro" from *A Lágrima*) or even entire books. And if knowing by heart implies an appropriation and has something of possession, it is also knowledge of the heart. It would, therefore,

be useless and demeaning to ascribe his seduction to specific audiences. The work of the poet served audiences differentiated by taste, social status, culture and political persuasion.

Words, like music, are liable to various modulations; they are, or can be, singing material. Guerra Junqueiro knew it and demanded that it was so: what "doesn't have music . . . is useless!" or what "doesn't sing, doesn't vibrate, is no good!" Based on this evaluation criterion, the verse of Junqueiro's work is stretched, loiters, stops, dances or sets off running like giddily rolling waves. Rich rhymes, vocal strata, ostensive use of metaphor, visually and aurally expressive images, pauses that are eloquent silences, sequences, reiterative structures, synaesthetic and onomatopoeic processes, a vast range of nuances, of technical-compositional, rhetorical-stylistic, ideothematic aspects, interwoven in a harmonious, seamless music, concur both for the modulation of what is the music of Junqueiro, and for direct communication with the reader or listener, easily consigning the verses to memory.

It is not, however, about reestablishing the prestige of Guerra Junqueiro, since this task would be impossible and an improbable anachronism, so intimate is the relation between text and society in the work of the poet and so disproportionately resounding was his reception. The academic project, Revisiting/Discovering Guerra Junqueiro, is rather an attempt to rescue him from the fog of oblivion and from the cultural ambiguity in which troubled circumstances, orthodoxies and conflicting powers bound him, freezing the reading and interpretation of his work and action.

Adapted from Pereira, H. (2010). From Singing Material to Intangible Poetry: The Music of Junqueiro. Journal of Science and Technology of the Arts, 2(1), 58-60. doi:http://dx.doi.org/10.7559/citarj.v2i1.195

A quick look at the first sentence indicates that this is likely an arts passage, since it begins by discussing poetry and music. The second sentence contains flowery language and abstract concepts and is followed by a question. The first sentences of the next two paragraphs discuss a poet named Guerra Junqueiro. As a result, we can assume that this passage will discuss music and poetry, specifically the work of Guerra Junqueiro. The first paragraph may seem intimidating, but it also looks like it's just an introduction to a more straightforward piece, so this passage is likely not as challenging as the first paragraph appears.

Since this passage contains advanced vocabulary and abstract ideas, this is certainly a passage that many would triage for later, likely assigning a difficulty of Hard or Medium.

Choose

The first paragraph is exceptionally verbose, with unnecessarily convoluted language. Subsequent paragraphs indicate that the passage is primarily focused on a single individual and his poetic work. Furthermore, the passage is filled with strongly-worded, abstract claims. This abstraction and complexity suggest that the Interrogation method is a great fit for this passage. As a further benefit, Interrogating will likely allow a reader to connect the abstractions of the first paragraph with the rest of the passage, promoting deeper understanding than the more detached Outlining and Highlighting methods.

Read and Distill

As you read, locate sentences that discuss or elaborate on a single theme. Ask yourself why the author chose to include that information and how it contributes to the author's overall purpose before moving to the next "chunk" of information. It's normal to take longer to work through a passage when you use this approach as compared to the other two, but be mindful of time and don't get lost in the minutiae. Remember, the goal is to think about the big picture, so avoid interrogating sentence by sentence. If you're struggling to generate questions in order to interrogate passages, consider going through this example by reading the questions on the right, then trying to answer them yourself. Follow up by reading the expert response and evaluating how well you did. Interrogating is a challenging method to learn and master, so if you're just starting out with it, focus on building one skill at a time!

INTERROGATING METHOD	
Passage	**Expert Thinking**
The poet lives by words as the musician by sounds. One or other of the threads with which both build, investigate, decipher and seek to express wonderment refers back to the order of the originary. In the beginning was what? The network of ancient cosmogonies and the imbricate lexical and semantic web of documents and sources that give shape to our concepts bear witness to the universal operative power of both sound and syllables. The word is the purest symbol of the manifestation of being, of the being that thinks himself and expresses himself or the being that is known of, and communicated by another. Expressing the person, the word is part of its dynamism. Even if by analogy, does the same claim not apply to music?	S1–7. Why does the author build an analogy between words and music? *The author uses a lot of obtuse language here, but the first sentence makes a comparison between words and music that is reiterated with the rhetorical question at the end of the paragraph. It's tough to pin down exactly what the author is saying, but there's a general theme about words and music both having a "universal operative power," a suggestion that both are means for people to express themselves and communicate with each other. Thus, the author draws this analogy between words and music in order to highlight their similarities, specifically the way that both are used for self-expression and communication.*
Nothing authorizes us to say that Guerra Junqueiro (1850–1923) knew music or was able to read music beyond the most elementary level—even though, from a certain point he lived in a house inhabited by music. Both his daughters played piano and the youngest, Julia, excelled at the violin, to the point that, for example, in a public concert in 1898, she played the allegro from Mozart's 3rd Quintet with Gilhermina Suggia on cello; and even if he openly, to write some poetry, sought inspiration in certain pieces by Beethoven, which reminded him of those "immortal melodic souls of great epic tales that have died . . . ."	S1–2. Why does the author shift to talking about one man's experience with music and poetry? *The first paragraph drew parallels between music and words, and now the author is describing an example of a poet who is familiar with, but untrained in, music. She also mentions specific pieces of music that inspired the poet. Thus, the author uses Junqueiro as an example to reinforce her analogy between music and language from paragraph 1. In addition, as you may recall from previewing the passage, the remaining paragraphs focus on Junqueiro, so we can see that this paragraph serves to introduce us to the main topic of the passage: Junqueiro's musically inspired poetry.*

INTERROGATING METHOD	
Passage	**Expert Thinking**
Guerra Junqueiro has been a unique case in Portuguese literary history. No other poet, whatever the period being analysed, attained such popularity. The effectiveness of the poetry-reader communication meant a mighty power of persuasion, able to seduce, charm, convince of its truth. This happened in a society marked by a high level of illiteracy in which music was the fundamental language and collective means of communicating feelings and meanings. Many, not knowing how to read, learned by ear and recited by rote extensive compositions of Guerra Junqueiro (such as "O Melro" from *A Lágrima*) or even entire books. And if knowing by heart implies an appropriation and has something of possession, it is also knowledge of the heart. It would, therefore, be useless and demeaning to ascribe his seduction to specific audiences. The work of the poet served audiences differentiated by taste, social status, culture and political persuasion.	S1–2. Why does the author emphasize Junqueiro's popularity? *So far, there have been a lot of words in this passage, but only a few key ideas: the analogy between music and words/poetry as modes of expression, and the fact that Junqueiro's poetry was inspired by music (despite no formal musical training). We also know from our preview that the passage primarily focuses on Junqueiro's work. From this, it is reasonable to conclude that the author emphasizes Junqueiro's popularity both to justify why she decided to write about him in particular and to give her claims about the connection between music and poetry more force. After all, if she spoke about some obscure poet that nobody read anymore, then her audience would be less likely to take what she has to say seriously.* S3–8. Why does the author talk about the audience of Junqueiro? *The author presents a few key facts about Junqueiro's audience: they came from a society with limited education and literacy, they primarily communicated with music, and they learned to recite Junqueiro's poetry by listening to it (instead of reading it). Based on these facts, we can conclude that the author is suggesting that a Junqueiro poem functions exactly like a piece of music (at least for his audience), providing more support for the analogy between words and music as forms of expression.*

INTERROGATING METHOD	
Passage	**Expert Thinking**
Words, like music, are liable to various modulations; they are, or can be, singing material. Guerra Junqueiro knew it and demanded that it was so: what "doesn't have music . . . is useless!" or what "doesn't sing, doesn't vibrate, is no good!" Based on this evaluation criterion, the verse of Junqueiro's work is stretched, loiters, stops, dances or sets off running like giddily rolling waves. Rich rhymes, vocal strata, ostensive use of metaphor, visually and aurally expressive images, pauses that are eloquent silences, sequences, reiterative structures, synaesthetic and onomatopoeic processes, a vast range of nuances, of technical-compositional, rhetorical-stylistic, ideothematic aspects, interwoven in a harmonious, seamless music, concur both for the modulation of what is the music of Junqueiro, and for direct communication with the reader or listener, easily consigning the verses to memory.	S1–4. Why does the author use such long strings of descriptive terms for Junqueiro's poetry? *This paragraph begins with a few shorter sentences, but its lengthy concluding sentence spans more than half of the paragraph and contains more than a dozen commas. Rather than attempting to make sense of every single descriptor, you should focus on the general theme: most of the descriptors use language associated with music or sound (as opposed to the written word). Thus, this long list is really just evidence for the claim that Junqueiro's poetry is musical by nature, which in turn supports the author's contention that words and music are closely related as means of expression.*
It is not, however, about reestablishing the prestige of Guerra Junqueiro, since this task would be impossible and an improbable anachronism, so intimate is the relation between text and society in the work of the poet and so disproportionately resounding was his reception. The academic project, Revisiting/Discovering Guerra Junqueiro, is rather an attempt to rescue him from the fog of oblivion and from the cultural ambiguity in which troubled circumstances, orthodoxies and conflicting powers bound him, freezing the reading and interpretation of his work and action.	S1–2. Why is the academic project NOT about reestablishing the prestige of Junqueiro? Why is the author writing about this project? *The author focused in previous paragraphs on the appeal of Junqueiro's work among those who primarily used music and spoken word to express themselves. However, contemporary audiences tend to inhabit different historical circumstances (e.g., they are more likely to be literate), so the author suggests that his work would not be received in the same way today. This serves to anticipate a potential criticism of the project, namely, that more literate and less musically-inclined audiences would not appreciate Junqueiro's poetry. Rather than restoring the former popularity of his work, the project instead aims to introduce this forgotten poet to contemporary audiences ("rescue him from the fog of oblivion") and help those audiences better interpret his work (namely, as a type of expression analogous to music, as established in previous paragraphs).*

Despite difficult language and complicated sentence structure, a deep interrogation of this passage shows us that the author actually has a fairly simple goal—one we honestly might have missed had we used a less in-depth passage method. The author uses the relationship between music and words to illustrate why Junqueiro was such a popular poet in his time and to suggest that his works deserve more attention today. Based on our analysis of the passage, we might expect to see Function questions asking why the author discussed the relationship between music and words, or an Inference question asking you why the author feels that Junqueiro's work is so important. The most challenging questions you might see with a passage of this type would ask you to expand on ideas from the passage: extending the analogy from the first paragraph or applying it to new examples.

7.3 Social Sciences Passages

Now that we have seen how the Kaplan Passage Method can be used to work through humanities passages, let's look at how it can also be applied to social sciences. The next section will discuss the unique challenges of these passages and walk you through how the Kaplan Passage Method can be applied to a social sciences passage.

Social Sciences Passages

As in the humanities, passages in the social sciences tend to fall into two broad categories. The first category, which includes most of the passages in anthropology, education, linguistics, population health, psychology, and sociology, has a **scientific** form.

Many scientific passages are characterized by the presence of numbers and reference to empirical studies as evidence. The author's opinion tends to be less obvious than in arts passages, but the presence of Author keywords can still be used to guide you to it. In some cases, a scientific passage may even look like a passage you might find in the *Psychological, Social, and Biological Foundations of Behavior* section, so it is especially important than you not bring any outside knowledge into your analysis of these types of passages.

MCAT EXPERTISE

While psychology and sociology can be tested in both the *Critical Analysis and Reasoning Skills* section and the *Psychological, Social, and Biological Foundations of Behavior* section, the former will not require outside knowledge. In fact, bringing in outside knowledge to answer any question in the CARS section can lead you astray, drawing you toward Out of Scope answer choices.

The remaining passages tend to fall into the second category, which includes archaeology, economics, geography, history, political science, and studies of diverse cultures. These types of passages are best categorized as **historical passages**. Note that these types of passages are far more variable than the clear-cut scientific passages and may even resemble humanities passages at first glance.

Historical passages tend to draw on historical events and quotations from sources alive at the time of the events discussed. They are rarely heavily opinionated, but the nature of the subject matter itself allows the author to express a strong viewpoint if he or she so desires. Passages dealing with history, economics, and political science often contain empirical studies, bringing in some elements of scientific passages, or heavy theoretical discussions, making them more like philosophical humanities passages.

Nevertheless, though some social sciences may appear to cross categorical boundaries, distinguishing them broadly as social science and more specifically as scientific or historical is useful for making judgments about what approach is best and what questions will likely be asked.

Applying the Method to Social Sciences Passages

Now, let's take a look at how to apply the Kaplan method to a social sciences passage, keeping the unique challenges of this passage type in mind.

Preview

Preview the passage below, then read the following analysis to assess what you are already noticing and what other factors you could have considered.

The issue of climate change is undoubtedly one of the most important that humankind has faced. Yet as important as it is, it is merely one of hundreds of news items that may or may not be regularly reported in the news media. How the issue is treated in the press is a relevant inquiry to make, as research clearly indicates that an issue's salience in the media can translate directly into relevance in people's minds. We also know that media coverage is also linked to more than just awareness of a topic; it can also have an effect on policy debates and attitudes among audiences, for example.

We know that climate change is difficult for one person to openly observe, and specific weather events are not always directly linked to the larger issue of climate change. The ways that media stories on climate change are written or produced vary considerably and depend largely on media workers making explicit connections and reporting such through their outlets. However, the business of mediated news tends to favor certain kinds of coverage over others in the media. News producers look for specific crises, or spectacular events, often utilizing an episodic lens through which stories may be relayed to audiences. If there is no specific event, then the topic will be less likely to receive coverage. The issue of climate change may be important in the long term, but it may not receive ample coverage in the immediate short term. As a result, regardless of Al Gore's and Leonardo DiCaprio's efforts, the salience of global climate change may be reduced for the public, who are likely dealing with more pressing issues or things that are more immediately impactful on their lives; even the shift of presidential attention to issues such as jobs and the economy over climate change highlights this situation.

The media's attention to the problem of climate change varies considerably. One study found that coverage was events-based in France, whereas in the United States, "conflicts between scientists and politicians" received more emphasis. Coinciding with this, other studies have suggested that over the years, scientists were quoted less, and politicians more, as the topic became politicized. Another suggested that the issue of climate change was too abstract, that journalists did not have the proper background to assess the details, that there was a false sense of balance in the reports, and that scientists were not providing jargon-free language so that journalists and the public might easily understand the issue.

The topic may become relevant to news outfits when addressed in some way by government or supranational body officials, who may be dealing with incidences or policies that are linked to climate change. Besides having actual events occurring naturally in the environment, which is likely to be covered, we may see elites or organizations raising the issue and thus drawing news attention (e.g., by commissioning and issuing an environmental impact report). Research indicated that international climate summits and efforts of NGOs had a stronger impact on issue attention than weather events. This is likely also the case in the Gulf region where newspapers engage in a type of protocol journalism practicing "non-adversarial forms of journalism common in Gulf states, relying heavily on protocol news and content supplied by government and corporations," which emphasizes the importance of covering official activities and government-driven pronouncements. At the same time, the public relations literature indicates that actors are recognizing that getting out in front of an issue allows potentially for greater control of the narrative surrounding any story. That is to say, if an official of the Gulf Cooperation Council issues a press release, they are able to address the issue, while at the same time encourage a specific kind of coverage on the subject.

While a given country may not see a need to address or take immediate action on climate change, another country may be suffering the consequences of inaction. As a result of the potential for one country's policies to impact another, supranational organizations are becoming increasingly relevant in dealing with matters of regional and international dealings, including climate change.

Adapted from Freeman, B. C. (2016). Protecting the Gulf: Climate change coverage in GCC print media. Cogent Arts & Humanities, 3(1), 1212690. https://doi.org/10.1080/23311983.2016.1212690

The reference to climate change and its representation in the media indicates that this is likely a social sciences passage. Since the first sentences of the next three paragraphs build on the idea of how climate change is portrayed in the media, we can further categorize this as a historical passage, perhaps political science.

Since this passage appears straightforward, with simple sentence structure and vocabulary, this would be an Easy passage for many readers.

Choose

Although the passage has a more straightforward topic and doesn't appear to contain significant changes in trajectory, the length of the passage, and the fact that most of its paragraphs are is long and cover multiple ideas, makes Highlighting an excellent method to use.

Read and Distill

As you read, try to identify big picture topics in each paragraph. Then, use your highlighting tool to highlight one or two words that summarize those main ideas or will help you remember key points when you refer back to them. Aim to work through the passage as quickly as possible so that you have more time to come back and perform a deep analysis of part of the passage if a question specifically requires you to. If used efficiently, highlights will help you quickly locate the details needed to answer these future questions.

HIGHLIGHTING METHOD

Passage	Expert Thinking
The issue of climate change is undoubtedly one of the most important that humankind has faced. Yet as important as it is, it is merely one of the hundreds of news items that may or may not be regularly reported in the news media. How the issue is treated in the press is a relevant inquiry to make, as research clearly indicates that an issue's salience in the media can translate directly into relevance in people's minds. We also know that media coverage is also linked to more than just awareness of a topic; it can also have an effect on policy debates and attitudes among audiences, for example.	What did the first paragraph reveal about the theme and the purpose of this passage? *The passage will likely talk about climate change and its relationship with the media. More specifically, the passage will investigate how the media treats, and perhaps should treat, climate change. We can highlight "treated in the press" to capture the main idea of this paragraph.*
We know that climate change is difficult for one person to openly observe, and specific weather events are not always directly linked to the larger issue of climate change. The ways that media stories on climate change are written or produced vary considerably and depend largely on media workers making explicit connections and reporting it through their outlets. However, the business of mediated news tends to favor certain kinds of coverage over others in the media. News producers look for specific crises, or spectacular events, often utilizing an episodic lens through which stories may be relayed to audiences. If there is no specific event, then the topic will be less likely to receive coverage. The issue of climate change may be important in the long term, but it may not receive ample coverage in the immediate short term. As a result, regardless of Al Gore's and Leonardo DiCaprio's efforts, the salience of global climate change may be reduced for the public, who are likely dealing with more pressing issues or things that are more immediately impactful on their lives; even the shift of presidential attention to issues such as jobs and the economy over climate change highlights this situation.	How can we use highlighting to tackle long paragraphs? What types of detail should we focus on? *We need our highlighting to break down this long paragraph. The first part of the paragraph seems to focus on how reports on climate change are heavily dependent on the media producing it. We can highlight "depend" to remind us of this, and to draw attention to the sentence surrounding it. We would not want to highlight the full phrase as it is wordy and would lower the efficiency of the highlight. The next idea is about how the media has a bias for immediate, singular events rather than focusing on the big picture. We should highlight "look for specific crises" to represent this idea. The last important idea is how this short-term focus lowers the salience of climate change significantly for the public. Here, we can highlight "reduced for the public" rather than "salience of global climate change" because the former better captures the action of the idea. The key is that the salience is reduced, and simply highlighting the word "salience" will not convey that information as well as "reduced for the public."*

HIGHLIGHTING METHOD	
Passage	**Expert Thinking**
The media's attention to the problem of climate change varies considerably. One study found that coverage was events-based in France, whereas in the United States, "conflicts between scientists and politicians" received more emphasis. Coinciding with this, other studies have suggested that over the years, scientists were quoted less, and politicians more, as the topic became politicized. Another suggested that the issue of climate change was too abstract, that journalists did not have the proper background to assess the details, that there was a false sense of balance in the reports, and that scientists were not providing jargon-free language so that journalists and the public might easily understand the issue.	In a paragraph with a lot of details and examples, what should we highlight? *Since we know highlighting every detail will result in over-highlighting, which is not very productive, we need to focus on the main theme of these details. The first line already captures the "variability" aspect, so we won't highlight terms like "many factors." It seems that this is due to the "conflicts between scientists and politicians," which can be quickly summarized by the word "politicized."*
The topic may become relevant to news outfits when addressed in some way by government or supranational body officials, who may be dealing with incidences or policies that are linked to climate change. Besides having actual events occurring naturally in the environment, which is likely to be covered, we may see elites or organizations raising the issue and thus drawing news attention (e.g., by commissioning and issuing an environmental impact report). Research indicated that international climate summits and efforts of NGOs had a stronger impact on issue attention than weather events. This is likely also the case in the Gulf region where newspapers engage in a type of protocol journalism practicing "non-adversarial forms of journalism common in Gulf states, relying heavily on protocol news and content supplied by government and corporations," which emphasizes the importance of covering official activities and government-driven pronouncements. At the same time, the public relations literature indicates that actors are recognizing that getting out in front of an issue allows potentially for greater control of the narrative surrounding any story. That is to say, if an official of the Gulf Cooperation Council issues a press release, they are able to address the issue, while at the same time encourage a specific kind of coverage on the subject.	Another long paragraph with a lot of detail and examples. How do we use good highlighting to quickly digest this paragraph? *This paragraph introduces new actors in the climate change conversation: government and supranational bodies. We need to highlight a phrase that shows how this new idea elaborates on previous ones. Since the paragraph seems to focus on how governments and other groups can create additional media attention on the topic of climate change, we can highlight "organizations raising the issue." The next part introduced the Gulf region's journalism habits, which certainly stands apart from the rest of the passage. However, the general message is the same: government actions impact media activity on climate change and therefore awareness. We can also highlight "protocol journalism" to capture the idea that the key nature of the Gulf states' media is more government driven. Finally, the end of the paragraph introduces a new idea—that specific actors can not only increase media exposure but also "control the narrative" about the action as well.*

HIGHLIGHTING METHOD	
Passage	**Expert Thinking**
While a given country may not see a need to address or take immediate action on climate change, another country may be suffering the consequences of inaction. As a result of the potentials for one country's policies to impact another, supranational organizations are becoming increasingly relevant in dealing with matters of regional and international dealings, including climate change.	What does this last paragraph have to conclude? What is the final message? *This paragraph shows how one country's actions can impact others. So, we can highlight "supranational organizations" to capture the idea that institutions that go beyond nation states (what "supranational" really means) will be necessary to adequately address climate change.*

The goal of the Highlighting method is to make it easy to locate the key points of the passage, while not becoming distracted by superfluous details, examples, and digressions. In this case, our highlighting allows us to easily identify the author's main points, the structure of the passage, and key details. If a question requires it, we should feel confident that we could quickly locate where in the passage to research an answer. As a final step before moving on to the questions, always stop to consider the overall goal of the author in writing his or her passage. In this case, the author uses the passage to discuss media attention to climate change and how that attention can be manipulated by various actors.

Conclusion

And so our discussion of how to approach passages draws to a close. The skills we've covered—understanding rhetorical elements, using keywords, reading critically, using the Kaplan Method for CARS Passages, analyzing argumentation and logic, recognizing varieties of passages, and anticipating questions—will serve you well not only on the MCAT but also in medical school and as a physician.

When you read a CARS passage, you use keywords to guide your reading and determine how information is put together. You then draw inferences and set expectations for where the author is likely to go with a given argument in order to anticipate the questions you'll be asked. Similarly, in medicine, you will listen actively to your patients, assessing whether various aspects of their chief complaint and background information are likely to help guide you to a diagnosis. By forming predictions about their diagnosis as you take their history, you can then set expectations for some of the other signs and symptoms the patient may exhibit, which then informs your physical exam. You can also anticipate the questions the patient will ask you—*What does this mean for me? Do I have to take medication for this? Does this put me at risk for anything else?*—and answer them in advance, to paint as comprehensive a picture as possible. Thus, it should come as no surprise that the *Critical Analysis and Reasoning Skills* section gets its name because it tests your ability not to comprehend dance theory, musicology, archaeology, or linguistics, but to understand how to analyze and reason through complex information.

Passages are only a part of the picture, though. While we need to read the passage to gain information, the real points come from the questions. In the next four chapters, we will shift our focus to these questions, starting with the Kaplan Method for CARS Questions and then focusing on the three major question type categories, as identified by the AAMC: *Foundations of Comprehension*, *Reasoning Within the Text*, and *Reasoning Beyond the Text.*

Concept and Strategy Summary

The Kaplan Passage Method

PREVIEW FOR DIFFICULTY

- Look for the big picture
- Assess the relative difficulty
- Decide to read *now* or *later*

CHOOSE YOUR APPROACH

- Highlighting: Best for high-difficulty passages or if low on time
- Outlining: Gives a moderate understanding of the passage and allows for more time to work on the questions
- Interrogating: Gives a strong understanding of passage but allows for less time to work on the questions

READ AND DISTILL THE MEANING OF EACH PARAGRAPH

- Recognize keywords to identify the most important and testable content in each paragraph
- Use your approach from the choose step to extract your major takeaways from each paragraph:
    - Highlighting—highlight 1–3 key terms and phrases per paragraph you can use to quickly locate information later
    - Outlining—create a brief label for each paragraph that summarizes the main idea of that paragraph
    - Interrogating—thoroughly examine each major idea presented in the paragraph and determine why and how the author is using the information to build an argument
- Identify the reason the passage was written before moving into the questions

Figure 7.2 The Kaplan Method for Passages in CARS

Humanities Passages

- **Humanities passages** include topics from architecture, art, dance, ethics, literature, music, philosophy, popular culture, religion, studies of diverse cultures, and theater. Many passages in the humanities can be considered arts passages or philosophical passages.
    - **Arts passages** tend to include strong opinions, quotations, and descriptive language to illustrate examples.
    - **Philosophical passages** tend to be abstract and heavy on logic, focusing heavily on concepts and relations between them; they often appeal to the reader's memory or imagination.

Social Sciences Passages

- **Social sciences passages** include topics from anthropology, archaeology, economics, education, geography, history, linguistics, political science, population health, psychology, sociology, and studies of diverse cultures. Many passages in the social sciences can be considered scientific passages or historical passages.
  - **Scientific passages** tend to include empirical studies and more subtle author opinions. Remember not to use any outside knowledge when working through these passages in the CARS section.
  - **Historical passages** tend to draw on historical events and quotations from sources alive at the time; they may include empirical studies or theoretical evidence, which can make them similar to the other passage varieties.

Worked Example

Use the following Worked Example, in tandem with the following practice passages, to internalize and apply the strategies described in this chapter. The Worked Example matches the specifications and style of a typical MCAT *Critical Analysis and Reasoning Skills* (CARS) passage.

This is a great passage for the Outlining method. There are only two arguments to keep track of, and both are clearly laid out in terms of paragraph structure. In addition, paragraph length, sentence, and language complexity make it possible to create a short outline and still capture the main idea of the passage.

OUTLINING METHOD

Passage	Expert Thinking
In *Prisoners of Men's Dreams*, published in 1992, Suzanne Gordon argues that American feminism has lost sight of its original goal of transforming the world into a kinder, gentler place. Gordon deplores the sort of feminism that has triumphed instead: a cold, ruthless, "equal-opportunity" feminism, which aims for women's entrance into the masculine public world and their achievement by male standards of excellence.	As with most passages, the first paragraph is where the author will give us at least a hint of the main idea. What is being introduced here? *The author mentions Suzanne Gordon, which we can abbreviate as S.G., and her argument on how American feminism has changed from kind and gentle to cold and ruthless.* Are there indications of opinions here? *Yes. The words used to describe S.G.'s position clearly indicate that she favors the "kind" form of feminism over the "cold" form. However, it's not yet clear whether the author endorses S.G.'s view.* P1: S.G.: Amer femin change = bad, kind → ruthless
The heart of the book consists of excerpts from a hundred interviews with career women, who do a lot of complaining about fatigue and disillusion. Gordon's subjects comprise an unsurprising lot, given her presupposition of modern feminism's focus on successful women as products of overcoming male-centric and male-infused social and business structures. At the end, Gordon calls for a National Care Agenda that would make "caregiving" rather than competition the ultimate American value.	Based on the previous paragraph, what are we expecting to logically follow? *Since we were introduced to a conclusion without any support, we would expect some evidence for that conclusion to come next. And this is exactly what this paragraph contains.* Is there any change in S.G.'s point of view or any indication of the author's opinion? *No. S.G. continues to emphasize the "caregiving" or kind idea of feminism, and the author's opinion remains unclear.* P2: Support for S.G.'s femin, Nat Care Agenda
Suzanne Gordon is obviously an intelligent, sympathetic, and well-meaning person, but *Prisoners of Men's Dreams* is a good example of the kind of sentimental, unlearned effusion that has become a staple of contemporary feminism and that most men rightly ignore. And who could blame them? Rallying for the propulsion of women in the public and private spheres through carefully played attempts at the pity point are bound to be met with stolid expression and silenced ears.	The last two paragraphs discussed S.G.'s conclusion and the supporting evidence for it, but we still have not seen where the author stands on the issue. What is the author's opinion on the issue and what can help us determine this? *The author starts out the paragraph with a positive tone toward S.G. However, the Contrast keyword "but" precedes a change in tone: the author harshly criticizes S.G.'s idea of feminism.* Does the author provide any evidence to reinforce this point of view? *The sentence following the question provides an answer to it, a justification for why S.G.'s feminism should be ignored. The language is dense, but the author suggests some kind of problem with appealing to "pity." However, this is a detail we don't need to include in our outline.* P3: Auth: S.G. = well-meaning but unlearned

OUTLINING METHOD	
Passage	**Expert Thinking**
Like so many American feminists, Gordon is completely out of her depth as a social analyst. Awkward, unintegrated quotes from Adam Smith and Woodrow Wilson are waved around to disguise her lack of familiarity with economics, history, and political science. Gordon's quote appropriations smack of the same short-sighted social phenomena that lead to "Keep Calm and Carry On" paraphernalia being plastered on the walls with complete disregard of manifest intent of the message.	The first sentence of this paragraph continues along the same line that the previous one did. This means that we are expecting more support for the author's stance unless we see a Contrast keyword. Does the author change the stance at any point? *No. The entire paragraph is dedicated to explaining why S.G. is a poor social analyst.* P4: Auth: S.G. = bad social analyst
We are presented with the usual three-handkerchief, tear-jerker scenario about Big, Bad, Ugly America—that corrupt, empty, greedy society that all those wonderful, warm, benevolent people around the world look at with disgust. This point of view is the essence of chic these days among know-nothing feminists and the preening pseudoleftists who crowd our university faculties.	This paragraph is a further continuation of supporting the author's stance. How can we label it in a concise form? P5: Auth: America = bad is a pop view now
Well, let me tell you: as a child of Italian immigrants, I happen to think that America is the most open, dynamic, creative nation on God's green earth. As a scholar, I also know that it is capitalist America that produced the modern, independent woman. Never in history have women had more freedom of choice in regard to dress, behavior, career, and sexual orientation.	In the first sentence, we see a colon (:), which indicates that we have an example coming up and a continuation of the discussion. We only need to be on the lookout for a Contrast keyword if there is one. Does the author shift direction here? *No, the author continues to refute a view that she ascribes to S.G. by arguing for American capitalism.* P6: Auth: America = best for women now
And yet, Gordon's insistence on defining women as nurturant and compassionate drove me up the wall. My entire rebellion as a child in the Fifties was against this unctuous, preachy stuff coming from teachers, nuns, and Girl Scout leaders. This drivel was not the path to supporting and empowering this woman.	In this paragraph, we actually see the author showing strong emotion, which is uncommon in MCAT passages. Even though we see a Contrast keyword in the first line, it is there to signify the difference between S.G.'s and the author's opinions, which is not new info. P7: Auth = upset with S.G.'s feminism
This "transformative feminism" is just as repressive and reactionary as the "patriarchy" it claims to attack. Minerva save us from the cloying syrup of coercive compassion! What feminism does not need, it seems to me, is an endless recycling of Doris Day Fifties clichés about noble womanhood.	In the last paragraph, we see further strong opinion from the author. Did that opinion change compared to previous paragraphs? *No, it did not.* P8: Auth: S.G.'s "transform fem" = bad

Here's a sample outline for this passage:

P1. S.G.: Amer femin change = bad, kind → ruthless

P2. Support for S.G.'s femin, Nat Care Agenda

P3. Auth: S.G. = well-meaning but unlearned

P4. Auth: S.G. = bad social analyst

P5. Auth: America = bad is a pop view now

P6. Auth: America = best for women now

P7. Auth = upset with S.G.'s feminism

P8. Auth: S.G.'s "transform fem" = bad

Question	Analysis
1. The author of the passage would most likely claim that someone who did NOT agree with her view of feminism was:	This is an Apply question because we are asked for the author's Response to a new scenario. Noticing the word *NOT* in the question stem, we could use elimination or rephrase it in order to get closer to a prediction. *Someone who [does] NOT agree with [the author's] view of feminism* is just a more convoluted way of asking for what the author thought about Gordon and similar feminists. In paragraph 3, the author derides these feminists as "sentimental" and "unlearned." Both of these are excellent predictions.
A. independent.	We can eliminate **(A)** because the author identifies with and is supportive of the "modern, independent woman" in paragraph 6.
B. sentimental.	**(B)** is a perfect match and is the correct answer.
C. rebellious.	**(C)** is an Opposite because the author confidently declared her own "rebellion as a child" in paragraph 7.
D. matriarchal.	**(D)** is a Faulty Use of Detail because while Gordon claims to attack patriarchy, according to paragraph 8, she does not necessarily advocate matriarchy; and further, the author is not against matriarchy—she is simply against the "cloying syrup of coercive compassion" that feminists might think is matriarchy.

Question	Analysis
2. In the passage, "transformative feminism" (last paragraph, first sentence) is used to mean:	Because this question gives a direct quote that needs interpreting, it must be a Definition-in-Context question. We can expect to read the sentence before and after to get the gist. Looking at the surrounding text, *transformative feminism* seems to be about "teachers, nuns, and Girl Scout leaders" applying Gordon's theory of womanhood—the one that the author has continually railed against as too "nurturant and compassionate." In this sense, these typical "womanly" virtues are what the author believes America has successfully gotten away from, and it would be regressive to be forced to act that way. The prediction should be something like: *encouraging women to act in a nurturing and compassionate way.*
A. a political agenda with caregiving as its guiding principle.	**(A)** is a Distortion that describes Gordon's National Care Agenda. While this is an extension of Gordon's basic argument, *transformative feminism* describes the changes in women's roles, not political policy.
B. the process of reinventing power structures to provide equal opportunities for women.	**(B)** is an Opposite of Gordon's description of feminism; in paragraph 1, the author mentions that Gordon is against "equal-opportunity feminism."
C. encouraging people to adopt "womanly" virtues as cultural standards.	**(C)** matches closely with the prediction and is the correct answer.
D. a goal of redefining feminine roles to include nurturing and compassion.	**(D)** is a tempting Distortion of the author's words. Gordon already sees the feminine role as nurturant, so the word *redefining* is flawed in this answer choice. In other words, Gordon claims that feminine roles already include nurturing and compassion—but that women should be more strongly encouraged to take on these roles.

Question	Analysis
3. Which of the following assumptions is NOT implicit in the author's view?	We notice the *NOT* in the question stem, which makes this a Scattered Inference question (specifically of the Assumption subtype). These questions are often solved faster by elimination rather than predicting because the one correct answer could be nearly anything, while the three wrong answers are statements that must be true in order for the author's conclusions to follow.
A. Stereotypes concerning female attributes are incorrect.	**(A)** can be eliminated because the author's central argument is anchored in her belief that Gordon's account of womanly virtue is inaccurate.
B. "Equal-opportunity" feminism is fundamentally ideal.	At first, **(B)** seems like it could be inferred because the author argues against Gordon's definition of women as caring and compassionate throughout the passage. However, this does not mean that the author necessarily views "equal-opportunity" feminism as ideal. Indeed, it would not be surprising if the author noted some potential flaws with "equal-opportunity" feminism as well. Because **(B)** might be true—but does not have to be—it is *NOT* an *assumption made by the author* and is therefore the correct answer.
C. Gordon's feminism is consistent with Fifties-style clichés of feminine roles.	**(C)** is suggested in the final paragraph when the author rails against Gordon's view, stating that "feminism does not need... an endless recycling of Doris Day Fifties clichés about noble womanhood."
D. American values and mores are considerably more liberal than those of other nations.	**(D)** is well-supported in paragraph 6, when the author states, "I happen to think that America is the most open, dynamic, creative nation on God's green earth."

Question	Analysis
4. The author's strongest objection to Gordon's thesis is that:	Because this question asks about the relative strength of the author's objections, this is a Function question. The word *strongest* stands out, because while there are many objections made, very few were actually backed up with any evidence besides the author's personal conviction. The *strongest objection* had better be defensible. The author refers to Gordon's work as "unlearned," questions her writing and social analysis ability, and associates her opinions with "know-nothing[s]." A prediction like *Gordon's thesis is uneducated rambling* fits here.
A. it lacks scholarly rigor.	**(A)** closely matches with the prediction and is the correct answer.
B. it offers political solutions for moral questions.	**(B)** is Out of Scope because the author never mentioned morality or *moral questions* in the passage.
C. it depicts modern feminism as cold and ruthless.	**(C)** is an accurate description of Gordon's view of feminism but it's not a significant objection of the author's.
D. the modern independent woman is free to choose her dress, career, and behavior.	The author would certainly agree with **(D)**, but it is not actually an objection against Gordon; in theory, Gordon wants those same things, just cast in a much different light.

Question	Analysis
5. The author's opinion of Gordon's view of American values can best be described as:	Be careful to unpack the distinct ideas in this question stem. Ultimately, this Inference question ask for a particular author opinion, specifically about Gordon's characterization of American values. Paragraph 5 is the place to go, as suggested by our outline. There, the author describes Gordon's negative view of America as something that "know-nothing feminists and preening pseudoleftists" endorse. The language in the paragraph is hostile and doesn't even seem to take Gordon's view seriously. We can thus predict that the author has a strongly negative opinion.
A. disinterested.	**(A)** can immediately be eliminated. *Disinterested* is not a synonym for uninterested; rather, *disinterested* means *unbiased*—which certainly does not describe the author's opinion of Gordon!
B. tolerant.	**(B)** is an Opposite; the author derides Gordon's views throughout the passage.
C. uninformed.	**(C)** is incorrect because the question stem asks for our description of the author's opinion—not the author's description of Gordon's view. The passage provides enough concrete details that it would seem inappropriate to describe the author's opinion as uninformed.
D. dismissive.	**(D)** matches our prediction perfectly. The author clearly dismisses Gordon's view of America in paragraph 5.

PRACTICE QUESTIONS

Passage 1 (Questions 1–5)

GO ONLINE

In recent years, extensive media attention has been given to enormous damages awarded in the US civil litigation tort system. In 1996, 79-year-old Stella Liebeck was awarded $2.7 million in punitive damages from McDonald's after sustaining third-degree burns from spilled coffee. The system awarded Michael Gore nearly $4 million in 1994 after BMW sold him a car that had been repainted and sold as new.

Awards such as these spurred businesses, insurance companies, and lobbyists to claim an "explosion" of legal liability. In response, many legislators called for tort system reform that included limiting the amount of damages, controlling legal fees, and redefining the concept of "fault" administered by the judges. Jury verdicts that appear, on superficial inquiry, to be blatantly excessive seem to challenge our system of compensation. Some claim that juries find negligence in order to provide compensation for victims who have large medical bills and lost wages, at the expense of "deep pocket" defendants.

In his seminal article in the *Maryland Law Journal*, "Real World Torts: An Antidote to Anecdote," Marc Galanter examines the issue. As the title suggests, in order to investigate the tort system, Galanter used empirical data to examine whether, on the whole, these "anecdotes" truly represented how the system compensates injured parties.

Galanter found that all tort claims form a dispute pyramid charting the progress from an injury to a jury verdict. Injuries form the broad base of the pyramid. On the next level, approximately 8 percent of injuries become grievances (events for which an injury was noticed). Of these grievances, 85 percent become claims (where the injured brings the problem to the alleged wrongdoer), and 23.5 percent of claims become disputes (having failed to reach an informal agreement). Next, 58 percent of plaintiffs with claims contact a lawyer, and 32.8 percent of these result in a court filing. Of all court filings, only 7 percent result in a verdict, and only 34.7 percent of these are decided in favor of the plaintiff. This means that an injured person gets a jury verdict in his favor only 0.007 percent of the time.

For example, medical malpractice results in approximately 100,000 deaths a year. At the tip of Galanter's pyramid, only 21 of the 100,000 deaths will result in a verdict. Finally, only 7 people will receive damage awards from a jury.

Galanter concludes that the system is hardly unbalanced in favor of plaintiffs. The proposed tort reform would actually increase insurance company profitability and reduce payments to the most seriously injured tort victims. Punitive damage awards are extremely rare, only applied in the most egregious cases, and always subject to judicial review. The awards discourage businesses from releasing harmful products into the stream of commerce.

Moreover, according to Galanter, court filings in the law division of the circuit court of Cook County have actually declined during the period from 1980 to 1994. His observations are consistent with a 1999 study by the National Center for State Courts, which found that tort filings have decreased by 9 percent since 1986. By looking at existing empirical data instead of isolated, inflammatory cases, legislators will be able to do a better job of deciding if the system is in need of reform and, if so, what type of reform is appropriate.

1. The author primarily mentions Liebeck's award (paragraph 1) in order to:
   A. give an example to support his overall claim.
   B. give an example of a verdict that is blatantly excessive.
   C. give an example of a verdict that has caused legislators to call for tort reform.
   D. introduce evidence for a conclusion made later in the passage.

2. Which of the following situations would be most analogous to the situation faced by a potential tort plaintiff, based on the information in paragraph 4?
   A. A young basketball prospect trying to make it to the NBA
   B. A group of children picking sides for a baseball game
   C. A young, qualified woman looking for a job
   D. An injured woman trying to reach an emergency room

3. Based on the information in the passage, the author believes that Galanter's pyramid:
   I. is applicable to medical claims.
   II. should compel legislators to change their views.
   III. is biased against "deep pocket" companies.

   A. I only
   B. II only
   C. I and II only
   D. I, II, and III

4. Which of the following, if true, would most WEAKEN the conclusion implied in paragraph 4?
   A. Galanter's article was published in 1995.
   B. A study showed that filings for divorce followed a much different pattern.
   C. Galanter's study dealt with only product liability cases.
   D. Most doctors carry medical malpractice insurance.

5. Based on the information in the passage, the author would argue for all of the following EXCEPT:
   A. juries should not be so compassionate toward victims at the expense of wealthy businesses.
   B. legislators should examine all data.
   C. the media spotlight does not necessarily clarify problems.
   D. courts might award damages as a way to ensure that businesses practice in the public's best interest.

Passage 2 (Question 6–12)

. . . [post-World-War-II Director of Policy Planning George F.] Kennan's strategy had been to try to bring about changes, over time, in the Soviet concept of international relations: to convince Russian leaders that their interests could be better served by learning to live with a diverse world rather than trying to remake it in their own image. Kennan had rejected both war and appeasement to accomplish this; it could only be done, he thought, through a long-term process of "behavior modification"—responding positively to whatever conciliatory initiatives emanated from the Kremlin, while firmly countering those that were not . . .

Kennan took the position that it was as important to reward the Kremlin for conciliatory gestures as it was to resist aggressive ones. This meant being prepared to engage in such negotiations that would produce mutually acceptable results. The [Truman] administration conveyed the appearance of being willing to discuss outstanding issues with Moscow, but Kennan regarded several of its major actions between 1948 and 1950 . . . [among them] the formation of the North Atlantic Treaty Organization (NATO) . . . would reinforce Soviet feelings of suspicion and insecurity, and hence, to narrow opportunities for negotiations . . .

The initiatives for the North Atlantic Treaty came from the Western Europeans themselves, and reflected the uneasiness they felt over the disparity in military power in Europe: the Russians had thirty divisions in Eastern and Central Europe alone; comparable combined US, British, and French forces came to fewer than ten divisions. Thus, the Western Union countries (Great Britain, France, and Benelux), together with the United States and Canada, agreed on the outlines of a treaty providing that an attack on any one of them would be regarded as an attack upon all.

Kennan had not been involved in the initial discussions, but he made clear his reservations about the course the administration chose to follow. These boiled down to three points: (1) that the Europeans had mistaken what was essentially a political threat for a military one, and that they consequently risked "a general preoccupation with military affairs, to the detriment of economic recovery"; (2) that outside the immediate North Atlantic, "which embraces a real community of defense interest firmly rooted in geography and tradition," any alliance extended to only some countries would render the rest more vulnerable . . . ; (3) that an alliance made up of [Western European] nations would amount to "a final militarization of the present dividing-line through Europe," and that "no alteration, or obliteration, of that line could take place without having an accentuated military significance." Such a development might be unavoidable, "but our present policy is still directed . . . toward the eventual peaceful withdrawal of both the United States and the U.S.S.R. from the heart of Europe . . ."

These were not isolated concerns. There was worry in Washington that emphasis on rearmament would delay recovery; indeed, one condition attached to the administration's military assistance program for Western Europe was that economic revival would continue to have first priority. The question of how to include some countries without appearing to write off others also caused a great deal of agonizing: in the end, the administration stretched the concept of "North Atlantic" to encompass Italy, but refused to extend it to Greece, Turkey, Iran, or to form a comparable pact with non-communist countries of the Western Pacific. There was less concern about Kennan's third point simply because most observers already regarded division, by mid-1948, as an accomplished fact . . . Despite its reservations, the administration went on to conclude a North Atlantic Treaty and initiate a program of military assistance to its members. Kennan came to see, regretfully, that [because of military insecurity of Europeans] there were few alternatives . . .

6. The passage suggests that Kennan's "behavior modification" approach to changing the Soviet concept of international relations was:
   - **A.** unlikely to be successful if the Kremlin always made conciliatory gestures.
   - **B.** moderate in comparison with the approaches he decided to reject.
   - **C.** a logical outgrowth of his extensive background in behavioral psychology.
   - **D.** an extension of American strategy during World War II.

7. The passage suggests that the impetus for the formation of NATO was:
   - **A.** information that a Russian attack on Western Europe was impending.
   - **B.** the understanding that no nation could withstand a Russian attack without assistance.
   - **C.** the desire to aid the Western European economic recovery as well as to guarantee military assistance.
   - **D.** the fear that the Soviets would try to capitalize on their military advantage.

8. Kennan assumed which of the following in making his first counterpoint against NATO?
   - **A.** The formation of the military alliance would spur economic growth.
   - **B.** The presence of the thirty Soviet divisions did not mean they were going to attack.
   - **C.** The economic recovery in Europe had been progressing slowly.
   - **D.** It's always a mistake to make military affairs a higher priority than economic affairs.

9. Kennan's reaction to the administration's refusal to extend NATO membership to Greece, Turkey, or Iran was most likely one of:
   - **A.** understanding, because these countries did not have the same geographic defense interests as the Europeans.
   - **B.** approval of the fact that the concept of "North Atlantic" was not overextended.
   - **C.** disappointment that those countries could not now be employed in anti-Soviet strategy.
   - **D.** trepidation that these countries were now more open to potential enemy aggression.

10. Which of the following explains why the Truman administration was not worried about Kennan's objection that NATO would amount to "a final militarization of the present dividing-line through Europe"?
    - **A.** They believed that it would be possible to alter the line through negotiations of peaceful withdrawal.
    - **B.** They wanted to maintain a strong American military presence in Europe.
    - **C.** They felt it was too late to prevent the solidification of the dividing line.
    - **D.** Neither of Kennan's other two objections to NATO had given them cause for concern.

11. The passage suggests that, with regard to the reservations expressed by Kennan about NATO, the administration was:
    - **A.** often in agreement but ultimately undeterred.
    - **B.** unresponsive to his proposals for improvement.
    - **C.** able to counter each of his criticisms.
    - **D.** forced to carefully reexamine its objectives.

12. Which of the following conclusions would be most in accord with the theme of the passage?
    - **A.** Military alliances invariably have drawbacks that render them ineffective.
    - **B.** Behavioral modification is the only way to change a government's concept of international relations.
    - **C.** Coherent international strategy can flounder because of the military situation.
    - **D.** Negotiations should be conducted between two powers once military equilibrium has been established.

PRACTICE ANSWERS

Passage 1 (Questions 1–5)

Sample Highlighting

P1. "litigation tort system"; **P2.** "call for tort system reform"; **P3.** "truly represent"; **P4.** "dispute pyramid" and "only 7 percent"; **P5.** "Medical malpractice"; **P6.** "damage awards"; **P7.** "declined" and "empirical data."

Sample Outlining

P1. Introduction to tort system; examples of extreme damages awarded

P2. Insurance/lobbyists: tort awards too excessive

P3. Galanter examines empirical evidence

P4. Galanter pyramid findings: tiny amount of successful tort claims

P5. Malpractice ex: 7 of 100,000 win damages

P6. Galanter: system not in favor of plaintiffs; awards = good: keeps businesses in line

P7. Overall, tort filings declining without legislation, Auth: legislators should use empirical data

1. **C**

This Function question directs us back to Paragraph 1 which mentions extreme examples the tort system awards. Even without rereading, predict that Liebeck must be one of those very high monetary awards that caused lobbyists to be so aggressive in fighting against the tort system in general and call for reform. This matches **(C)**. **(A)** is not possible simply because the author does not overtly make any claims. The author is largely neutral and simply sets the facts in front of us to make our own decisions. **(B)** distorts the author's mention of "jury verdicts that appear, on superficial inquiry, to be blatantly excessive"—the use of the phrase *on superficial inquiry* implies that the author may not agree that these damages actually are excessive. Finally, **(D)** is vague enough that it could sound plausible, but there is no later argument or conclusion that requires Liebeck's award specifically. Generally, there isn't any conclusion about coffee, McDonald's, burns, or the elderly, that would depend on this example either.

2. **A**

The best description of "the situation faced by a potential tort plaintiff" in paragraph 4 is as follows: Galanter's pyramid findings demonstrated that only a tiny number of tort claims actually result in decisions in favor of the plaintiff. The main point is that there must be a tiny number of success stories from a much larger pool of individuals. This matches **(A)**. In each of the other cases, close to 100 percent success would be expected (or at least a much higher percentage than the number of young basketball prospects who make it to the NBA).

3. **C**

The combination of *Based on the information in the passage* and Roman numerals in the question stem tells us that this is a Scattered Detail question. In paragraphs 4 and 5, the author brings in the medical example to illustrate the extremely low rate of success that actually is seen in tort cases. This is not a blatant endorsement of Galanter, but coincides with Galanter's pyramid findings. Statement I is exactly the example that is being used, so eliminate **(B)**. Statement II is true because it is the recommended course of action given at the end of the passage, which is primarily supported by Galanter's pyramid argument. With **(A)** eliminated, we can investigate Statement III. This claim is actually made by lobbyists and insurance companies in paragraph 2—and is part of the larger claim that Galanter ultimately refutes. Thus, Statement III is untrue, making **(C)** the correct answer.

4. **C**

This is a Strengthen–Weaken (Beyond the Text) question, so let's start by determining the conclusion implied by paragraph 4. The main point of Galanter's argument stems from the assumption that analyzing all of the empirical data will give the fullest picture and not allow anecdotal bias. If it were possible that the data Galanter used was false or incomplete for some reason, this would seriously weaken his argument overall. **(C)** reflects that prediction; if Galanter's study was only specific to product liability, then it can't be generalized to other similar cases or other tort suits. **(A)** reflects on the dates given in the passage—Galanter's study seems to investigate data until 1994. As long as Galanter's study was published after this point, there is no negative effect on his argument, eliminating this answer. The pattern of divorce cases, as described in **(B)**, has no effect on the argument because there is no reason to believe that divorce would (or wouldn't) follow the same patterns as tort cases. Finally, whether or not physicians have malpractice insurance does not appear to be related to the number of cases brought to court or decided in favor of the plaintiff, so **(D)** would also have no effect on Galanter's argument.

5. **A**

We are asked for arguments the author would make *based on the . . . passage*; with the word *EXCEPT* included, this must be a Scattered Inference question. While the author does not overtly state an opinion, it can be inferred that the author sides overall with Galanter's thesis. With that as a general prediction, we can eliminate any answer choice that fits with Galanter (and, by extension, the author)—and the one answer that does not fit is correct. This means that **(A)** is immediately correct. The author would not openly support being more lax to wealthy businesses at the potential detriment to victims. This position is exactly Opposite Galanter's, making it the correct answer. The other claims are all made in the passage: **(B)** is supported by the claim "By looking at existing empirical data, . . . legislators will be able to do a better job of deciding if the system is in need of reform." **(C)** is supported by the "extensive media attention" described in paragraph 1. **(D)** is supported in paragraph 6, where the author states that "awards discourage businesses from releasing harmful products into the stream of commerce."

Passage 2 (Questions 6–12)

Sample Highlighting

P1. "learning to live with" and "behavior modification"; **P2.** "willing to discuss" and "feelings of suspicion"; **P3.** "disparity in military"; **P4.** "mistaken," "vulnerable," "dividing-line," and "peaceful withdrawal"; **P5.** "economic revival," "stretched," and "less concern."

Sample Outlining

P1. Kennan: use behavior modification to improve Soviet international relations

P2. Behavior modification requires positive and negative reinforcement, ex: Truman administration

P3. NATO = Western nations uneasy, think Soviet military too big: attack on 1 = attack on all

P4. Kennan's reservations about NATO: (1) political *vs.* military threat, (2) other countries vulnerable, (3) solidifies military line in Europe

P5. Administration agrees with Kennan's 1 and 2, assumes 3 is already fact; NATO still goes through

6. **B**

Because this question is asking about a conclusion the author implies, this is an Inference question of the Implication subtype. According to paragraph 1, behavior modifications were Kennan's advocated method to improve international relations with Soviets. It relied on consistent positive and negative responses, and not on military or extreme actions. This method is in contrast to the military strategy presented later in the passage, NATO—which Kennan opposed. Predict that Kennan's approach is more subtle and less aggressive than the other option. This matches closely with **(B)**. **(A)** isn't well-supported by the passage, and it comes across as an Opposite. *Conciliatory gestures* would mean that the Soviets are taking actions that the Western nations approve of, so this would theoretically mean that great progress (success) was underway. **(C)** ascribes an extensive background to Kennan that was never mentioned (or even hinted at) in the passage. Finally, **(D)** has two issues: first, American World-War-II strategy was not explicitly described in detail (to determine if this could be an offshoot from that), and second, it appears that Kennan's *"behavior modification"* strategy wasn't actually used by America—against Kennan's reservations.

7. **D**

This is another Inference question, so we start by identifying the relevant text in the passage. The most sensible place to go is the paragraph where NATO is first mentioned, paragraph 3. Here, we learned that Western nations were uneasy and felt that the Soviet military was too big. Specifically, Russia had more divisions than combined Western forces, and the West feared a possible (but entirely theoretical) attack. This matches closely with **(D)**. **(A)** intones a similar idea, but is Extreme and distorts the information. The attack was possible, but there was no evidence that *information* was available that showed the attack was *impending* (about to happen). **(B)** is similarly Extreme; even though certain Western European countries feared an invasion (and the passage hints at the dire fate of even smaller countries), the idea that *no nation could withstand a Russian attack* is simply too broad. Finally, **(C)** contradicts both Kennan and the administration, who say in the passage that *emphasis on rearmament would delay recovery*. Thus, while NATO did guarantee military assistance, it went directly against the principle of *aid[ing] Western European economic recovery*.

8. **B**

The word *assumed* tells us that this is an Inference question of the Assumption subtype. Our task is to find a missing piece of evidence for Kennan's first conclusion in paragraph 4. Kennan accused "the Europeans [of] mistak[ing] what was essentially a political threat for a military one." One way to attack an Assumption question is the Denial Test, described further in Chapter 9. In the Denial Test, one takes the opposite of each answer choice—whichever answer choice, when turned into its opposite, undermines Kennan's argument will be the correct answer. The opposite of **(A)** would be: *The formation of the military alliance would not spur economic growth.* This is a statement Kennan is likely to agree with, given the statement that "emphasis on rearmament would delay recovery" in paragraph 5, so it can be eliminated. The opposite of **(B)** would be: *The presence of the thirty Soviet divisions meant they were going to attack.* This undermines Kennan's argument because it means that the Europeans were not just responding to a political threat—there was a very real military threat as well. Because this negated statement ruins Kennan's argument, this must be the correct answer. Negating **(C)** would yield: *The economic recovery in Europe had been progressing quickly.*

This assumption does not impact Kennan's argument and can be eliminated. Finally, **(D)** if negated would suggest that it is not always a mistake to favor military over economic affairs, but this doesn't negatively impact Kennan's argument: Kennan never suggests such an extreme position.

9. **D**

This is yet another Inference question, asking for Kennan's likely response. Kennan's second point of contention was that other countries are vulnerable. Kennan is worried that "any alliance extended to only some countries would render the rest more vulnerable." **(D)** paraphrases that prediction. Both **(A)** and **(B)** express a neutral to positive reaction, which does not match Kennan's actual response. **(C)** might sound reasonable in a real-world scenario, but Kennan was generally against uniting formally against Russia. It's doubtful, then, that he would be *disappoint[ed] that those countries could not now be employed in anti-Soviet strategy.*

10. **C**

This is a Detail question, asking for the Truman administration's thoughts on Kennan's final point of contention in paragraph 4. Paragraph 5 is where we learn what the administration's feelings actually are. The end of that paragraph serves as an excellent prediction: "most observers already regarded division, by mid-1948, as an accomplished fact." This matches **(C)**. **(A)** is a Faulty Use of Detail because it ascribes one of Kennan's opinions from the end of paragraph 4 to the Truman administration. **(B)** distorts the description of the troops in Europe. The passage discusses the Western European troops that are present, but doesn't mention how many—or whether there even are—American ones there. Further, there is no evidence in the passage that the Truman administration desired *a strong American military presence* in Europe at all. **(D)** is an Opposite: the administration agreed with Kennan on the first two points (mostly), giving them *cause for concern.*

11. **A**

This Inference question is quite similar to the Detail question that preceded it. We know from paragraph 5 that the Truman administration agreed with Kennan on two of his three points, but that they still decided to go ahead with the treaty "despite [his and their] reservations." This prediction matches nicely with answer **(A)**: the administration was *often in agreement* with Kennan but *ultimately undeterred* by the reservations that he had. **(B)** indicates that Kennan had specific *proposals*, but they were never mentioned or implied in the passage. Kennan had general strategies like *"behavior modification,"* but not specific proposals. **(C)** is patently untrue because the administration actually agreed with the majority (two out of three) of Kennan's points. Finally, **(D)** was also never directly discussed, and—if anything—the administration went ahead with actions to further *military assistance* and *economic revival* (the only two objectives that we're told about in the passage).

12. **C**

This is an excellent chance to turn the main idea of the passage into a prediction. Kennan begins by providing general suggestions on how to improve relations with Russia, before moving on to a more focused discussion on his reservations about NATO and his expectations that it would negatively impact Soviet relations. Kennan's *"behavior modification"* strategy, first introduced in paragraph 1, was hampered by the formation of NATO, and NATO was formed in response to the military imbalance in Europe. Thus, overall, Kennan clearly believed that the development of NATO was not beneficial to international relations with Russia. This matches closely with **(C)**. **(A)** and **(B)** are Extreme: NATO had *drawbacks*, sure, but was not *render[ed] . . . ineffective* by them. And behavioral modification is one way—but not the *only way—to change a government's concept of international relations.* **(D)** is what the members of NATO clearly thought, but it is neither a main theme of the passage nor what Kennan—the primary voice in the passage—thought.

CHAPTER 8

Question and Answer Strategy

CHAPTER 8

Question and Answer Strategy

In This Chapter

Introduction

LEARNING OBJECTIVES

After Chapter 8, you will be able to:

- Solve questions by applying the Kaplan Method for CARS Questions: Type, Rephrase, Investigate, and Match
- Recognize and avoid common wrong answer pathologies
- Identify correct answers by focusing on scope, author agreement, and tone

Thus far, this book has examined the multifarious aspects of *Critical Analysis and Reasoning Skills* (CARS) passages. You've learned about their rhetorical and logical characteristics in Chapter 3. You've seen how to approach them with thoughtful Previewing in Chapter 4 and Reading in Chapter 5. Chapter 6 built on these foundations and focused on Distilling the meaning of the passage. Finally in Chapter 7, these threads were brought together as we discussed the Kaplan CARS Passage Strategy.

Now, we turn to question stems and answer choices. We'll start by outlining the Kaplan Method for CARS Questions. Subsequently, we'll look at the recurring traps that the test makers set for the unwary student, which we call Wrong Answer Pathologies. In the final portion, we'll consider the counterpart to pathologies: patterns common in correct answers.

8.1 The Kaplan Method for CARS Questions

In Chapter 7 of *MCAT CARS Review*, we introduced the Kaplan Method for CARS Passages and explored the ways in which it can be applied to a variety of passage types. In this section, we'll introduce the Kaplan CARS question method, which takes the basic form shown in Figure 8.1.

TYPE THE QUESTION

- Read the question, **NOT** the answers
- Identify the question type and difficulty
- Decide to attack *now* or *later* in the same passage

REPHRASE THE STEM

- Determine the task to be accomplished based on the question type
- Simplify the phrasing of the original question stem
- Translate the question into a specific piece of information you can either locate or infer

INVESTIGATE POTENTIAL SOLUTIONS

- Search for the answer in your interrogation, your outline, or the passage
- Predict what you can about the answer
- Be flexible if your initial approach fails: when in doubt, refer back to the passage itself

MATCH YOUR PREDICTION TO AN ANSWER CHOICE

- Search the answer choices for a response that is synonymous with your prediction
- Eliminate answer choices that diverge from the passage
- Select an answer and move on

Figure 8.1 The Kaplan Method for CARS Questions

Note: The Kaplan Method for CARS Questions, as well as the Kaplan Method for CARS Passages, CARS Question Types, and Wrong Answer Pathologies, are included as tear-out sheets in the back of this book.

Type

You might notice that the first step of the question method is similar to the Preview step of the Kaplan Method for CARS Passages. This is not a coincidence, but rather a consequence of the timing constraints posed by the section. Every question is worth the same number of points, which means you want to avoid being derailed by any given question. It is important to approach the MCAT by being realistic about the nature of the exam: at least one or two questions within each section are so difficult that you're likely to get them wrong no matter how many minutes you spend on them. Though it may be tempting to spend a long time on a challenging question, it is better to recognize which questions those are right away, so you can instead use that time to gain more points by answering a larger number of more straightforward questions.

To that end, your first task with any question will be to read the stem, and *only* the stem, for the sake of deciding either to work on it *now* or to triage (to use an apt medical metaphor) and save it for *later in the passage*. In Chapter 4, we discussed the variety of factors used to assess the difficulty of passages during the Preview step of the Kaplan CARS Passage Method. For questions, however, difficulty is best assessed by identifying the **question type**. Because question type is the main variable to be focused on, this step of the Question Method is referred to as the Type step. The three broad categories of question that the AAMC presents in CARS were introduced in Chapter 1, and these categories can be further separated into more granular question types. Almost all *Foundations of Comprehension* questions fall into one of four types, while *Reasoning Within the Text* and *Reasoning Beyond the Text* categories can each be split into two predominant types, along with a handful of other types that only appear rarely. All of these question types are discussed in greater depth in Chapters 9–11. Starting each question with the Typing step not only allows for proper assessment of difficulty but also makes Rephrasing and Investigating a question much easier. We will discuss how question type indicates difficulty and how to apply the Type step most effectively to each question type in Chapters 9–11.

REAL WORLD

Think of question typing as equivalent to diagnosing a question. In medical school, you will learn the appropriate steps to undertake if a patient presents with chest pain, abdominal pain, a severe headache, confusion, and so on. The ten question types should be thought of as the "presentation" of the question, which determines the appropriate steps to take through the rest of the Kaplan Method for CARS Questions.

Why avoid looking at the answer choices when determining question difficulty? The primary reason is that most of them are wrong. If you glance at just one of them, for instance, it's three times more likely to be incorrect than correct and could seriously mislead you about the question. Inexperienced test takers immediately jump to the answers, and the AAMC punishes them for it by wording wrong options seductively. Selecting the first answer that looks good without first thinking through the question itself is a recipe for failure. Thus, in the next steps, Rephrase and Investigate, we will cover how to think through the question before jumping to the answer choices in the Match step.

MCAT EXPERTISE

To avoid getting seduced by wrong answer choices, cover all answer choices with your hand or your noteboard; leave them out of your field of view until you're ready for them in the Match step. Reading the answer choices can lead to the misinformation effect, discussed in Chapter 3 of *MCAT Behavioral Sciences Review*.

Rephrase

Once you have decided to attack a question, it's time to rephrase it. The **Rephrase** step goes beyond simply restating the question stem. The purpose of the Rephrase step is to provide you with a clear **task** and direction that you can use to attack the question. To this end, rephrase the question stem, focusing on the task itself and any relevant context given in the stem, to clearly identify what the question is demanding of you. Simpler question types like Main Idea and Definition-in-Context always involve one specific task (recognizing the big picture and explaining the meaning of part of the text as used in the passage, respectively). Even the most complex question types will have one major task to accomplish, though it may involve multiple steps. For example, almost all questions of the Apply type involve one of three tasks: gauging the author's response, predicting a likely outcome, or finding a good example (as described in Chapter 11 of *MCAT CARS Review*). Even if you feel you understand the task as phrased in the original question stem, the Rephrase step is an important check to ensure that you have fully read the question and know what you have to do to answer it. One of the most common causes of missed questions in the CARS section is misreading of question stems; taking the additional time to Rephrase will earn you valuable points on Test Day.

Sometimes rephrasing is more difficult to accomplish, such as when the task is obscured by unclear language or by extraneous information in the stem. In either of these cases, try working your way backwards, starting with the part of the question directly before the question mark or colon (in a question that requires completing a sentence). Try to simplify the phrasing of that part first, and then connect it to any other relevant information in the stem. If you still struggle to identify the task of the question, consider the question type you found in the Type step, because different instances of the same question type often have similar tasks. Finally, don't waste time on Test Day writing out the rephrased question; simplifying the question in your head is usually sufficient.

It's worth pointing out that Rephrasing the stem into a task can be done even without any passage knowledge due to the standardization of CARS questions. Taking notice of common patterns will allow you to devise a plan of attack even if you had difficulty with the passage itself. This skill of Rephrasing will require practice to master, but it's worth it. To help you master this step, here are some example question stems from the passage at the end of the chapter, along with sample rephrasing and identification of the task:

The author mentions Knorozov in the third paragraph in order to:

- Simplify: Why was Knorozov mentioned (P3)?
- Task: Identify how Knorozov was used to build the author's argument.

Which of the following would most call into question the author's argument about the complexity of Mayan writing?

- Simplify: What would weaken the author's argument about the complexity of Mayan writing?
- Task: Recall or refer to the author's argument on Mayan writing complexity and consider how each answer would affect that argument.

The author of the passage would be LEAST likely to agree with which of the following?

- Simplify: What would the author disagree with?
- Task: Recall or refer to the author's central ideas and arguments and identify which answer choice the author would disagree with.

MCAT EXPERTISE

If you are unable to Rephrase a question and arrive at a series of actionable steps, review the question types and their tasks in Chapters 9, 10, and 11.

Notice that the rephrased question stems are fairly similar to the original wordings. This is often the case, particularly for more straightforward questions. Remember, simplifying the question stem is a stepping stone to allow you to Rephrase the question stem as a task to be accomplished. You should not need to simplify the phrasing of each question, but it is an option if you need to. Notice that the third question stem is simplified by converting the clunky "LEAST likely to agree" to *disagree*. Other troublesome phrases, such as double negatives, should also be simplified to avoiding careless misreading errors.

After the Rephrase step you should have a clear understanding of the task(s) to be completed in order to reach the correct answer. *How* you go about accomplishing those objectives is the focus of the Investigate step.

Investigate

The next step in the question and answer strategy is to **Investigate** potential solutions using your rephrasing of the question. Specifically, you will follow the directives in your Rephrase to **predict** what the correct answer should look like. How you use the passage to make this prediction depends on your passage approach (**Highlight**, **Outline**, or **Interrogate**).

Highlighting

A **Highlighting** approach to the passage should provide the location of important pieces of information and central ideas, but will typically lack details. Thus, the corresponding Investigate step should include rereading specific portions of the passage, as directed by the Rephrase step and the ideas highlighted. Ensure that you are not scanning the entirety of the passage during this step; instead consider the task of the question and your highlighting to determine where you should look. For example, consider the following rephrasing of the second question from above:

Rephrase: Understand the author's argument on the complexity of Mayan writing and consider how the answer choices affect that argument.

Investigate: Review what you've highlighted to locate where in the passage the complexity of Mayan writing was mentioned, and reread it for deeper understanding in order to form a prediction about what would weaken it.

Outlining

An **Outlining** approach to the passage should not only note the location of important pieces of information, but also summarize the central ideas of each paragraph. Thus, you should expect your written outline to be sufficient to form predictions for questions that require understanding of the main ideas. However, questions that ask about a specific detail or require inferences from passage information will often require returning to the passage. Again, ensure you are not scanning the entirety of the passage to make your prediction—use your outline to determine where to look. Now, consider the same example from before with outlining:

Rephrase: Understand the author's argument on the complexity of Mayan writing and consider how the answer choices affect that argument.

Investigate: Review your outline to find the relevant paragraph(s), as well as the conclusion of the argument if you've included it. Use that to predict what would weaken the argument or, if that's insufficient, return to the relevant paragraph(s) and review the argument's evidence to aid you.

Interrogating

An **Interrogating** approach to the passage will usually be in-depth enough to leave you with a solid understanding of the central ideas of the passage and their interconnections. As a result, you'll find you can answer many questions without even referring back to the passage. Of course, a quick double check of the passage is always allowed, but keep in mind that with an Interrogating approach there will not be much time left for passage research during the questions. Let's look at the same example one more time, now with a mind toward interrogating:

Rephrase: Understand the author's argument on the complexity of Mayan writing and consider how the answer choices affect that argument.

Investigate: Recall from your interrogation the conclusion of the argument and the type of evidence the author used to support it, and then consider how that argument might be weakened. Use that to make a prediction and only return to the passage if you get stuck.

While investigating your predictions, it's important to be mindful of how specific or in-depth your prediction should be. Predictions lay along a spectrum from focused to general. Focused predictions should be specific enough to allow you to directly match your prediction to an answer choice. In contrast, general predictions just set broad expectations about what the correct answer should include or exclude. Where your prediction falls on this spectrum primarily depends on the question type and question task. Your mantra for predicting should be *predict what you can*. Here are some examples:

The author mentions Knorozov in the third paragraph in order to:

- Focused prediction: This prediction should refer specifically to the role Knorozov plays in paragraph three.
- The correct answer choice will likely be a paraphrase of your prediction.

Which of the following would most call into question the author's argument about the complexity of Mayan writing?

- General prediction: This prediction begins with a summary of the author's conclusion about the complexity of Mayan writing and the kind of evidence used to support it. You'll be looking for something that weakens this argument.
- The correct answer will be a statement that weakens the argument by attacking its conclusion, its evidence, or an unstated assumption.

The author of the passage would be LEAST likely to agree with which of the following?

- General prediction: This prediction should consist of a brief recounting of the major ideas of the passage. You'll be looking for something inconsistent with one or more of these ideas.
- The correct answer will often be a statement that conflicts with one of the major ideas. However, in some cases, the disagreement may be with one of the more specific details of the passage, so you may need to use process of elimination, ruling out answer choices that are consistent with your summary of the major ideas.

Armed with these predictions, you will be better prepared to evaluate the gauntlet of misleading answer choices. But perhaps you're wondering: why bother with this Investigation step and with making predictions at all? After all, you've taken multiple choice tests before that included reading comprehension sections, and perhaps you've never needed much strategy to do well on them. Why change when it comes

MCAT EXPERTISE

If you ever find yourself searching the entirety of the passage during your Investigate step, you could be in trouble! When returning to the passage, you should know what you are looking for and its general location in the passage. If you don't know where in the passage to look, stop and save the question for later in the passage. It's likely that while answering the other questions in the passage, you will become more familiar with the passage and may even stumble upon the information you needed. This problem is most likely to occur when using the Highlighting method for the passage, but fortunately this quick approach should leave you with some extra time to investigate the questions.

to CARS? There are several reasons why Investigating is key to success in CARS. For one, the CARS section is quite different from postsecondary reading comprehension exams. CARS passages are more likely to be on topics that you lack familiarity with, so you're less likely to have an intuitive sense of what the answers to questions will be. In addition, the answer choices in the CARS section are designed to lure testers who do not use the passage to answer the question. They often sound like something you read in the passage, or they appeal to outside information that seems right but is irrelevant. Without a thoughtful prediction you are more likely to fall for these alluring yet incorrect answer choices.

An apt analogy would be going to the grocery store hungry and without a shopping list. As many of us have experienced in this very situation, you are more likely to succumb to the influence of professional marketers, which may be peddling unhealthy food choices. If you had a grocery list, you'd be more likely to leave with healthier options and the items you actually require. Plus, you'd likely finish your grocery shopping faster, as you'll avoid aimlessly walking down the aisles trying to figure out what it was you were intending to bring home.

Finally, what should you do if you can't locate relevant information in the passage or don't know where to look? Typically such questions are best to try later, at the very end of a passage set, after you've researched the other questions and already reread some of the text. You may find that by the time you return to it, the effort you put into other questions ended up revealing an unhelpful question stem's correct answer. When you do attempt these questions, a general prediction and process of elimination will usually end up being the best plan. As a final note here, you will always want to return to these questions and at least guess before you leave the passage, since it is more effective in CARS to complete all questions when you work their relevant passages, rather than returning at the very end of the section.

Match

The final step of the question method is to **Match** your prediction to the correct answer. First evaluate the choices, and if you see an item that closely resembles what you expected, reread every word of that answer choice carefully to make sure it says precisely what you think it does. Rarely will you find a word-for-word match for your prediction. Instead, your best bet is to look for a correspondence between ideas, by searching for a choice that shares a similar meaning with your prediction but uses different words. Once you have found such an answer, select it and move on to the next question. At that point, *reading the other choices will not be worth your time*—be confident that you've answered the question when you find that conceptual match.

If you aren't able to find a choice that is synonymous with your prediction, don't feel that you immediately need to resort to process of elimination (although that is a valid strategy). Part of being flexible is being able to revise your initial prediction; to set new expectations if the answer choices point you in that direction. The answer choices could technically be considered an additional source of information for arriving at the correct answer, but keep in mind that they include a lot of misinformation and so should be treated with caution.

MCAT EXPERTISE

If you read a question stem and it doesn't give you very much to work with, don't just say *I don't know* and jump straight to the answer choices. Take a moment to remember the main themes of the passage and then use those themes to help with the process of elimination. This will help you avoid being distracted by answer choices that are seductive but that do not fit with the passage.

Sometimes the question stem just doesn't give you very much to work with, and on other occasions you'll search through the answers for your prediction but find no likely match. In these cases, you will have to use the process of elimination, which may require multiple returns to the text as you research each choice individually. If you were able to set expectations during the Investigate step for wrong choices, however, less additional research will be required. Keep in mind that an answer requires only one major flaw for elimination, so the Wrong Answer Pathologies described in the next section can greatly expedite the process.

When all else fails, you can fall back on educated guessing. Eliminate whatever you can and then go with your gut among the remaining options. Never make a blind guess unless you're completely out of time and need to fill in an answer choice. Even crossing off just one wrong answer will increase your chances of randomly choosing the correct one by 33 percent, while crossing off two doubles your chances. If time allows, try working on any unanswered questions for the passage and see if that effort allows you to return to rule out additional incorrect options.

MCAT EXPERTISE

Should I compare answer choices?
Your default assumption should be that only one answer choice is flawless and that the others contain at least one flaw each, sufficient for ruling them out. However, you may occasionally find questions containing superlatives (*strongest challenge*, *most supported*, *best example*, and so on) in which you need to compare two or more answers that have the same effect but to different degrees. When making such comparisons, don't assume that an Extreme answer is necessarily wrong, especially if the question stem includes the words *if true* or similar language. A stronger answer that nevertheless produces the desired outcome would be the correct choice. Thankfully, questions that actually require distinctions of this type are rare.

8.2 Wrong Answer Pathologies

The AAMC has designed the CARS section to be a fair test of critical thinking skills. The focus on fairness is great news for test takers because it means that the questions are not designed to unfairly mislead you. There will never be a question with two correct answer choices or one in which all of the options are wrong. Each question you encounter on Test Day will have one right answer and three that are incorrect for at least one reason. Even better, there are only so many ways an answer can be incorrect; in fact, a few of them are found so frequently that you can treat them like recurring signs and symptoms of answer choice "illness." Naturally, we call them **Wrong Answer Pathologies**.

A choice only needs one fatal flaw to be worth eliminating, but often wrong answer options have many issues, so don't necessarily be alarmed if you ruled out a wrong answer for a different reason than the one mentioned in a practice question's explanation. In addition to having some occasional overlap, the following list of pathologies is not meant to be exhaustive; it includes only the four patterns we've identified as the most common through researching all of the released MCAT material. In the Kaplan Method for CARS Questions just detailed, pathologies function as recurring expectations for wrong answers, which you can assume fit for most of the questions you encounter (with a few significant departures noted below).

Note: The Wrong Answer Pathologies, as well as the Kaplan Method for CARS Passages, Kaplan Method for CARS Questions, and CARS Question Types, are included as tear-out sheets in the back of this book.

KEY CONCEPT

Wrong Answer Pathologies are the most frequent patterns found in incorrect answer choices. They are so common that you'll find at least one in just about every CARS question and even in many of the questions in the three science sections!

Faulty Use of Detail (FUD)

The test makers will often include accurate reflections of passage details in wrong answers, primarily to appeal to those students who jump at the familiar language. What makes the use of a detail "faulty" is that it simply doesn't answer the question posed. It may be too specific for a question that requires a general answer, or it may be that the detail comes from the wrong part of the passage. Even if a choice comes from the right paragraph, the detail cited might not be relevant to the question posed, which is often the case in Strengthen–Weaken (Within the Passage) questions. A thorough prediction made in your Investigate step makes catching these FUDs much easier. Remember, the correct answer must be true to the passage and must answer the question!

Out of Scope (OS)

With the noteworthy exception of *Reasoning Beyond the Text* questions (for which this pathology does not usually apply), an answer choice that is outside the scope of the passage will inevitably be wrong. Typically, such answers will be on topic but will bring in some element that the text does not discuss. For instance, if an author never makes comparisons when discussing different ideas, an Out of Scope answer choice might involve the author ranking two or more concepts. Another common OS pattern is the suggestion that an entity or idea was the first of its kind or the most influential, when the author entirely avoids discussing its historical origins or impact. Remember that information that is unstated but strongly suggested (such as assumptions and implications) does not count as Out of Scope as will be the case with the correct answers to many *Reasoning Within the Text* questions, so don't be too quick to reject a choice as OS just because the author does not explicitly say it.

Opposite (OPP)

Whenever an answer choice contains information that directly conflicts with the passage, we call it an Opposite. Often the difference is due simply to the presence (or absence) of a single word like *not* or *except*, a prefix like *un–* or *a–*, or even a suffix like *–less* or *–free*. Be especially careful when stems or choices involve double (or triple) negatives; they're much less difficult to understand if you Rephrase them with fewer negations. Moreover, don't assume that just because two answer choices contradict each other that one of them has to be correct. For example, suppose an author argues that it is impossible to prove whether or not a divine being exists, a variant of the religious view known as *agnosticism*. If a question accompanying the passage were to ask for a claim the author agreed with, *God exists* and *There is no God* would both be Opposites of the correct answer.

REAL WORLD

A Faulty Use of Detail answer choice is like a politician who dodges a question during a debate:

- Moderator: *Candidate X, what will you do to improve the economy?*
- Candidate X: *You know, that's a wonderful question. My platform is to stand for all Americans' rights and to represent them fairly. If elected, I will help lead our country with integrity.*

Even if the candidate is being honest, the response just doesn't answer the question posed!

BRIDGE

The answers to many *Reasoning Within the Text* questions will be claims unstated by the passage that can nonetheless be inferred. Remember that inferences include assumptions (unstated evidence) and implications (unstated conclusions). Inferences are discussed in Chapter 3 of *MCAT CARS Review*.

Distortion (DIST)

Extreme answers and other answers that "twist" the ideas in the passage further than the author would prefer are what we call Distortions. Although they do not automatically make a choice incorrect, the following are common signals of distorted claims:

- Strong words like *all*, *always*, *none*, *never*, *impossible*, or *only*
- A prefix like *any–* or *every–*
- A suffix like *–est* or *–less*

MCAT authors typically do not take radical positions on issues, so it's worth noting whenever they do. In those rare cases, extreme choices would not actually be Distortions of the author's view and might be correct. The other major case in which extreme answer choices should not be immediately ruled out is when the question stem tells you that you can treat the answer choices as true and your task is only to gauge which would have the greatest impact on a particular argument. This is often the case with Strengthen–Weaken (Beyond the Passage) questions.

8.3 Signs of a Healthy Answer

If you're like most students prepping for the CARS section, you've had a dispute with at least one question explanation. *Hey, what about what the author says in the first paragraph?* you may have wondered, or perhaps you've said to yourself (or aloud!), *But couldn't you think of it like* this *instead?* While you may be in the habit of arguing for points with college professors, it does you no good to try to argue with the MCAT. The AAMC is extremely deliberate about how they word correct answers, always taking care to include exactly one per question.

Correct answer choices can vary widely in appearance, but there are patterns in how they are written as well. If the traps that can lead you astray on Test Day are appropriately called Wrong Answer Pathologies, then these corresponding traits can be thought of as indicators of good health. While the following signs are not enough by themselves to make an answer right, you can generally expect them to correspond to the correct choices in most types of questions.

Appropriate Scope

You might say correct answers follow the "Goldilocks principle" when it comes to scope: not too broad, not too specific, but just right. The **scope** defines the limits of the discussion, or the particular aspects of the larger topic that the author really cares about. Consideration of the purpose of the passage as you Read and Distill should give you an idea of the scope of the passage overall. As a general rule (with an important exception), correct answers to MCAT questions will remain within the scope of the passage, but you can formulate a more precise expectation of what scope the correct answer needs to have by identifying the question's type and task.

MCAT EXPERTISE

The scope of a text refers to the particular aspects of a topic that the author addresses. Every paragraph in a CARS passage has its own scope, and together you can think of them as constituting the scope of the whole passage. Similarly, each answer choice will have its own scope, which could mimic any part of the author's discussion or depart from the passage entirely. It is essential to note that having the same scope doesn't necessarily mean having identical content. For instance, unstated assumptions in an argument are definitely within the scope of the passage, even though the information they contain is left unsaid by the author.

Main Idea questions will always have correct answers that match the scope of the entire passage. They will typically include at least one wrong answer that is too focused (Faulty Use of Detail) and at least one that goes outside the passage entirely (Out of Scope). In contrast, Detail and Definition-in-Context questions usually require more refined scopes to identify their correct answer choices. If a clue directs you to a particular portion of the passage, the correct answer, more often than not, will have the same scope as the referenced text (or what immediately surrounds it).

The important exception to the rule that answers must remain within the scope of the author's discussion applies to the category of *Reasoning Beyond the Text* questions, addressed in Chapter 11. As their name suggests, these broaden the scope to new contexts, sometimes appearing to have no connection to the passage whatsoever. Note, however, that some *Reasoning Beyond the Text* questions will present new information in the stem but have answers that stick to the scope of the passage anyway. So be savvy with the answer choices in *Reasoning Beyond the Text* questions: while the correct answer choice will tend to lie slightly outside the scope of the passage, don't automatically rule out an answer choice just because it *happens* to be in scope.

Author Agreement

Unless a question stem explicitly asks about an alternative viewpoint or a challenge to the information presented in the passage, a correct answer choice will be consistent with what the author says. This is one reason why considerations of **tone** (most clearly reflected by Author keywords) are usually important enough to be worth including as you Read and Distill the passage (Highlight, Outline, Interrogate), as was recommended in Chapter 5. Generally, a correct answer should not contradict anything that the author says elsewhere in the passage, with the possible exception of sentences that speak in a different voice than the author's (such as quotes or references to others' opinions). In short, if it doesn't sound like something the author would say, you'll most likely want to rule it out.

Synonymous Phrasing

The correct answer should match your prediction made in your Investigate step, be it a focused or general prediction. When evaluating answer choices using your prediction, keep in mind that the AAMC often phrases their correct answers with different terms than those presented in the passage. This can lead students to mistakenly rule out correct answers as Out of Scope, simply because they contain unfamiliar language. Thus, when evaluating answer choices, it is key to remember that consistency of meaning is more important than consistency of phrasing.

Weaker Is Usually Better

One final consideration is a consequence of the fact that the AAMC tends to select passages by authors who do not take extreme views. You may find one or two passages on Test Day with more radical writers; for them, a stronger claim in the answer choices may actually be a good sign. However, for most of the passages you'll encounter, authors tend to use numerous Moderating keywords to limit the strength of their claims. Because a stronger claim has a higher burden of proof (that is, stronger evidence must be provided to support the claim), most authors avoid them to make what they write seem more plausible. Thus, you should generally give preference to answer choices that use weaker language such as *can*, *could*, *may*, *might*, *is possible*, *sometimes*, *often*, *likely*, *probably*, and *in some sense*. Exceptions to this tendency are questions that instruct you to consider the answer choices as true and gauge their effect on an argument. These were addressed earlier in the discussion of the Distortion Wrong Answer Pathology.

Conclusion

This chapter is only an introduction to the question method; the three chapters that follow are a necessary supplement for seeing how the method functions in practice. Specific strategy suggestions and worked examples are included for each of the most common question types and tasks, together constituting well more than 90 percent of what you'll encounter on Test Day. The explanations accompanying these sample questions will also identify their Wrong Answer Pathologies, giving you some concrete examples to go with the explanations provided here.

Concept and Strategy Summary

Kaplan Method for CARS Questions

- **Type** the question
  - Read the question, **NOT** the answers
  - Identify the question type and difficulty
  - Decide to attack *now* or *later in the same passage*
- **Rephrase** the stem
  - Determine the task to be accomplished based on the question type
  - Simplify the phrasing of the original question stem
  - Translate the question into a specific piece of information you can either locate or infer
- **Investigate** potential solutions
  - Search for the answer in your interrogation, your outline, or the passage
  - Predict what you can about the answer
  - Be flexible if your initial approach fails: when in doubt, refer back to the passage itself
- **Match** your prediction to an answer choice
  - Search the answer choices for a response that is synonymous with your prediction
  - Eliminate answer choices that diverge from the passage
  - Select an answer and move on

Wrong Answer Pathologies

- **Faulty Use of Detail** (**FUD**) answer choices may be accurate statements, but they fail to answer the question posed.
  - The answer choice may be too specific for a question that requires a general answer.
  - The answer choice may use a detail from the wrong part of the passage.
  - The answer choice may be from the right paragraph but still not be relevant to the question posed.
- **Out of Scope** (**OS**) answer choices usually bring in some element that the passage does not discuss (and that cannot be inferred from the passage).
  - The answer choice may make connections or comparisons that the author did not discuss.
  - The answer choice may make a statement about the significance or history of an idea that the author did not.
  - The answer choice may otherwise bring in information that does not fall within the constraints of the passage.

- **Opposite** answer choices contain information that directly conflicts with the passage.
  - The answer choice may contain (or omit) a single word like *not* or *except*.
  - The answer choice may contain a prefix like *un–* or *a–* or a suffix like *–less* or *–free*.
  - The answer choice may say that a given claim is true, when the author is ambivalent.
- **Distortion** answer choices are extreme or twist the ideas in the passage further than the author would prefer.
  - The answer choice may use a strong word like *all*, *always*, *none*, *never*, *impossible*, or *only*.
  - The answer choice may contain a prefix like *any–* or *every–* or a suffix like *–est* or *–less*.
  - The answer choice is usually more radical than the author because radical positions are hard to support and are rare in MCAT passages.

Signs of a Healthy Answer

- Correct answers tend to have the right **scope**—not too broad, not too specific, but just right.
- Correct answers tend to be consistent with the author's statements and opinions.
- Correct answers use language that differs from the passage but is still consistent with the ideas discussed.
- Correct answers tend to use Moderating keywords, such as *can*, *could*, *may*, *might*, *is possible*, *sometimes*, *often*, *likely*, *probably*, and *in some sense*.

Worked Example

Use the Worked Example on the next page in tandem with the subsequent practice passages, to internalize and apply the strategies described in this chapter. The Worked Example matches the specifications and style of a typical MCAT *Critical Analysis and Reasoning Skills* (CARS) passage.

The following passage is well suited for the Interrogating method. With a simple structure and four medium-length paragraphs, you will be able to spend more time thinking about the connection between paragraphs and why the author includes specific details. The passage is informative and relatively straightforward, and while there are a lot of details, they are strongly tied to the central argument of the passage, making Interrogating useful for identifying the unique contribution of each major new detail added.

INTERROGATING METHOD

Passage	Expert Thinking
Mayan signs are by nature highly pictorial, often representing in considerable detail animals, people, body parts, and objects of daily life. The pictorial principle is taken to the extreme in inscriptions composed of "full-figure" glyphs, in which individual signs and numbers become animated and are shown interacting with one another. None of this should be taken to mean that the Mayans had simple picture writing. The Mayans wrote both logographically and phonetically, and within its phonetic system alone, the Mayans had multiple options. All English words are formed from various combinations of only 26 phonetic signs. By contrast, all Mayan words can be formed from various combinations of nearly 800 consonant–vowel glyphs, each representing a full syllable. Sounds are formed by combining a particular consonant with one of the five vowels (hence a syllabary, rather than alphabet).	S1–2. Why did the author emphasize that Mayan signs are highly detailed? *The details in the images provide a lot of information, in contrast to the reader's possible preconception that signs are often simple and straightforward.* S3–7. Why is it important to note that the Mayans had two writing systems? Why does the author bring up the English language? *The author distinguishes between logographic (or pictorial) signs and phonetic (or syllabic) ones, explaining that the Mayans used both. She uses the English language as a contrast, demonstrating that its phonetic alphabet is simpler than the Mayan phonetic syllabary, just one of their two writings systems. This is consistent with the first two sentences, providing further evidence that the Mayans had a complex system of writing.*
The combination of consonant–vowel syllabic glyphs and logographs enabled the scribes a variety of choices to write the words of their texts in detail. For example, one very common honorific title in Mayan texts is *ahaw*, meaning "lord" or "noble." *Ahaw* may be written in logographic form as a head in profile, with the distinctive headband or scarf that marked the highest nobility in Mayan society. But it is also possible to write the word as a combination of three phonetic, syllabic signs: *a–ha–wa*. Likewise, the word *pakal* ("shield") can be indicated by a depiction of a shield or by the combination of syllabic elements *pa–ka–la*.	S1–5. Why does the author describe how specific words like *ahaw* work in the Mayan language? *These words are used as examples of how logographs and phonetic language can describe the same word. This provides further evidence for the argument the author introduced in the first paragraph: having a variety of ways to depict the same concept adds to the overall complexity and utility of the Mayan writing system.*

INTERROGATING METHOD	
Passage	**Expert Thinking**
Because many Mayan signs remain undeciphered, it's not possible to state precisely the relative proportions of logographic and syllabic signs. But a significant number of the logograms have been deciphered, and the number of deciphered syllabic signs keeps growing. Epigraphers have filled more than half of the syllabic grid (which plots the consonants of the spoken Mayan language against its vowels and thus represents the totality of signs needed to write the language). Half of the grid may seem a meager proportion, but it must be remembered that the discovery of the structure of the syllabic elements—Knorozov's main contribution—was made only a little more than 30 years ago. Furthermore, the consonant-vowel syllables that are already understood are the common ones. Many of the empty spaces in the syllabic grid remain so because they are linguistically rare; rare signs are more difficult to translate than common ones.	S1–5. Why has the author turned to discussing undeciphered Mayan signs? Why is it reasonable that approximately half of the possible Mayan signs have not been deciphered? *Keep in mind that the author is trying to build an argument that the Mayans had a particularly complex system of writing. The fact that the system is not completely understood could serve as a potential objection to this conclusion, so the author tries to set the record straight here. While a lot of the syllabary is unknown, researchers can at least identify what pieces are missing based on their knowledge of the Mayan language. It makes sense that only half have been deciphered because most of the missing pieces are uncommonly used syllables anyway and because the process of deciphering has only been going on for about 30 years (as the reference to Knorozov serves to highlight).*
Nonetheless, the pace of phonetic decipherment is bound to increase in the coming years as more resources are trained on it. One aspect of Mayan writing that may complicate this progress is the fact that different signs can have the same value. Two signs that share a value are known as allographs. Such equivalences are common in Mayan texts (there are at least five different signs that could be chosen to represent the Mayan syllable *ba*). Each scribe chose from several different signs to convey the sounds. In evaluating a particular phonetic interpretation of a syllable, it's helpful to identify as many of the variant forms as possible; the process of recognizing allographs depends on the slow work of comparing many texts in order to find variant spellings of the same word.	S1–5. Why do allographs possibly hinder deciphering Mayan writing? How does this relate to the bigger idea introduced in P1? *Allographs are two signs that share the same value, and the researchers need to compare many texts to find the different variants, slowing down the process. Ultimately, allographs relate to the central argument introduced in P1 because they indicate the complexity of Mayan writing.* *After reading the whole passage, the goal should be evident: the author writes in order to argue that the Mayan system of writing is distinctively complex and to explain some of this complexity.*

Question	Analysis
1. The author mentions Knorozov in the third paragraph in order to:	This is a Function question, which is a *Foundations of Comprehension* question. Because we read the passage noticing its rhetoric, this is a question that we can do right away. The question stem even tells us to go to paragraph 3, which deals with progress in completing the syllabic grid. Knorozov was used to justify why only about half of the syllabic grid has been filled in—he discovered the syllable structure only "a little more than 30 years ago."
A. prove that the recent discovery of Mayan signs has led to its lack of decipherment.	**(A)** is a Distortion on two counts: first, the recent discovery is of the syllable structure of the language, not the *Mayan signs* themselves; second, *prove* is far too strong a word—the author is not nearly so extreme.
B. offer an explanation for what may appear to be a relative paucity in the completion of the Mayan syllabic grid.	**(B)** fits and is the correct answer. Knorozov and his recent discovery is part of the justification for why there is so little filled in on the syllabic grid. This also fits with our interrogation of paragraph 3.
C. argue that expert linguists have been unsuccessful in their attempts to decipher and understand allographic Mayan signs.	**(C)** is a Faulty Use of Detail; the challenge presented by allographs is described in paragraph 4 and is not associated with Knorozov.
D. show how the understanding of other linguistic structures may improve the comprehension of Mayan syllabic signs.	**(D)** is a classic Out of Scope answer choice as *other linguistic structures* are never discussed in paragraph 3.

Question	Analysis
2. As used in the passage, the term "logographic" most closely refers to:	This is a Definition-in-Context question. As another *Foundations of Comprehension* question, we can do this right away. We saw "logographic" in paragraphs 1 and 2. When introduced in paragraph 1, the gist of the word was that it was a picture. Paragraph 2 then describes the writing of the word *Ahaw*: "*Ahaw* may be written in logographic form as a head in profile." Because a picture is used to represent the word in *logographic* form, they must be pictorial symbols.
A. a written phonetic representation of a word.	**(A)** and **(B)** are both Opposites that describe the other form of Mayan writing presented in the passage.
B. syllabic division of an individual word.	
C. representation of an idea as an image.	**(C)** simply paraphrases the idea of pictorial representation, so it's the correct answer.
D. a visual picture of an idiomatic phrase.	**(D)** is a Distortion; although logographs are visual representations, the author suggests that they represent single words, not entire *idiomatic phrase*[s].

Question	Analysis
3. The author of the passage would be LEAST likely to agree with which of the following?	This is an Apply question. Because this is a *Reasoning Beyond the Text* question type, this is one that we should save for later. The question asks what *the author . . . would be LEAST likely to agree with,* so the answer should be inconsistent with the author's beliefs. The author believes that Mayan writing is complex for several reasons but is steadily being deciphered. The correct answer will contradict this or some other point from the passage.
A. Languages with writing composed of pictorial signs can demonstrate a remarkable degree of complexity and detail.	**(A)** is an Opposite; the author would certainly agree with this statement given her belief that the language is more than just "simple picture writing"; the description of the "800 . . . glyphs," multiple ways of writing the same word, and allographs support this idea.
B. Linguistic signs based on syllabic or phonetic coding may be easier to decipher than those based on visual images.	**(B)** is Out of Scope; this comparison is never made, so it's impossible to say whether the author would agree with this statement or not.
C. Pictorial languages are restricted to the expression of simple ideas because of their emphasis on images.	The author would certainly disagree with **(C)**, making it the correct answer. Mayan writing doesn't convey only *simple ideas*, the author would argue, because it's more than just "simple picture writing."
D. The existence of allographs in Mayan signs indicates the complexity of this linguistic system.	**(D)** is an Opposite; this summarizes the point of paragraph 4: allographs are a complication that makes deciphering Mayan writing more difficult.

Question	Analysis
4. The author discusses the words *ahaw* and *pakal* in order to:	This is another Function question, so we should do it right away. Where did we see the words *ahaw* and *pakal*? The words were in paragraph 2. The purpose of paragraph 2 is to explain that a logograph (picture) and a combination of syllables can both be used to represent the same word.
A. estimate the number of meanings that some common Mayan words may possess.	**(A)** Distorts the author's point: the same word can have multiple ways of being represented, not multiple *meanings*.
B. compare the flexibility of Mayan logographs to that of consonant–vowel syllables.	**(B)** is Out of Scope, as the author does not suggest a comparison between the *flexibility* of phonetic and logographic writing—just that they can both be used to write certain words.
C. illustrate the difficulty of filling in the syllabic grid due to words being linguistically rare.	**(C)** is a Faulty Use of Detail; even though *the syllabic grid* has been slow to fill due to *linguistically rare* words, this is a point from paragraph 3 rather than paragraph 2, where *ahaw* and *pakal* are introduced.
D. demonstrate that Mayan words may appear in both logographic and syllabic form.	**(D)** matches cleanly with the prediction and is the correct answer.

Question	Analysis
5. According to the author, which of the following would best address some of the decipherment problems associated with Mayan signs?	This is an Inference question, asking for a conclusion the author did not explicitly state, or an implication. Because the example is not explicitly stated in the passage, we should save this question until later. Problems with *decipherment* are mentioned in paragraphs 3 and 4. In paragraph 3, the author noted that the remaining unknown syllables are rarely used and thus harder to translate. Paragraph 4 continued the idea by talking about allographs. The author does say in the last lines of the passage that deciphering allographs depends on *comparing many texts*. Look for an answer that sounds like the points in either of these two paragraphs.
A. Additional financial and scholarly resources should be directed toward this linguistic effort.	**(A)** is a Distortion; the author argues at the beginning of paragraph 4 that the "pace of phonetic decipherment is bound to increase" as more resources are trained on it, but stops short of saying that more resources *should* be trained on it, or suggesting what *form* those resources might take. This answer choice changes a statement of fact to a recommendation, as well as interpreting *resources* far more specifically than can be justified by the passage.
B. More attention should be focused on identifying logographic signs than on categorizing syllabic signs.	**(B)** is an Opposite; because allographs are phonetic representations, the author would argue that if anything, there should be more attention given to syllabic signs than logographic signs—many of which have already been deciphered.
C. Scholars should prioritize the completion of the syllabic grid.	**(C)** Distorts the author and is extreme; the author actually seems content with the current, incomplete state of the grid and its slow but steady progress toward being filled in.
D. Careful study of comparative texts should continue in order to evaluate phonetic interpretation of each syllable.	**(D)** closely reflects statements at the end of paragraph 4 that "the process of recognizing allographs depends on the slow work of comparing many texts in order to find variant spellings of the same word." That makes it correct.

Question	Analysis
6. The author implies which of the following about the ratio of logographic to syllabic signs in Mayan writing?	This is an Inference question and best saved for later. The question stem references the ratio between two types of signs. Refer back to paragraph 3 where this is discussed. The author says that "it's not possible to state precisely the relative proportions of logographic and syllabic signs" because many of the signs are undeciphered. The clear implication is that as these signs are deciphered, the ratio will become clearer.
A. Researchers disagree about the correct way to determine it.	**(A)** is Out of Scope; the author suggests that the lack of a ratio is due to a lack of information, not because of *disagree*[ment].
B. Its practical value has failed to attract serious attention.	**(B)** is also Out of Scope, as the author doesn't indicate anything about the *practical value* of knowing this ratio.
C. A meaningful ratio will likely never be established.	**(C)** is a Distortion; the author mentions that the number of deciphered signs is growing. So, if anything, the author would argue that a ratio may be established some day in the future.
D. More work must be done before the ratio can be determined.	**(D)** matches perfectly. Once all logographs and syllables are translated, we would know the ratio—but that will require *more work*.

Question	Analysis
7. Which of the following would most call into question the author's argument about the complexity of Mayan writing?	This is a Strengthen–Weaken (Beyond the Passage) question, so we should save it until last. We need to *call* the *complexity of Mayan writing into question*; this means we need to find that answer that would make the author's conclusion less likely. Where do we see the author's argument about the complexity of Mayan writing? In paragraph 3, the author states that Mayan writing is complex due to the fact that it includes both logographic and syllabic systems. In paragraph 4, the author goes on to state that part of the problem in deciphering Mayan writing is the presence of so many allographs that scribes had to choose from when writing. To make the conclusion about complexity less likely, we need to find a statement that challenges the conclusion, a piece of evidence used to support it, or an unstated assumption.
A. It has recently been discovered that allographs are actually just stylistic differences between scribes.	**(A)** is exactly what we are looking for: if allographs are actually just stylistic differences—and not "different signs that could be chosen to represent" the same syllable, then one key piece of evidence for the author's conclusion is refuted. While this isn't a particularly strong challenge to the argument (because there's a lot more evidence provided for it in the passage), it's stronger than any of the other choices, which makes it correct.
B. The syllabic grid is only partially complete.	**(B)** is a statement made by the author in paragraph 3. The author explains in that paragraph why this lack of completion is reasonable, so this choice doesn't challenge the author's argument.
C. Other pictorial languages, such as Egyptian, have been deciphered.	**(C)** is incorrect because the relative extent of decipherment was shown in paragraph 3 not to be relevant to the author's conclusion.
D. Languages based on logographs are less complicated than modern syllabic languages.	**(D)** is incorrect because Mayan is not a purely logographic language, but uses both logographs and a phonetic syllabary. The comparison here is irrelevant to the author's argument.

PRACTICE QUESTIONS

Passage 1 (Questions 1–7)

GO ONLINE

The rich analysis of Fernand Braudel and his fellow *Annales* historians have made significant contributions to historical theory and research, not the least of which is a broad expansion of potential routes of scholastic analysis. In a departure from traditional approaches, the *Annales* historians assume that history cannot be limited to a simple recounting of conscious human actions, but must be understood in the context of forces and material conditions that underlie human behavior. Braudel was the first *Annales* historian to gain widespread support for the idea that history should synthesize data from various social sciences, especially economics, in order to provide a broader view of human societies over time (although Febvre and Bloch, founders of the *Annales* school, originated this approach).

Braudel conceived of history as the dynamic interaction of three temporalities. The first of these, the *evenementielle*, involved short-lived dramatic "events," such as battles, revolutions, and the actions of great men, which had preoccupied traditional historians like Carlyle. *Conjonctures* was Braudel's term for larger cyclical processes that might last up to half a century. The *longue durée*, a historical wave of great length, was for Braudel the most fascinating of the three temporalities. Here he focused on those aspects of everyday life that might remain relatively unchanged for centuries. What people ate, what they wore, their means and routes of travel—for Braudel, these things create "structures" that define the limits of potential social change for hundreds of years at a time.

Braudel's concept of the *longue durée* extended the perspective of historical physical space as well as time. Until the *Annales* school, historians had generally taken the juridical political unit—the nation-state, duchy, or whatever—as their starting point. Yet, when such enormous timespans are considered, geographical features may well have more significance for human populations than national borders. In his doctoral thesis, a seminal work on the Mediterranean during the reign of Philip II, Braudel treated the geohistory of the entire region as a "structure" that had exerted myriad influences on human lifeways since the first settlements on the shores of the Mediterranean Sea. So, the reader is given such esoteric information as the list of products that came to Spanish shores from North Africa, the seasonal routes followed by Mediterranean sheep and their shepherds, and the cities where the best ship timber could be bought.

Braudel has been faulted for the impression of his approach. With his Rabelaisian delight in concrete detail, Braudel vastly extended the realm of relevant phenomena; but this very achievement made it difficult to delimit the boundaries of observation, a task necessary to beginning any social analysis. Indeed, to identify an appropriate scope of work when all options and directions for academic inquiry are available before one's eyes is a nearly impossible task. Further, Braudel and other *Annales* historians minimize the differences among the social sciences. Nevertheless, the many similarly designed studies aimed at both professional and popular audiences indicate that Braudel asked significant questions that traditional historians had overlooked.

1. The author refers to the work of Febvre and Bloch in order to:
   - **A.** illustrate the need to delimit the boundaries of observation.
   - **B.** suggest the relevance of economics to historical investigation.
   - **C.** debate the need for combining various social science approaches.
   - **D.** show that previous *Annales* historians anticipated Braudel's focus on economics.

2. In the third paragraph, the author is primarily concerned with discussing:
   - **A.** Braudel's fascination with obscure facts.
   - **B.** Braudel's depiction of the role of geography in human history.
   - **C.** the geography of the Mediterranean region.
   - **D.** the irrelevance of national borders.

3. The passage suggests that, compared with traditional historians, *Annales* historians are:
   - **A.** more interested in other social sciences than in history.
   - **B.** critical of the achievements of famous historical figures.
   - **C.** skeptical of the validity of most economic research.
   - **D.** more interested in the underlying context of human behavior.

4. Which of the following statements would be most likely to follow the last sentence of the passage?
   - **A.** Few such studies, however, have been written by trained economists.
   - **B.** It is time, perhaps, for a revival of the Carlylean emphasis on personalities.
   - **C.** Many historians believe that Braudel's conception of three distinct "temporalities" is an oversimplification.
   - **D.** Such diverse works as Gascon's study of the Lyon and Barbara Tuchman's *A Distant Mirror* testify to his relevance.

5. According to the passage, some historians are critical of Braudel's perspective for which of the following reasons?
   - **A.** It seeks structures that underlie all forms of social activity.
   - **B.** It assumes a greater similarity among the social sciences than actually exists.
   - **C.** It fails to consider the relationship between short-term events and long-term social activity.
   - **D.** It rigidly defines boundaries for social analysis.

6. Which of the following historical phenomena would the author most likely consider an example of the *longue durée*?
   - **A.** Julius Caesar's crossing of the Rubicon, which led to a four-year civil war in Rome
   - **B.** The occurrence in ancient Rome of devastating malaria outbreaks roughly every half-century
   - **C.** A gradual shift toward a drier Mediterranean climate that lasted from 250 to 600 c.e. and impacted food availability and daily life
   - **D.** The battle of Carrhae in 53 b.c.e., in which the Parthians exterminated a large Roman army

7. Suppose Braudel was once quoted as saying, "For me, the most interesting aspects of history are those in which one man's decisions turned the tides of history." What relevance would this new information have for the passage?
   - **A.** It would weaken the author's claim that Braudel was most interested in the *longue durée* temporality.
   - **B.** It would weaken the author's claim that Braudel was not interested in the actions of great men.
   - **C.** It would strengthen the author's claim that Braudel was most interested in the *evenementielle* temporality.
   - **D.** It would strengthen the author's claim that Braudel considered the best historical analysis to focus on the actions of great men.

Passage 2 (Questions 8–12)

A balance of power arises when a group of neighboring countries enters a state of economic and military equilibrium. In a balance of power system, a nation-state cannot violate the independence or the essential rights of another without incurring reprisals from neighboring states acting to restore balance. Conflict ought to be avoided, but may be necessary when a potential hegemon takes aggressive action. This is because peace or the safety of an individual nation is less important in balance of power than preserving equilibrium in the system. Diplomacy and trade in a balance of power is thus a continuum of action and reaction, rather than a series of attempts at independent policy-making.

What conditions are necessary for a balance of power to occur? First, states must be aligned in a "state system." The states involved must be independent, close in proximity (often possessing shared borders), and near equals in power. When one state far outstrips its close neighbors in power, it dictates economic and military policy for the region. When neighbor states are commensurate, however, interaction on issues of shared concern occurs on a level playing field of policy making.

A second factor in the formation of a balance of power is the framework of the system itself. To maintain an effective balance of power, a system must include a minimum of three states. A true equilibrium cannot exist between two states because one state inevitably gains ascendancy over the other. A balance of power is also generally characterized by a common ground of culture in the state system. A group of states is more likely to align in a cooperative manner if constituent states perceive a degree of cultural similarity with their neighbors. Added to this, the mechanism of diplomacy, skilled diplomats, and economic alliance structures must be in place for a balance of power to thrive. If this framework exists, then the system will be sufficiently flexible to survive short-lived economic fluctuations and military aggression.

Inherent in this need for a framework is the third precondition for the evolution of a balance of power: rational estimation. Countries involved in a balance of power must have a rational means of estimating the power of individual and combined states within the system. The flow of economic information between countries is crucial for an economic equilibrium to persist; a nation's economic planning should consider the dynamics of the entire state system in addition to its own agriculture and industry. A similar diffusion of information must occur on national security issues. This means that the stability of a state system depends on the development of sophisticated intelligence agencies for estimating the military capability and activity of other states.

The importance of military intelligence is perhaps best illustrated on a smaller scale. Suppose that your neighbor has stolen your lawnmower and you are considering taking retribution by picking a fight with him. For your survival and for the continued survival of the balance of power in your neighborhood, you must first be aware of all relevant personal defense issues. In particular, it is critical that you find out whether your neighbor possesses a gun and whether he might use it under such circumstances. Without the gathering of such information, the balance of power in the neighborhood cannot be maintained.

8. Which of the following helps explain why a balance of power is generally characterized by a "common ground of culture" (paragraph 3)?

- **A.** Countries with similar cultures often have mechanisms of diplomacy in place.
- **B.** Cultural differences between two nations are often the source of military conflict.
- **C.** A hegemon often influences the cultures of all nations surrounding it.
- **D.** Alliances among nations are more common if there are shared cultural beliefs.

9. The author's use of the phrase "level playing field" (paragraph 2) is probably meant to indicate that in a balance of power:

- **A.** states coordinate policy in a democratic manner.
- **B.** diplomatic initiatives generally work to the advantage of all member states.
- **C.** each state has some influence on economic and military events.
- **D.** member states form alliances against countries outside the balance of power.

10. The passage suggests that in order to participate in a balance of power, a country should:

I. plan its agricultural and industrial development.
II. possess information on its neighbors' economies.
III. disclose its military secrets to its neighbors.

A. II only
B. I and II only
C. I and III only
D. I, II, and III

11. The analogy between an international balance of power and the interrelations between neighbors is based on which of the following assumptions?

A. Both neighbors and neighbor states must avoid conflict whenever possible.
B. Power relations between neighbors are dependent on similar factors to those between neighbor states.
C. A degree of cultural similarity is required for peaceful coexistence.
D. The use of force for self-preservation is always justifiable.

12. According to the passage, which of the following would be LEAST critical to the survival of a neighborhood's balance of power?

A. An alliance network of a minimum of five or more neighbors
B. The presence of a number of skilled negotiators
C. Information about the location of dangerous weapons
D. Knowledge of neighbors' typical behavior in times of conflict

PRACTICE ANSWERS

Passage 1 (Questions 1–7)

Sample Highlighting

P1. "synthesize data"; **P2.** "dramatic event," "cyclical processes," and "aspects of everyday life"; **P3.** "geographical features"; **P4.** "delimit the boundaries" and "minimizes difference."

Sample Outlining

P1. *Annales vs.* traditional historians; Braudel popularized

P2. Braudel's three temporalities

P3. Expansion on *longue durée* as geohistory

P4. Braudel critics: "scope too broad, differences of social sciences blurred"; Author: "but asked new questions"

1. **D**

A classic Function question: Why did the author refer to Febvre and Bloch? Go back to the end of paragraph 1 where Febvre and Bloch are mentioned and read the surrounding text: "Braudel was the first *Annales* historian to gain widespread support of the idea that history should synthesize data from various social sciences … (although Febvre and Bloch, founders of the *Annales* school, originated this approach)." In other words, while Braudel popularized this approach, it was actually Febvre and Bloch who came up with it in the first place. This prediction matches closely with **(D)**, which has Febvre and Bloch *anticipat*[ing] Braudel's approach. **(A)** and **(C)** both suggest that Febvre and Bloch could be used to argue against the *Annales* approach, which is inconsistent with their roles as originators of this same approach. **(B)** tries to Distort the author's use of Febvre and Bloch: economics is clearly relevant for the *Annales* approach to historical investigation, but Febvre and Bloch are not used in the context of supporting the relevance of economics—they are mentioned only because they used economics before Braudel.

2. **B**

This is another Function question that essentially asks, *What is the role of the third paragraph?* Refer back to paragraph 3 and summarize its purpose: *Expansion on* longue durée *as geohistory.* This prediction matches closely with **(B)**, which reflects on *the role of geography in human history.* **(A)** is a Faulty Use of Detail wrong answer; the author does mention Braudel's attention to *esoteric information*, but this is a minor detail from the paragraph that misses the full purpose of expanding on the *longue durée.* **(C)** is also a Faulty Use of Detail; the *geography of the Mediterranean region* is used to support the use of geography in Braudel's approach, but the primary concern is not to discuss *the geography of the Mediterranean* specifically—rather, it is to show how geography can influence human history. **(D)** Distorts the author's words. Braudel does shy away from national borders to focus on geographic borders, but that does not mean that all national borders are *irrelevan*[t].

3. **D**

This is an Inference question. Based on the evidence presented in the passage, we should be able to infer a difference between traditional and *Annales* historians. From our initial analysis, we know that the two are contrasted in paragraph 1, so this is where we will look for clues. The key to the answer is given in the second sentence: "In a departure from traditional approaches, the *Annales* historians assume that history . . . must be understood in the context of forces and material conditions that underlie human behavior." In other words, the *Annales* historians are *more interested in the underlying context of human behavior* than traditional historians, as **(D)** states. **(A)** Distorts the author's words: the *Annales* are interested in incorporating social sciences into historical analysis, but that does not mean that they are more interested in other social sciences than in history—they are historians, after all! **(B)** is also a Distortion. The *Annales* historians propose that history is more than just the actions of famous figures, but this doesn't mean that they are *critical of* those figure's *achievements*. Finally, **(C)** is an Opposite. *Annales* historians want to incorporate economic research findings into historical analysis and so should not be *skeptical of the validity* of such approaches.

4. **D**

We want to predict what direction the author would go in if the passage were continued in this *Reasoning Beyond the Text* question. The purpose of the last paragraph was to mention criticisms of Braudel's approach and respond to those criticisms. The paragraph ends by responding to criticism by citing the contribution of Braudel's work: "studies . . . indicate that Braudel asked significant questions that traditional historians had overlooked." If another sentence were added, it should continue along the same lines of highlighting the influence or merits of Braudel. This prediction matches perfectly with **(D)**. **(A)** would backtrack on the author's support by questioning the professionalism of the authors of such studies. **(B)** is both Out of Scope and Opposite. Incorporation of the *Carlylean approach* mentioned in paragraph 2 would come out of nowhere at the end of paragraph 4, and it also represents the traditional approach to historical analysis; the author is unlikely to support a revival of this approach. Finally, **(C)** is a criticism of Braudel, and because the author is defending Braudel in the last part of the paragraph, this answer choice is an Opposite.

5. **B**

This is a Detail question, so we simply need to find the appropriate information in the passage. From our initial analysis, we know that the beginning of paragraph 4 is where the criticisms of Braudel are brought up. In fact, Braudel's critics believed his *scope* was *too broad*, and that the *differences* among *social sciences* were *blurred*. This prediction matches **(B)**. **(A)** is a Faulty Use of Detail. Braudel seeks structures like geohistory that underlie social activity in history; however, this is not in paragraph 4 and is not a criticism of Braudel. **(C)** is Out of Scope; while Braudel's "three temporalities" are discussed, the *relationship between them* is never addressed. Finally, **(D)** is an Opposite answer choice. According to paragraph 4, critics thought Braudel's approach "made it difficult to delimit the boundaries of observation." This statement is at odds with *rigidly define*[d] *boundaries* in this answer choice.

6. **C**

This is an Apply question of the Example subtype. We must use the author's description of the *longue durée* to identify a similar example. Paragraph 2 defines the three temporalities and defines the *longue durée* as being an "historical wave of great length" on the order of hundreds of years. Only **(C)** has even remotely that long of a duration, making it the correct answer. **(A)** and **(D)** are short-lived events in history just as the *evenementielle* temporality is defined in the passage. **(B)** describes a cyclical event that *occur[s] . . . roughly every half-century*, which would fit the definition of the *conjonctures* temporality described in the passage.

7. **A**

The question asks us to *suppose* new information. Thus, this is a *Reasoning Beyond the Text* question. If we glance at the answer choices, we can see that this is a Strengthen–Weaken question. The question stem introduces a quote from Braudel that emphasizes his interest in the influence of *one man's decisions* in history. The focus on one man's influence is most similar to the *evenementielle* temporality discussed in paragraph 2. We also see in this paragraph that Braudel considered the *longue durée* the "most fascinating of the three temporalities." The quote given in the question stem would go against the information in the passage, so we can immediately cross off **(C)** and **(D)**, which claim that this quote would *strengthen* the author's claim. The logic presented here matches with **(A)**. **(B)** is a Distortion because the author claims that Braudel is most fascinated by the *longue durée* temporality, but that is not the same thing as saying Braudel is *not interested* at all in other aspects of history. In fact, Braudel gives the actions of great men their own temporality, the *evenementielle*, so Braudel must be at least somewhat interested in these actions.

Passage 2 (Questions 8–12)

Sample Highlighting

P1. "cannot violate"; **P2.** "States involved must"; **P3.** "minimum of three" and "diplomacy"; **P4.** "estimating the power"; **P5.** "smaller scale."

Sample Outlining

P1. Balance of power: key is maintaining equilibrium

P2. Needs independent nations, proximity, and level playing field

P3. Factors needed in framework: $\geq$3 states, common culture, diplomacy

P4. Also need shared information on other states (power, economy, security)

P5. Neighbor analogy

8. **D**

In this Detail question, we need to pick the choice that matches the explanation for this claim in the passage. In paragraph 3, the author clarifies the phrase "common ground of culture" by pointing out that states with similar culture are "more likely to align in a cooperative manner." This prediction is almost identical to **(D)**. **(A)** is a Faulty Use of Detail; while *mechanisms of diplomacy* are needed for a balance of power according to the author, they are a separate, third necessary factor mentioned in this paragraph (in addition to having at least three states and common culture) and are not part of the "common ground of culture" described. **(B)** is Out of Scope because the passage does not have any discussion of cultural differences causing military conflict—it just states that cultural similarities foster cooperative behavior. **(C)** is similarly Out of Scope because the author does not discuss the influence of a hegemon on the culture of surrounding states.

9. **C**

This Inference question asks us to interpret a phrase used by the author. The phrase is found in the last sentence of paragraph 2. Here, the author claims that states "commensurate," or nearly equal in power, will decide policy on a "level playing field." This sentence contrasts with the previous one, which claims that "When one state far outstrips its close neighbors in power, it dictates economic and military policy for the region." Taking these together, we can infer that roughly equal states will all play a role in "dictat[ing] economic and military policy for the region." This prediction matches closely with **(C)**. **(A)** and **(B)** are subtle Distortions. While the author says that states will decide policy on a "level playing field," he or she does not state that this occurs *democratic*[ally] or *to the advantage of all member states*. Notice that these first three answer choices are all very similar in that they suggest that each member state gets a say in policy making—but **(C)** is the mildest answer choice that strays the least from what is stated in the passage. Generally, an answer choice that stays closer to what is written in the passage is a safer bet on the MCAT. Finally, **(D)** is definitely Out of Scope, as the passage does not address the interaction between states within the balance of power and those *outside the balance of power*.

10. **B**

This is a Scattered Inference question, so find the related information in the passage. Statements I and II are both brought up in the same sentence in paragraph 4: "a nation's economic planning should consider the dynamics of the entire state system in addition to its own agriculture and industry." In other words, a nation should *plan its agricultural and industrial development*, Statement I, while also taking into consideration the other nations in the balance of power, which would necessarily require *possess*[ing] *information on* [their] *economies*, Statement II. Thus, we can eliminate **(A)**. Statement III is not supported by the passage, however. The author clearly believes military intelligence is an important aspect of the balance of power and that states should seek out military information about other member states, but the author does not say that states ought to willingly give away military secrets to other states. This idea is a Distortion. With Statement III eliminated, the answer is **(B)**.

11. **B**

The word *assumption* immediately lets us know this is an Inference question of the Assumption subtype. Here we want to identify an assumption the author makes while forming the analogy in paragraph 5. One assumption in any analogy is *always* that the two entities are similar enough to be compared. Therefore, we should look for an answer that, were it not true, would mean that neighbors and states are actually not so similar. **(B)** matches this assumption—if *power relations between neighbors* were not *dependent on similar factors to those between neighbor states*, then the comparison made in the last paragraph would make no sense. **(A)** is a Distortion; while it makes sense that both neighbors and states would aim to *avoid conflict whenever possible*, the word *must* makes this answer extreme. Even if neighbors occasionally quarreled, and neighbor states occasionally fought, the analogy would still hold. In fact, the analogy even suggests that a conflict may occur over the lawnmower. **(C)** is also extreme; while the author suggests that cultural similarity facilitates cooperation, there is no suggestion that this similarity is absolutely *required for peaceful coexistence*. Finally, **(D)** is Out of Scope. The author seems to suggest that force may be needed to maintain a balance of power but does not make any reference to what is *justifiable*. Note the use of *always* also makes this answer choice extreme.

12. **A**

This is a *Reasoning Within the Text* question that does not neatly fit into any category because it is asking us to extend the analogy given in paragraph 5 a bit more. In this case, we will have to take the characteristics of balances of power described in paragraphs 2–4 and translate them to the neighbor example. Note that we want the *LEAST* important characteristic, implying that three answer choices are qualities the author would deem important for a neighborhood's balance of power. Consider your initial analysis of the passage: in paragraph 3, the author states that *≥3 states* are required in the balance of power—which is less than the number given in **(A)**. This makes it the correct answer; an alliance network could have fewer than five neighbors (specifically, three or four neighbors) and still have a balance of power. The other answer choices are all stated in the passage. **(B)** is mentioned in paragraph 3, as "*skilled diplomats*"—a sensible correlate to *skilled negotiators*—are needed in the balance of power. **(C)** and **(D)** are actually stated directly in the passage: "for the continued survival of the balance of power . . . it is critical that you find out whether your neighbor possesses a gun and whether he might use it under such circumstances [of conflict]."

CHAPTER 9

Question Types I: *Foundations of Comprehension* Questions

CHAPTER 9

Question Types I: *Foundations of Comprehension* Questions

In This Chapter

Introduction

LEARNING OBJECTIVES

After Chapter 9, you will be able to:

- Identify Main Idea, Detail, Inference, and Definition-in-Context questions
- Solve *Foundations of Comprehension* questions with strategies based on question type
- Recognize common features of *Foundations of Comprehension* questions

In order to get into college, you likely had to take an exam like the SAT® or ACT®, both of which feature sections that test reading comprehension. Most of the questions in those sections were straightforward, requiring you merely to search the text for a key fact, to define the meaning of a term used in a passage, or to identify the author's thesis. Some of the more challenging questions may have required you to imagine things from the writer's perspective or to explain why she used a certain word or phrase. Such questions can also be found in the *Critical Analysis and Reasoning Skills* (CARS) section of the MCAT, though you will find that they are generally more difficult than the analogous questions found on your precollege exams.

Chapters 9, 10, and 11 of this textbook will follow the same general pattern. For each question type, we will briefly describe distinguishing features before examining some sample question stems. Then, after discussing strategic approaches to the question type, we'll attack an example passage excerpt, providing at least one sample question of each type. The question types in this chapter all fall under the *Foundations of Comprehension* category and will be examined in the following order: Main Idea, Detail, Inference, and Definition-in-Context.

Note: The Question Types, as well as the Kaplan Method for CARS Passages, Kaplan Method for CARS Questions, and Wrong Answer Pathologies, are all included as tear-out sheets in the back of this book.

9.1 Main Idea Questions

Questions that ask about the big picture or major themes of the passage are what we call **Main Idea questions**. Only a small number of the questions in the CARS section fall into this type, but they are easy to recognize. The Kaplan Method for CARS Passages will arm you with everything you need to attack Main Idea questions, typically without referring back to the passage.

Sample Question Stems

- The author's central thesis is that:
- Which of the following best characterizes the main idea of the passage?
- The primary purpose of the passage is to:
- The author of the passage is primarily interested in:
- Which of the following titles best captures the main theme of the passage?
- The author can best be viewed as a proponent of:
- The language used in the passage makes it clear that the intended audience is:
- What is the author's central concern?

BRIDGE

Main Idea questions that ask about the audience or the medium are checking your rhetorical analysis skills. Analysis of the likely author, her tone, and her voice can reveal the intended audience and most likely medium. Rhetoric was discussed in Chapter 3 of *MCAT CARS Review*.

Main Idea questions will often use some variant of the phrase *central thesis*, *primary purpose*, or—of course—*main idea*, or they might make some kind of general reference to the *author*. On rare occasions, Main Idea questions will bring in a more challenging aspect of the rhetorical situation, such as the *audience* or the *medium*.

Strategy

You should decide to work on a Main Idea question as soon as you encounter it, because these questions can get you some quick points. Your best bet for your Investigate step is simply to *Go for the goal!* As discussed in Chapter 6, no matter which Distill approach you take, you will consider the overarching purpose or goal of the passage before moving on to the questions. If you encounter a Main Idea question, the purpose that you considered during your passage distillation will usually echo the correct answer. When searching for a Match in the answer choices, both the verb and the content of each answer choice should be carefully considered. For Main Idea answer choices, the verb used to describe the author's purpose in writing the passage is just as important as the content of the answer.

In the event that none of the choices come close to matching what you thought the author's purpose was, you should use a systematic process of elimination to remove Faulty Use of Detail answers that are too narrow, Out of Scope options that go too far afield, and any answer choice that has the wrong tone (positive, negative, ambivalent, or impartial) or degree (extreme *vs.* moderate) as discussed in Chapter 8 *of Critical Analysis and Reasoning Skills Review.*

MCAT EXPERTISE

The wrong answer choices in Main Idea questions are very predictable. One or more tend to be too narrow, reflecting the ideas from only one paragraph. One or more tend to be too broad, becoming Out of Scope. One or more tend to embody the wrong tone (positive, negative, ambivalent, or impartial) or degree (extreme *vs.* moderate).

Main Idea Questions—Applied Example

There is no shortage of evidence for the existence of systemic biases in ordinary human reasoning. For instance, Kahneman and Tversky in their groundbreaking 1974 work proposed the existence of a heuristic—an error-prone shortcut in reasoning—known as "anchoring." In one of their most notable experiments, participants were exposed to the spin of a roulette wheel (specially rigged to land randomly on one of only two possible results) before being asked to guess what percentage of United Nations member states were African. The half of the sample who had the roulette wheel stop at 65 guessed, on average, that 45 percent of the UN was African, while those with a result of 10 guessed only 25 percent, demonstrating that prior presentation of a random number otherwise unconnected to a quantitative judgment can still influence that judgment.

The anchoring effect has been observed on repeated other occasions, such as in Dan Ariely's experiment that used digits in Social Security numbers as an anchor for bids at an auction, and in the 1996 study by Wilson et al. that showed even awareness of the existence of anchoring bias is insufficient to mitigate its effects. The advertising industry has long been aware of this bias, the rationale for its frequent practice of featuring an "original" price before showing a "sale" price that is invariably reduced. Of course, anchoring is hardly alone among the defective tendencies in human reasoning; other systemic biases have also been experimentally identified, including loss aversion, the availability heuristic, and optimism bias.

Example:

1. The author's primary task in the passage is to:

    A. search for evidence of systemic biases in normal human thinking.
    B. discuss empirical findings on anchoring and other reasoning biases.
    C. show that anchoring is the most commonly occurring error in reasoning.
    D. demonstrate that knowledge of anchoring bias is insufficient to prevent it.

Solution: Utilizing the Kaplan Question Strategy discussed in Chapter 8, start by identifying the question type. This question asks you to identify the *primary task* of the author, which is indicative of a Main Idea question. During your Read and Distill of the passage, you should consider the overall goal of the passage, meaning that these questions can be answered quickly. Attack and answer questions like this one as soon as you encounter them to earn quick points on Test Day.

The second step is to Rephrase the question. A simplification in this case might be *Why did the author write this passage?* or *What was the author trying to convey to the reader?* The task in Main Idea questions is usually well defined by the question stem, so the Rephrase step should be minimal.

The Investigate step will remain the same for Main Idea questions no matter which method you selected in your Choose step for the passage. Main Idea questions ask you to synthesize all of the information presented into an overall purpose. To solve, you should recall your reasoning about the purpose of the passage from the end of your Distill step, then find a Match in the answer choices. In this case, the author was attempting to present evidence for systemic biases in reasoning, especially anchoring.

That prediction from your Investigate step should immediately lead you to **(B)**. The phrase *discuss empirical findings* is roughly equivalent in meaning to our *present evidence*. Also, the scope is right, with the answer focusing on anchoring but mentioning other biases because the author does introduce anchoring as an example with the phrase "For instance."

On Test Day, whenever you find a match for a prediction or a choice that fulfills your expectation for correctness, go with that answer and move on to the next question. For these examples, though, it's worth discussing what's wrong with the other options. **(A)** comes close and has the appropriate scope, but the verb is wrong: in the very first sentence, the author says that "There is no shortage of evidence," so why would the author's task be to *search for* something so readily available? Moving on to **(C)**, we find an Out of Scope choice: while the author says that systemic biases are common to thinking, there is never any comparison among the types mentioned. Just because the author focuses on anchoring does *not* mean that anchoring is necessarily the most frequently occurring—the author could choose to discuss it for any number of reasons other than its commonness. Finally, **(D)** is a Faulty Use of Detail because it is too specific. The referenced claim is made in the second paragraph, but it applies only to the study conducted by Wilson and others. This answer does not encompass the focus of the whole passage.

9.2 Detail Questions

Detail questions ask about what is stated explicitly in the passage. These are probably what you typically think of when you imagine a "reading comprehension" question, as they tend to require searching the text to find the missing piece. While Main Idea questions focus on the big picture, Detail questions zoom in on some of the finer points of the passage. They are by far the most common question type in the *Foundations of Comprehension* category, constituting at least half of the questions that fall under this heading.

Sample Question Stems

- According to the author's account of [topic], [concept] is:
- The author states that [person] holds the view that:
- Which of the following, according to the passage, does the author associate with the idea of [concept]?
- The author's apparent attitude toward [alternative position] is:
- Based on the discussion in [paragraph reference], the work of [artist/writer] was widely regarded as:
- The passage suggests which of the following about [topic]? [list of Roman numerals]
- The author asserts all of the following EXCEPT:
- Which of the following claims does NOT appear in the passage?

Detail questions tend to contain simple, declarative language (*is* and *are*) rather than the subjunctive mood (*would* and *could*), often include phrases like *the author states* and *according to the passage*, and can take the form of incomplete sentences ending with a colon.

The last three examples listed above fall into what is known as the Scattered subtype. A **Scattered** question is one that either employs a set of Roman numeral options or uses a word like EXCEPT, NOT, or LEAST. While just about any type of question can be Scattered—from Scattered Function to Scattered Inference—Scattered Detail is perhaps the most common example of the Scattered subtype.

MCAT EXPERTISE

It is easy to forget that you are working on a LEAST/NOT/EXCEPT question when you're analyzing the answer choices. Highlighting those words in the question stem can provide a visual clue that you are looking for a false statement as opposed to the three true wrong answers.

Strategy

The only trick to working with Detail questions is that sometimes a seemingly straightforward question can actually require making an inference (which is, notably, another type of question in the *Foundations of Comprehension* category). If you are dealing with a true Detail question, though, you should follow the clues in the question stem, especially the content **buzzwords**, and keep these terms as a focus during your Rephrase of the question stem. You should then Investigate by using your Distill step to help you home in on the relevant portion of the passage. If you Highlighted or Outlined, the main idea of each paragraph should help you find the relevant detail. If you Interrogated, you should return to the chunk that contained the detail being asked about in the question stem. Once you find the precise sentence referenced, remember to read at least the sentence before and the sentence after (unless, of course, you're

BRIDGE

Sometimes, what appears to be a Detail question will actually require you to make an inference. Inference questions include both Assumption and Implication questions, the strategies for which are discussed later in this chapter.

looking at the first or last sentence of the entire passage!) to get a bit more context. Once you've read the relevant text, complete your investigation by putting the sentence into your own words, and then look for the best match to answer the question.

You will likely want to save the Scattered Detail questions you encounter for later because these will often require researching three or four different parts of the text, rather than just one. When working with these questions, your approach will vary quite a bit depending on which passage strategy you've chosen. Process of elimination is almost inevitable with questions of the Scattered subtype, but that doesn't mean you should immediately jump blindly into the answer choices. As with any question, take what you can from the stem to set some basic expectations.

MCAT EXPERTISE

For Detail questions, make sure to paraphrase the relevant text in a "short and sweet" format that will be easy to repeat to yourself while reading the answer choices. Much of the challenge of this type of question can be trying to figure out which answer choice actually matches your prediction—so make your prediction something that's easy to remember!

Detail Questions—Applied Example

One of the first examples of the ascendance of abstraction in 20th-century art is the Dada movement, which Lowenthal dubbed "the groundwork to abstract art and sound poetry, a starting point for performance art, a prelude to postmodernism, an influence on pop art . . . and the movement that laid the foundation for surrealism." Dadaism was ultimately premised on a philosophical rejection of the dominant culture, which is to say the dominating culture of colonialist Europe. Not content with the violent exploitation of other peoples, Europe's ruling factions once again turned inward, reigniting provincial disputes into the conflagration that came to be known by the Eurocentric epithet "World War I"—the European subcontinent apparently being the only part of the world that mattered.

The absurd destructiveness of the Great War was a natural prelude to the creative absurdity of Dada. Is it any wonder that the rejection of reason made manifest by senseless atrocities should lead to the embrace of irrationality and disorder among the West's subaltern artistic communities? Marcel Janco, one of the first Dadaists, cited this rationale: "We had lost confidence in our culture. Everything had to be demolished. We would begin again after the *tabula rasa*." Thus, we find the overturning of what was once considered art: a urinal becomes the *Fountain* after Marcel Duchamp signs it "R. Mutt" in 1917, the nonsense syllables of Hugo Ball and Kurt Schwitters transform into "sound poems," and dancers in cardboard cubist costumes accompanied by foghorns and typewriters metamorphosize into the ballet *Parade*. Unsurprisingly, many commentators, including founding members, have described Dada as an "anti-art" movement. Notwithstanding such a designation, Dadaism has left a lasting imprint on modern Western art.

Example:

2. As stated in the passage, prior to the Great War, the leaders of Europe were primarily focused on:

   A. fighting one another in World War I.
   B. colonizing other parts of the globe.
   C. gazing inward at local problems.
   D. rejecting the dominant culture.

Solution: The first step to the Kaplan Method for CARS Questions is the Type step, and when the stem says *as stated in the passage*, it usually signifies a Detail question (with some exceptions). You'll most likely want to work on Detail questions right away, rather than saving them for later.

The Rephrase step is next, and this question could be rephrased into a specific task: *Find in the passage what the leaders of Europe were focused on prior to the war.* There are two buzzwords in the question stem: *the Great War* and *the leaders of Europe*. The first comes directly from the start of paragraph 2, where you can judge from the context that it must be another name for what we now call World War I. (They of course didn't know there was going to be a second one at the time!) The second buzzword does not appear verbatim, but it shows up in the synonymous phrase "Europe's ruling factions" in the last sentence of paragraph 1. In addition to reading these two sentences to refresh yourself if necessary as part of your investigation, it may be worth reading one before and one after.

When we Investigate, the preceding sentence offers a key bit of context, illustrating that the author regards European culture quite negatively, with the phrase "the dominating culture of colonialist Europe." More of the same follows with the phrase "violent exploitation of other peoples," as this author highlights Europe's past as a global colonizer. The phrase "once again turned inward" is noteworthy for mirroring the language of *focused on* featured in the question stem. Putting this all together sets a thorough expectation for the correct answer: if the turning inward happened with the Great War, then prior to that, Europe's leaders must have been looking outward, focusing on their colonial acquisitions. This prediction finds a match in **(B)**.

Among the wrong options, **(A)** is based on a misunderstanding of the terminology. Though World War I is mentioned in the prior sentence, the author calls attention to the fact that it only later came to be known by that "epithet" (name) and then in the following sentence uses another name for it. Because the two terms refer to the same event, European leaders could not be focused on it before it happened. In contrast, **(C)** is an Opposite because Europeans were actually turning outward at their colonies. Finally, **(D)** is a Faulty Use of Detail: that phrase appears almost exactly, but it actually describes the founders of the Dada movement, not the European leaders.

MCAT EXPERTISE

One of the unique critical thinking challenges on the MCAT is based on the use of synonym. Test questions and answer choices are likely to use synonymous terms, rather than directly quoting the associated passage. Keep this in mind and remain flexible in your thinking when you Investigate a question stem or search for a Match in the answer choices to a particular question.

Example:

3. Based on the passage, which of the following is not a characteristic associated with Dadaism?

    A. A renunciation of European culture
    B. A reputation as an "anti-art" movement
    C. Importance for later 20th-century art
    D. The embrace of irrational atrocities

Solution: Although the test maker will often put the word *not* in italics or all caps, sometimes it will appear plainly in a question stem, as seen here. Don't be fooled: this is still a Scattered Detail question. Because Scattered Detail questions are often time-consuming, you'd likely want to make it the last question you work on in the question set. When you do commit to working through a Scattered Detail question, don't forget to Rephrase the stem and set some basic expectations. The phrasing *a characteristic associated with* is vague enough that it could mean something the author says, or it could potentially reflect another view identified in the passage. That means you can't rule out an answer choice just because it doesn't sound like something the author would say, so long as it sounds like something someone else in the passage would. Because both paragraphs are chock-full of claims about Dadaism, there's no point in searching the text to set additional expectations—you'll have to go with a more minimal prediction as the output of your Investigate step.

(A) comes directly from the author's second sentence: "Dadaism was ultimately premised on a philosophical rejection of the . . . culture of colonialist Europe." The word *renunciation* is a synonym of *rejection*—so cross off this option. Moving on to the next choice, while the author explicitly rejected the idea that Dadaism is "anti-art," that view is still reflected in the passage in the next-to-last sentence, where the author writes "many commentators, including founding members, have described Dada as an 'anti-art' movement." Another way of saying that many commentators describe it that way is to say that it has a reputation. Hence, **(B)** should also be eliminated. The next characteristic is found at both the beginning of the passage, with the quotation from Lowenthal, and at the end, with the author's statement that "Dadaism has left a lasting imprint on modern Western art." With **(C)** now off the table, we know that **(D)** must be the answer—but let's check it anyway.

This choice is very close to being a characteristic; however, the word *atrocities* prevents it from being associated with Dadaism and makes it the correct choice. The key sentence is worth repeating in full in order to untangle it: "*Is* it any wonder that the rejection of reason made manifest by senseless atrocities should lead to the embrace of irrationality and disorder among the West's subaltern artistic communities?" The author is saying that Dadaism embraces irrationality, although not in the same way that irrationality (*rejection of reason*) manifests itself in warfare (*senseless atrocities*). The preceding sentence offers some useful clarification: "The absurd destructiveness of the Great War was a natural prelude to the creative absurdity of Dada." Notice how the word *absurd* is used to join two concepts that are typically viewed as opposites: creation and destruction. The author is suggesting that, even though both the Great War and Dadaism defied reason, they did so in dramatically different ways. This irrationality or absurdness in Dadaism stood in opposition to the atrocities of World War I; the word *embrace* in the answer choice makes this statement *not* a characteristic of Dadaism. That's why **(D)** is correct.

9.3 Inference Questions

What makes **Inference questions** distinctive is that they deal with unstated parts of arguments: information that is not explicitly written by the author but that *must* be true given what is claimed in the passage. It is crucial to understand that the correct answers to Inference questions are not simply assertions that are *possibly* true or that *could* be accurate; rather, they are necessary assumptions or at least highly probable implications. We call these Inference questions because, as first noted in Chapter 3 of *MCAT CARS Review*, it is the common name of the process used to arrive at both **assumptions** (unstated evidence) and **implications** (unstated conclusions), collectively known as inferences.

Sample Question Stems

- In [paragraph reference], it is reasonable to believe that the author assumes:
- Implicit in the discussion of [topic] is the underlying assumption that:
- The passage suggests [claim] because:
- On the basis of the author's account of [topic], which of the following might reasonably be inferred?
- The author implies that [concept] is NOT:
- It is reasonable to conclude that the author regards [person or position] as:
- The author says [quotation], but also [paraphrase of different claim]. These beliefs imply:
- Which of the following inferences is most justified by information provided in the passage? [list of Roman numerals]

As these examples indicate, during the first step of the Kaplan Method for CARS Questions—in which you Type the question —you can recognize that you might be dealing with an Inference question if you see one or more of the following words (or variations of them): *assume*, *because*, *conclude*, *imply*, *infer*, *justify*, *reasonable*, or *suggest*. However, if a question stem contains any of these words and new information of some kind, then it will fall into the *Reasoning Beyond the Text* category instead (which we will discuss in Chapter 11). If there are no new elements, you are likely dealing with an Inference question.

As previously alluded to, there is actually something of a continuum between Detail and Inference questions; the position of questions on that continuum varies based on the complexity of the reasoning used to solve them. A question that tells you to make an inference might require one relatively simple step, such as canceling out a double negative or identifying a paraphrase of lines from the passage. Sometimes you'll even find you were told to look for an item that the *passage suggests* or the *author implies* but, as you Investigate, you may discover that the answer was in the text virtually word for word. In cases such as these, consider yourself fortunate, because you've uncovered a Detail question in disguise! The predictions you make for these questions will likely be more than adequate to match to the correct answer.

The downside to this ambiguity is that sometimes a question that uses simple declarative language will require a lot more critical thinking than you expected. Just because a stem includes a phrase like *the author asserts* or *as stated in the passage* (common question stems for Detail questions) does not necessarily mean that the answer is stated in a straightforward manner. Notwithstanding such trickery on the part of the AAMC, as long as you recognize the possibility that apparent Detail questions can be disguised as Inference questions and are prepared to apply the Denial Test strategy to those questions, you can avoid being fooled!

Strategy

There are only two basic tasks with the Inference question type. When you recognize that you need to make an inference, first ask yourself whether you are looking for a missing but essential piece of *evidence* (an **assumption**) or for a *conclusion* that is unstated but highly probable given what is said (an **implication**).

Assumptions

If you Rephrase the stem and your task is to identify an assumption, your Investigation should begin by determining what claim the assumption is supposed to underlie. Sometimes the question stem will say this explicitly, or it may use quoted text or paragraph numbers to refer to a particular claim. If not, depending on which passage strategy you used, you may need to go back and do research in the passage. If you need to go back to the passage as you investigate, you should reread the relevant passage sentences and isolate the specific statement, taking care to see whether there's any existing evidence in the surrounding text that's used to support the statement. Logic keywords are your best friends here, but keep in mind that a lot of authors use them sparsely, so they won't always be there to help. The Match to assumption questions might include words or ideas that are similar to a piece of evidence actually stated, so you can use any that you locate for your prediction, or just go with whatever links the evidence provided to the conclusion when you reread the text—*now I see what the author is taking for granted!*

KEY CONCEPT

Inferences are unstated parts of arguments that must be true based on what the passage says. Assumptions are unstated pieces of evidence, while implications are unstated conclusions.

Implications

When the task you identify as you rephrase the question stem is identifying an implication, center your investigation of the solution on any particular sentences referred to in the question stem. Watch out for Logic keywords, just as with assumptions.

Now, however, you'll be looking to see whether the particular statements referenced are used to support anything else. If so, use those explicitly supported conclusions to set your expectations for the correct answer. Alternatively, if another implication occurs to you when you reread, you can use that as a prediction.

Whenever you can't find a Match for your prediction—if you thought it was a Detail question, if you didn't know where to look because the stem had no paragraph reference clues, or if the answer choices just turned out very differently than you expected—plan B is to use a special version of process of elimination known as the **Denial Test**. Let's take a look at how it works in practice.

The Denial Test

For each answer choice, negate (take the opposite of) what is being said. In question stems that end with a colon (:), you may need to take part of the text from the end of the stem and combine it with the answer choice to create a sentence that can then be denied. Be careful with sentences that already contain negative words, prefixes, and suffixes, because sometimes just removing that text is not enough to change the meaning in the proper way. If nothing else, any claim can be denied by adding *It is not the case that* to the beginning of the sentence. Once you've denied the claim, think about the effect the denied claim has on the passage. If you're not sure what effect it has, look for clues in the question stem or the answer choice itself to see if it refers to a particular part of the passage that you could reread. If it's clear that the negated claim has no significant impact, then eliminate that answer.

When you come across an answer choice that logically conflicts with the text once you've negated it, you've likely found the correct answer. However, sometimes multiple answer choices will have denials that cause problems for the text, so when executing the Denial Test you should *always test every answer choice* and pick the one for which negation has the most detrimental impact on arguments or assertions from the passage. Keep in mind that this can be time-consuming, which is why it's generally a good idea to save a question for later if you believe the Denial test will be necessary to answer it.

MCAT EXPERTISE

While the Denial Test will always reveal the correct answer in an Inference question, it's very time-consuming. If you cannot set good expectations for the right answer during the Execute step, triage the question and return to it later with the Denial Test.

It is hard to understand how the Denial Test works simply by reading about it. Check out the Worked Example below and make sure to practice the Denial Test on Inference questions to get used to using it.

Inference Questions—Applied Example

In 1941, an exuberant nationalist wrote: "We must accept wholeheartedly our duty and our opportunity as the most powerful and vital nation . . . to exert upon the world the full impact of our influence, for such purposes as we see fit and by such means as we see fit." If forced to guess the identity of the writer, many US citizens would likely suspect a German jingoist advocating for *Lebensraum*. In actuality, the sentiment was expressed by one of America's own: Henry Luce, the highly influential publisher of the magazines *Life*, *Time*, and *Fortune*. Luce sought to dub the 1900s the "American Century," calling upon the nation to pursue global hegemony as it slipped from the grasp of warring Old World empires. As a forecast of world history, Luce's pronouncement seems prescient—but is it justifiable as a normative stance?

Not all of Luce's contemporaries bought into his exceptionalist creed. Only a year later, Henry Wallace, vice president under FDR, insisted that no country had the "right to exploit other nations" and that "military [and] economic imperialism" were invariably immoral. It is a foundational assumption in ethics that the wrongness of an act is independent of the particular identity of the actor—individuals who pay no heed to moral consistency are justly condemned as hypocrites. So why should it be any different for nation-states? In accord with this principle, Wallace proselytized for "the century of the common man," for the furtherance of a "great revolution of the people," and for bringing justice and prosperity to all persons irrespective of accidents of birth. Sadly, Wallace never had the chance to lead the United States in this cosmopolitan direction; prior to Roosevelt's demise at the beginning of his fourth term, the vice presidency was handed to Harry Truman, a man whose narrow provincialism ensconced him firmly in Luce's camp. And with Truman came the ghastly atomic eradication of two Japanese cities, the dangerous precedent set by military action without congressional approval in Korea, and a Cold War with the Soviet Union that brought the world to the brink of nuclear destruction.

Example:

4. One can most justifiably conclude on the basis of the author's discussion that Henry Luce assumed that:

   A. the United States did not have the right to create a military or economic empire.

   B. nation-states are never bound by the same ethical principles that persons are.

   C. the same normative standards should apply to both Americans and Germans.

   D. moral rules that govern individual behavior do not necessarily apply to countries.

Solution: The *assumed* at the end of the question stem makes the Type step for this question simple: it must be an Inference question. Rephrasing the question stem should give you something along the lines of "Based on the passage, what did Henry Luce assume?" Using this rephrase, we know we're looking for an assumption from a portion of the passage relevant to Luce's argument, and we'll want to rely on our Distill technique, possibly in combination with referring back to the passage itself, in order to Investigate. Luce advocated for what the author calls the *exceptionalist* stance that the United States, as the most powerful country at the time, was free to do as it pleased. This is in contrast to Henry Wallace, who explicitly rebuffed Luce's view, so watch out for Faulty Uses of Detail answer choices that would actually describe Wallace.

Denying **(A)** yields *the United States* ***did*** *have the right to create a military or economic empire*, which is completely consistent with Luce's view. Eliminate it. As written, **(A)** is actually a view attributed to Wallace by the author, so it's both a FUD and an Opposite as this was the major point of disagreement between the two.

(B) contains *never*, which is an Extreme keyword, but because Luce seems to have a fairly strong position, don't jump to dismissing the answer as a Distortion quite yet. It's a bit trickier to negate: to say that it is *not* the case that some event *never* happens is the same as saying the event *sometimes* occurs. Thus, the proper negation of **(B)** is *nation-states are* ***sometimes*** *bound by the same ethical principles that persons are*, which is completely consistent with Luce, who presumably believes that ethical principles do apply to nations sometimes, say on those occasions when the nation in question is not the United States. Eliminate it.

For **(C)**, the contradiction would be *the same normative standards should* ***not*** *apply to both Americans and Germans*, which is again consistent with Luce's *"American Century"* idea. Hence, we can definitely cross off **(C)**, which, as written, is actually another one of Wallace's beliefs—a second Opposite/FUD combo.

All that remains is **(D)**, which does indeed destroy the argument if rejected. Countering it by saying *moral rules that govern individual behavior* ***do*** *necessarily apply to countries* would mean that they *always* apply. But this supports the anti-hypocrisy argument that the author makes in favor of Wallace and against Luce at the beginning of paragraph 2. Thus, denying **(D)** would make Henry Luce's argument fall apart, and so it is an assumption Luce has made. **(B)** was simply a more extreme version of this claim, rightly rejected as a Distortion because an author is more likely to assume a weak form of a statement than a strong one.

BRIDGE

Remember that the opposite of an Extreme Negative keyword (like *never*) is a Moderating Positive keyword (like *sometimes*). The circular view of Author keywords presented in Chapter 5 of *MCAT CARS Review* is a great way to visualize this change.

Example:

5. It is reasonable to infer that the author believes that:

   A. Harry Truman was the worst US president of the 20th century.
   B. Franklin Roosevelt did not endorse the idea of the "American Century."
   C. Henry Wallace would not have approved of the use of atomic weapons.
   D. Henry Luce did not provide an accurate historical assessment of the 1900s.

Solution: We can Type this question as an Inference question given the phrase *reasonable to infer*. With a stem like this, which has no specific clues or references, you may find Rephrasing barely changes anything, and thus choose to dive straight into the Investigate step. With a stem that gives almost no information, we have little choice but to proceed with the Denial Test. It's not entirely clear what kind of inference this is, because an author "believes" both assumptions and implications. While you can guess based on the author's descriptive language that he or she does not approve of Harry Truman, you do not know the writer's feelings toward any of the other American presidents in the 1900s, so denying that Truman was the worst president would not have too much of an impact. Therefore **(A)** is a Distortion, too extreme to attribute to the author.

For **(B)**, you do not really know what the author believes about Franklin Delano Roosevelt (FDR). We only know that he was the US president serving over both Wallace and Truman. Therefore, we can make no safe inferences about FDR's attitude toward Luce's point of view: this Out of Scope option should be discarded.

Negate **(C)** and you arrive at *Henry Wallace **would** have approved of the use of atomic weapons*. If this were so, had he become president, he would have been guilty of the very misdeeds for which the author blames Truman—and the author would be utterly inconsistent in praising the one and condemning the other. In fact, because the author rails against hypocrisy, it is clear that he or she would have to hold the two men to the same ethical standards. Thus, denying **(C)** would considerably undermine the author's argument in paragraph 2, and this is almost certainly the answer.

To make sure, finish executing the Denial Test: rejecting **(D)** would have no negative impact—in fact, it would support the author's claim that Luce's view "seem[ed] prescient" as an historical prediction. This assumption suggests that the author indeed views Luce's view as historically accurate, even if ethically questionable. You can now be confident that **(C)** is the answer.

9.4 Definition-in-Context Questions

Definition-in-Context questions are the final question type to fall into the *Foundations of Comprehension* category. The task generated from the Rephrase of these questions is always the same: define the word or phrase as it is used in the passage. These questions are infrequent on Test Day, but are straightforward and make for easy points when they do occur.

Sample Question Stems

- As used by the author, the word [term] most nearly means:
- In [paragraph reference], what is the author's most likely meaning when stating [quotation]?
- The author's choice of the phrase [term] is probably intended to suggest:
- As used in the passage, [term] refers to:
- In [paragraph reference], the author asserts that [claim]. What the author most likely means by this is:
- Which of the following is most synonymous with [concept] as discussed in the text?

This list of stems makes it clear that Definition-in-Context questions always feature a reference to a word, phrase, or an entire claim from the passage, the meaning of which you are tasked with identifying. Quotation marks and italics are common features used to call attention to the terms, but on occasion a Definition-in-Context question stem may lack these.

Strategy

Although these questions ask about the meanings of words, a dictionary will not help you here, and in some cases it could even lead you astray. Trap answers in these questions are often the common definitions of the word, which are tempting Out of Scope choices that fail to match the context. These questions tend to be relatively fast to solve because they refer only to small portions of the text. Thus, you should generally decide to work on these questions as soon as you see them.

Your Investigation with a Definition-in-Context question will be to go to the text and surrounding context, if necessary, to see how the word or phrase is used in the passage. With this question type, the question stems will usually contain a paragraph reference, but use your recall of your Interrogation, your Outlining, or your Highlighting if this reference is not given to locate the relevant sentence. For this question type in particular, even if you interrogated the passage, it's likely you will need to go back into the passage and reread the surrounding materials. If reading that sentence doesn't give you enough to work with, look at the sentences before and after the target term or phrase. Wrap up your Investigation with a clear prediction: define the term or phrase in your own words based on how it was used in the passage. Author keywords may be especially helpful, because answer choices with the wrong tone can immediately be ruled out.

MCAT EXPERTISE

An author may imbue common words with a special meaning in the passage. Therefore, be sure to check how the author actually uses the word in a Definition-in-Context question, rather than looking for a dictionary definition of the term. Wrong answers in these questions are often accurate definitions of the term that do not match how the term was used in the passage.

Definition-in-Context Questions—Applied Example

In 1941, an exuberant nationalist wrote: "We must accept wholeheartedly our duty and our opportunity as the most powerful and vital nation . . . to exert upon the world the full impact of our influence, for such purposes as we see fit and by such means as we see fit." If forced to guess the identity of the writer, many US citizens would likely suspect a German jingoist advocating for *Lebensraum*. In actuality, the sentiment was expressed by one of America's own: Henry Luce, the highly influential publisher of the magazines *Life*, *Time*, and *Fortune*. Luce sought to dub the 1900s the "American Century," calling upon the nation to pursue global hegemony as it slipped from the grasp of warring Old World empires. As a forecast of world history, Luce's pronouncement seems prescient—but is it justifiable as a normative stance?

Not all of Luce's contemporaries bought into his exceptionalist creed. Only a year later, Henry Wallace, vice president under FDR, insisted that no country had the "right to exploit other nations" and that "military [and] economic imperialism" were invariably immoral. It is a foundational assumption in ethics that the wrongness of an act is independent of the particular identity of the actor—individuals who pay no heed to moral consistency are justly condemned as hypocrites. So why should it be any different for nation-states? In accord with this principle, Wallace proselytized for "the century of the common man," for the furtherance of a "great revolution of the people," and for bringing justice and prosperity to all persons irrespective of accidents of birth. Sadly, Wallace never had the chance to lead the United States in this cosmopolitan direction; prior to Roosevelt's demise at the beginning of his fourth term, the vice presidency was handed to Harry Truman, a man whose narrow provincialism ensconced him firmly in Luce's camp. And with Truman came the ghastly atomic eradication of two Japanese cities, the dangerous precedent set by military action without congressional approval in Korea, and a Cold War with the Soviet Union that brought the world to the brink of nuclear destruction.

Example:

6. The author's use of the term "provincialism" in paragraph 2 comes closest in meaning to:

   A. German jingoism.
   B. economic imperialism.
   C. nationalistic exceptionalism.
   D. exuberant cosmopolitanism.

Solution: With the stereotypical Definition-in-Context question structure, Typing this question as Definition-in-Context should be a quick and easy step, with little to no Rephrase needed prior to proceeding with the Investigate step. The clue in the question stem points to the second paragraph, but the reference might still be difficult to find in a paragraph full of *-isms*. The key sentence is: "Sadly, Wallace never had the chance to lead the United States in this cosmopolitan direction; prior to Roosevelt's demise at the beginning of his fourth term, the vice presidency was handed to Harry Truman, a man whose narrow provincialism ensconced him firmly in Luce's camp." This sentence gives you a lot to work with. You can see that *provincialism* is a view attributed to Truman and Luce, who are contraposed to the "cosmopolitan" Wallace. In a passage that features a few different perspectives, it's not surprising to see a question like this that requires keeping straight who holds which view. This passage presents two major sides, and it's clear that *provincialism* represents a view that belongs to Truman, whose views align with those of Luce.

With these expectations established, look for a match in the answers. Because both nationalism (at the start of paragraph 1) and exceptionalism (at the start of paragraph 2) are views attributed to Luce, it is evident that **(C)** must be correct. **(A)** is a Faulty Use of Detail; *a German jingoist* was mentioned in the first paragraph, but this phrase is no good because Luce and Truman are both Americans. **(B)** echoes the quotation from Wallace as he criticizes Luce—"military [and] economic imperialism"—but all of the examples given for Truman show evidence of military action, not economic: "the ghastly atomic eradication of two Japanese cities, the dangerous precedent set by military action without congressional approval in Korea, and a Cold War with the Soviet Union that brought the world to the brink of nuclear destruction." The emphasis in **(B)** is therefore misplaced. Lastly, **(D)** uses a variant of *cosmopolitan*, a characteristic that was attributed to Wallace, so it's an Opposite.

Conclusion

Reading comprehension is a skill you've been honing your whole life—most exams you've had in literature and English classes have most likely centered on your ability to understand the text you read. On the MCAT, reading comprehension is important not only for answering *Foundations of Comprehension* questions, but also for understanding the passage itself. While they take many forms, including Main Idea, Detail, Inference, and Definition-in-Context, all *Foundations of Comprehension* questions share a few common features. Answering all of these question types is facilitated by a solid understanding of the structure of the passage, which can be used to locate the relevant text for the answer. The answers to all of these questions must be stated, paraphrased, or implied directly in the passage. That said, not all of these questions are asking for the same thing: these question types differ in significant ways. For example, Main Idea and Definition-in-Context questions should usually be answered as soon as you see them, whereas Detail (especially Scattered Detail) and some of the more challenging Inference questions may be more time-consuming and should therefore be saved until the end of the question set. Each of these questions can be Typed, Rephrased, and Investigated in predictable ways. We'll continue discussing question types in the next two chapters as we explore *Reasoning Within the Text* and *Reasoning Beyond the Text* questions.

Concept and Strategy Summary

Main Idea Questions

- **Type: Main Idea questions** ask for the author's primary goal or purpose in writing the passage.
  - These questions often contain words like *central thesis*, *primary purpose*, or *main idea*.
  - Less commonly, these questions may ask about a different aspect of the rhetorical situation, such as the *audience* or the *medium*.
- **Rephrase:** Phrase the question in such a way that you have a task for your Investigate step, such as *Why did the author write this passage* or *What clues in the passage tell me who the author's audience is?*
- **Investigate:** Recall the purpose that you constructed during the Distillation step of the Kaplan Passage Strategy, and predict that the correct answer will be synonymous with that purpose.
- **Match:** Your prediction should closely correlate to the right answer. If there is no clear match, or if you cannot perform any of the earlier steps of the *Kaplan Method for CARS Questions*, use the process of elimination.
  - Wrong answer choices may be too narrow (Faulty Use of Detail) or too broad (Out of Scope).
  - Wrong answer choices may have the wrong tone (positive, negative, ambivalent, or impartial) or degree (extreme *vs.* moderate).

Detail Questions

- **Type: Detail questions** ask about what is stated explicitly in the passage.
  - Unstated parts of arguments *must* be true given what is claimed in the passage.
  - Detail questions are the most likely to use the **scattered** format, which uses Roman numeral options or words like *EXCEPT*, *NOT*, or *LEAST*.
- **Rephrase:** Phrase the question in a way that maintains any **buzzwords** from the stem that will guide you to where in the passage you need to Investigate.
- **Investigate:** If you do not recall enough information to make a prediction, reread the relevant sentence, as well as the sentences before and after.
  - Make a concise prediction so you can repeat it to yourself between answer choices.
  - For Scattered Detail questions, locate all three of the wrong answers in the passage so you can eliminate them as options.
- **Match:** Your prediction should match the right answer. If there is no clear match, or if you cannot perform any of the earlier steps of the Kaplan Method for CARS Questions, use the process of elimination.

Inference Questions

- **Type: Inference questions** look for unstated parts of arguments.
  - These questions are similar to Main Idea questions, although they focus on the purpose of only one portion of the passage (usually one sentence or one paragraph).
  - **Assumptions** are unstated evidence.
  - **Implications** are unstated conclusions.
  - These question stems often contain words like *assume*, *because*, *conclude*, *imply*, *infer*, *justify*, *reasonable*, or *suggest*.
- **Rephrase:** Determine whether you are looking for an assumption (evidence) or implication (conclusion). Then, rephrase the claim the answer is supposed to support (assumptions) or be supported by (implications). It is helpful to build the directionality of the support into your Rephrase step.
- **Investigate:** Reread the relevant sentence if necessary, noting the explicit evidence and conclusions given.
  - For assumption questions, the answer is either similar to the evidence given or links the evidence to the conclusions.
  - For implication questions, the answer is either similar to the conclusions given or is another logical conclusion one could draw from the evidence.
- **Match:** Your prediction to an answer choice. If there is no clear match, or if you cannot perform any of the earlier steps of the Kaplan Method for CARS Questions, use the process of elimination, removing any answer that conflicts with the author's main argument or the paragraph's purpose.
  - Apply the Denial Test: negate each answer choice.
  - Whichever answer choice—when negated—has the most detrimental effect on the argument made in the passage is the correct answer choice.

Definition-in-Context Questions

- **Type: Definition-in-Context questions** ask you to define a word or phrase as it is used in the passage.
  - These questions often call attention to the term to be defined using quotation marks or italics, but not always.
  - Definition-in-Context questions always reference a word, phrase, or an entire claim from the passage.

- **Rephrase:** For Definition-in-Context questions, the rephrase will always look the same: *What does this word mean in this passage?*
- **Investigate:** Reread the sentence with the word or phrase, and perhaps the surrounding context, as necessary. Reword the author's definition of the term in your own voice as your prediction.
- **Match:** Your expectations should match the right answer. If there is no clear match, or if you cannot perform any of the earlier steps of the Kaplan Method for CARS Questions, use the process of elimination.

Worked Example

Use the Worked Example below, in tandem with the subsequent practice passages, to internalize and apply the strategies described in this chapter. The Worked Example matches the specifications and style of a typical MCAT *Critical Analysis and Reasoning Skills* (CARS) passage.

Highlighting can be beneficial when a passage is complex and the author jumps and pivots from one perspective to the next. The following passage has those qualities, and it also has unusual structural features that don't lend themselves well to Outlining or Interrogating. Thus, the Highlighting method should be a good choice as a way to approach this passage, specifically because it can be used to clearly indicate the points where major ideas are introduced and the points where the author shifts her focus. This will speed up efforts to find relevant information while answering questions later.

HIGHLIGHTING METHOD	
Passage	**Expert Thinking**
. . . Until last year, many people—but not most economists—thought that economic data told a simple tale. On one side, productivity—the average output of an average worker—was rising. And although the rate of productivity increase was very slow during the 1970s and early 1980s, the official numbers said that it had accelerated significantly in the 1990s. By 1994, an average worker was producing about 20 percent more than 1978.	What is the main theme of the passage? What words could be highlighted to demonstrate what this passage and paragraph are about? *The first sentence hints that the author is about to dispute conventional wisdom. However, this isn't the best phrase to highlight, as it's not the focus of the paragraph. A more informative highlight would be "productivity increase," as it captures what this paragraph is about.*
On the other hand, other statistics said that real, inflation-adjusted wages had not been rising at the same rate. Some commonly cited numbers showed real wages actually falling over the last 25 years. Those who did their homework knew that the gloomiest numbers overstated the case . . . Still, even the most optimistic measure, the total hourly compensation of the average worker, rose only 3 percent between 1978 and 1994 . . .	What parallel structure did you notice in this paragraph? What phrase can be highlighted to indicate the new concept introduced in this paragraph? *This paragraph's first line is similar to the "On one side" sentence in the previous paragraph. More accurately, this paragraph is a continuation of the first paragraph and completes the thought. Highlight the phrase "real wages actually falling" because it represents the conclusion of the author's first point. That is, the simple tale is how productivity went up, but real wages went down.*
. . . But now, experts tell us it may have been a figment of our statistical imaginations . . . a blue-ribbon panel of economists headed by Michael Boskin of Stanford declared that the Consumer Price Index [C.P.I.] had been systematically overstating inflation, probably by more than 1 percent per year for the last two decades, mainly failing to take account of changes in consumption patterns and product quality improvements . . .	What pivot point did you notice in this paragraph? What phrase should you highlight to capture this shift? *The first word of this paragraph indicates that the author is about to pivot into his second point. However, highlighting the "But now" phrase in the first sentence won't actually indicate what the author's point is when referring back to the passage. Instead, highlight "overstating inflation" to capture the new focus of this paragraph.* *When using the Highlighting method, do not highlight terms like C.P.I.: although it is a new term, the term is isolated and doesn't match the theme of this passage up to this point. This is the kind of term you might highlight if you were Outlining or Interrogating, as something a Definition-in-Context or Detail question might ask about.*

HIGHLIGHTING METHOD

Passage	Expert Thinking
. . . The Boskin Report, in particular, is not an official document—it will be quite a while before the government actually issues a revised C.P.I., and the eventual revision may be smaller than Boskin proposed. Still, the general outline of the resolution is pretty clear. When revisions are taken into account, productivity growth will probably look somewhat higher than before because some of the revisions will also affect how we calculate growth. But the rate of growth of real wages will look much higher—roughly in line with productivity. In other words, the whole story about workers not sharing in productivity gains will turn out to have been based on a statistical illusion . . .	How is this paragraph linked to the previous paragraph? What is the most important information in this paragraph? *A new term, "Boskin Report," is introduced, which relates to the mention of Michael Boskin in the previous paragraph. The first half of this paragraph mostly speaks about the technical aspect of the Boskin Report, but the actual main idea is about disproving that real wages are falling while productivity is climbing. Highlight "in line with productivity" or a related phrase to capture this idea. With this theme identified, this paragraph can be identified as a conceptual extension of the previous paragraph.*
It is important not to go overboard on this point. There are real problems in America, and our previous concerns were not pure hypochondriasis. For one, economic progress over the past 25 years has been much slower than in the previous 25. Even if Boskin's numbers are right, median family income—which officially has experienced virtually no gain since 1973—has risen by only about 35 percent over the past 25 years, compared with 100 percent over the previous 25. Furthermore, it is likely that if we "Boskinized" the old data—that is, if we tried to adjust the C.P.I. for the 50s and 60s to take account of changing consumption patterns and rising product quality—we would find that official numbers understated the rate of progress just as much if not more than they did in recent decades . . .	What pivots are in this paragraph? What phrases should you highlight to track the author's perspective? *In this paragraph, the author emphasizes that the claims discussed in the first two paragraphs about the "simple tale" are not totally unfounded. This can be represented by highlighting "not pure hypochondriasis." The following sentences dive into the details surrounding concerns about the economy. You could additionally highlight "understated the rate" because it parallels with the previous notion that the Boskin report is used to adjust the old data.*
. . . Moreover, while workers as a group have shared fully in national productivity gains, they have not done so equally. The overwhelming evidence of a huge increase in income inequality in America has nothing to do with price indices and is therefore unaffected by recent statistical revelations. Families in the bottom fifth, who had 5.4 percent of total income in 1970, had only 4.2 percent in 1994; over the same period, the top 5 percent went from 15.6 to 20.1. Corporate CEOs, who used to make about 35 times as much as their employees, now make 120 times as much or more . . .	What is the main idea for this paragraph? What are some low-yield details not worth highlighting? *The main idea in this paragraph is "income inequality," extending the previous paragraph's idea that the "simple tale" has some merit to it. The second half of this paragraph talks about the details of income inequality in American society. These details do not offer anything remarkably new.*

HIGHLIGHTING METHOD	
Passage	**Expert Thinking**
. . . While these are real and serious problems, however, one thing is now clear: the truth about what is happening in America is more subtle than the simplistic morality play about greedy capitalists and oppressed workers that so many would-be sophisticates accepted only a few months ago.	What final conclusion is drawn in the last paragraph? *The last paragraph brings the passage full circle by restating a conclusion found in the first paragraph. The author is arguing against the "simple tale" that real wages are falling, and claims that the actual picture is more complicated. "More subtle" captures that idea. For details and to answer most questions, we will have to refer back to previous paragraphs.*

Question	Analysis
1. According to the passage, "Boskinization" adjusts the C.P.I. by:	The words *according to the passage* tell us this is a Detail question, which should be quick points. The term "Boskinized" appears in paragraph 5, where we find that it means that *Boskin adjusted the C.P.I. to take account of changing consumption patterns and rising product quality.*
A. increasing wages and decreasing productivity to reconcile the present disparity.	**(A)** may be tempting because Boskin's model did, in the end, increase apparent wages, but the passage makes no mention of *decreasing productivity* measures, making this choice Out of Scope. This choice also does not match with how the term "Boskinized" is used in the passage.
B. taking into account technology's role in an improved efficiency.	**(B)** is also Out of Scope because there is no mention of *technology's role* in the passage.
C. reassessing patterns of consumption and quality of product.	This choice is a spot-on match with the passage, making **(C)** the correct answer.
D. evaluating the inequalities in various levels of incomes.	**(D)** gives us an option that discusses wage inequality. However, this idea was a facet of the author's argument, not of Boskin's revisions. This is a Faulty Use of Detail answer choice.

Question	Analysis
2. The Boskin Report does all of the following EXCEPT:	The word *EXCEPT* shows us that this is a Scattered Detail question—one that might be worth skipping on Test Day until more time is available. To investigate, we must find relevant details and eliminate them systematically, keeping in mind that the correct answer is the one NOT included in the passage.
A. reveals that the C.P.I. was inaccurate.	Paragraph 3 tells us that the Boskin Report demonstrated that the C.P.I. "had been systematically overstating inflation," eliminating **(A)**.
B. reconciles the present disparity between productivity and wage levels.	Paragraph 4 shows us that Boskin did reconcile wages and productivity, eliminating **(B)**.
C. reveals the reasons for the increasing disparity between the highest and lowest income earners.	**(C)** is not present in the passage. While the income disparity was discussed, no mention was made of its causes—making this the correct answer.
D. provides possible clarification for economic progress in the 1950s and 1960s.	In paragraph 5, we see that Boskin's work, if applied to the 1950s and 1960s, "could find that official numbers understated the rate of progress," eliminating **(D)**.

Question	Analysis
3. In the first paragraph the author assumes:	The word *assumes* shows us that this is an Inference question, meaning our first Investigate action is to look back at paragraph one and its themes.
A. the total productivity of America has not seen a significant increase since the 1970s.	**(A)** can be eliminated because this is discussed in paragraph 6, rather than paragraph 1.
B. productivity is a measure of economic growth.	**(B)** The unstated evidence in paragraph 1 is that productivity is a measure of economic growth—making this the correct answer.
C. each American worker's productivity is directly proportional to overall national productivity gains.	**(C)** is Opposite, because the author states that "while workers . . . have shared . . . in national productivity gains, they have not done so equally." It is also Out of Scope because it comes from paragraph 2.
D. Boskin's report is unable to explain the discrepancy between productivity growth and wage increases.	**(D)** This paragraph focuses on *income inequality*, not *the discrepancy between productivity growth and wage increases*. Faulty Use of Detail.

Question	Analysis
4. The author's primary purpose in presenting this passage is to:	The words *primary purpose* identify this as a Main Idea question, which usually can be answered quickly. The goal in this passage that we identified when highlighting is *to argue that the C.P.I., with or without Boskin's work, understated real wages; and that the productivity/wage disparity is more complicated than it first appears.*
A. argue that overreliance on the C.P.I. is insufficient for explaining the current state of the American worker.	**(A)** is correct; the passage primarily addresses the idea that the "C.P.I. . . . understated real wages" (and therefore does not fully explain the current state of the American worker).
B. argue that wages actually increased from 1978 to 1994.	In **(B)**, the answer is far too specific as it applies only to paragraph 2 and not the entire passage.
C. argue that a capitalistic oppression of the worker is the primary cause of the current economic climate.	The *capitalist oppression of the worker* is part of the *simplistic* model dismissed in paragraph 7. Because the author does not agree with this model, **(C)** can be eliminated.
D. suggest that partisan division in Congress would be more adequate for explaining the current economic climate.	Congressional divide, while perhaps present in real life, was never mentioned in the passage, making **(D)** Out of Scope.

Question	Analysis
5. The author's use of the term "statistical imagination" in paragraph 3 most nearly indicates:	This is a Definition-in-Context question. A quick scan of paragraph 3 shows us that "statistical imagination" refers to a shortcoming of the traditional C.P.I. model, due to failure to "take account of changes in consumption patterns and product quality improvements." Be on the lookout for any wrong answers that sound like a standard definition for *imagination*.
A. wage data for the last 25 years has been falsified.	**(A)** is Out of Scope because we're never told that data was *falsified*.
B. the pessimistic view of the economy indicated by the C.P.I. is overstated due to underestimation of the significance of key variables.	**(B)** matches closely with the prediction and is the correct answer.
C. mathematical models of the economy are less accurate than anecdotal reports.	**(C)** is Out of Scope because this passage never compares *mathematical models* to *anecdotal reports* in terms of validity.
D. the C.P.I. is a completely unreliable tool for explaining the economic climate.	The C.P.I. certainly has some issues but is not the *completely unreliable tool* mentioned in **(D)**—this is Extreme.

PRACTICE QUESTIONS

Passage 1 (Questions 1–6)

GO ONLINE

The United States has less than half of the 215 million acres of wetlands that existed at the time of European settlement. Wetland conversion began upon the arrival of European immigrants with their traditional antipathy to wetlands and with the will and technology to dry them out. In the mid-19th century, the federal government awarded nearly 65 million acres of wetlands to 15 states in a series of Swamp Land Acts. But the most rapid conversion occurred between the mid-1950s and mid-1970s, when an estimated 450,000 acres per year were lost, primarily to agriculture.

This conversion has meant the loss of a wide range of important wetland functions. Wetlands inhibit downstream flooding, prevent erosion along coasts and rivers, and help remove or assimilate pollutants. They support scores of endangered birds, mammals, amphibians, plants, and fishes. Wetlands provide aesthetic and open-space benefits, and some are critical groundwater exchange areas. These and other public benefits have been lost to agricultural forestry and development enterprises of all kinds, despite the fact that most of the conversion goals might have been obtained with far less wetland loss through regional planning, stronger regulation, and greater public understanding of wetland values.

At best, existing wetland laws and programs only slow the rate of loss. Despite the growing willingness of government to respond, wetland protection faces significant obstacles. Acquisition as a remedy will always be limited by severe budget constraints. The Emergency Wetlands Resources Act allocates only $40 million per year in federal funds, supplemented by relatively modest state funds, for wetland purchase. Ultimately, the wetlands that are protected will be a small percentage of the approximately 95 million acres remaining today. Wetland acquisition by private environmental groups and land trusts adds qualitatively important, but quantitatively limited, protection. Government incentives to induce wetland conservation through private initiatives are limited and poorly funded. Some private developers have recognized that business can protect selected wetlands and still profit. Recreational developments in Florida have benefited from wetland and habitat protection that preserves visual amenities. It is doubtful, however, that these business decisions to save wetlands would have occurred without strong government regulation; the marketplace does not generally recognize the public benefits of wetlands for flood control, fish and wildlife, and other long-term values.

One possible strategy (and the one presently being implemented) is to protect each and every wetland in threatened areas according to stringent permit guidelines that do not distinguish by wetland types or values. This approach may be environmentally desirable, but it has not worked. About 300,000 acres of wetlands are lost each year. An alternative strategy is to develop a regional management approach focused on valuable wetlands in selected areas that are under intense pressure. Broad regional wetland evaluations could identify critical wetland systems that meet particular local and national needs and avoid abandonment of any wetlands without careful review of the trade-offs. Cooperating federal, state, and local interests can then anticipate and seek ways to prevent wetland losses and can guide future development in areas where alternative options exist. There is no general federal authority to conduct such planning for wetland system protection. But there are several authorities under which a program to anticipate and prevent wetland losses on an area-wide basis can be developed.

1. In paragraph 1, it is reasonable to believe that the author assumes:

- **A.** wetland conservation efforts are cost prohibitive.
- **B.** wetland conservation began with the European immigration to the United States.
- **C.** wetlands are a source of tourism income for many areas in the United States.
- **D.** wetlands can be converted to agricultural purposes through currently available means.

2. According to the passage, all of the following contributed to the rapid loss of wetlands in the United States EXCEPT:

- **A.** technological innovations implemented by European settlers.
- **B.** development of commercial and residential real estate complexes.
- **C.** increased rezoning for the purposes of agricultural and industrial operations.
- **D.** conversion of wetlands for agricultural uses.

3. As used by the author, the word "authority" (paragraph 4) most nearly means:

- **A.** a person with extensive or specialized knowledge of a subject.
- **B.** having ascertained something from a reliable source.
- **C.** the power to influence others due to one's knowledge about a topic.
- **D.** a person or organization having power or control in particular.

4. Implicit in the discussion in this passage is the underlying assumption that the author believes:

- **A.** Wetlands under intense pressure should be focused on first.
- **B.** Using current policies, wetland losses cannot be fully combated, only minimized.
- **C.** The loss of wildlife due to wetland losses is the only reason the government should be concerned.
- **D.** Advocating for increased governmental spending in wetlands conservation will solve the problem of wetland losses.

5. Which of the following does the author state as beneficial functions of the wetlands?

- **I.** Supporting communities of endangered amphibians
- **II.** Providing renewable forestry options
- **III.** Erosion and flood control

- **A.** III only
- **B.** I and II only
- **C.** I and III only
- **D.** I, II, and III

6. The author can best be viewed as a proponent of:

- **A.** developing new strategies and improving current efforts to prevent wetland losses.
- **B.** maintaining the current stringent permit guidelines to protect conservation areas.
- **C.** promoting development of wetlands for visual amenities and aesthetic benefits.
- **D.** current government incentives to induce wetland conservation through private business funding.

Passage 2 (Questions 7–12)

Although many may argue with my stress on the continuity of the essential traits of American character and religion, few would question the thesis that our business institutions have reflected the constant emphasis in the American value system on individual achievement. From the earliest comments of foreign travelers down to the present, individuals have identified a strong materialistic bent as a characteristic American trait.

The worship of the dollar, the desire to make a profit, the effort to get ahead through the accumulation of possessions, all have been credited to the egalitarian character of the society. As Tocqueville noted in his discussion of the consequences of a democracy's destruction of aristocracy, "They have swept away the privileges of some of their fellow creatures which stood in their way, but they have opened the door to universal competition."

A study of the comments of various 19th-century foreign travelers on American workers reveals that most of these European writers, among whom were a number of socialists, concluded that social and economic democracy in America has an effect contrary to mitigating compensation for social status. American secular and religious values both have facilitated the "triumph of American capitalism" and fostered status striving.

The focus on egalitarianism and individual opportunity has also prevented the emergence of class consciousness among the lower classes. The absence of a socialist or labor party, and the historic weakness of American trade-unionism, appear to attest to the strength of values that depreciated a concern with class.

Although the American labor movement is similar to others in many respects, it differs from those of other stable democracies in ideology, class solidarity, tactics, organizational structure, and patterns of leadership behavior. American unions are more conservative; they are more narrowly self-interested; their tactics are more militant; they are more decentralized in their collective bargaining; and they have more full-time salaried officials, who are on the whole much more highly paid. American unions have also organized a smaller proportion of the labor force than have unions in these other nations.

The growth of a large trade-union movement during the 1930s, together with the greater political involvement of labor organizations in the Democratic party, suggested to some that the day—long predicted by Marxists—was arriving in which the American working class would finally follow in the footsteps of its European brethren. Such changes in the structure of class relations seemed to these observers to reflect stagnancy and limitations on social mobility—the decline of opportunity and the petrifaction of class lines. To them, such changes could not occur without modification of the traditional value system.

A close examination of the character of the American labor movement suggests that it, like American religious institutions, may be perceived as reflecting the basic values of the larger society. Although unions, like all other American institutions, have changed in various ways consistent with the growth of an urban industrial civilization, the essential traits of American trade unions, as of business corporations, may still be derived from key elements in the American value system.

7. In the context of the passage, the phrase "strong materialistic bent," as used in the sentence "From the earliest comments of foreign travelers down to the present, individuals have identified a strong materialistic bent as a characteristic American trait," (paragraph 1) refers to:

- **A.** European socialists' view of aristocrats.
- **B.** European travelers' concern with democracy.
- **C.** American society's emphasis on acquiring wealth.
- **D.** American religion's criticism of secular values.

8. Based on the information given in the passage, which of the following is/are NOT true?
   - **I.** American society emphasizes class solidarity over individual achievement.
   - **II.** American unions are less interested in nonlabor issues than unions in other democracies.
   - **III.** American labor organizations and American religious institutions share some of the same values.

   A. I only
   B. II only
   C. II and III only
   D. I, II, and III

9. According to the passage, all of the following have influenced the outlook of the American labor movement EXCEPT:
   A. secular values.
   B. religious values.
   C. urban industrial civilization.
   D. foreign labor movements.

10. According to the passage, which of the following is a part of the "traditional value system"?
    A. Class solidarity
    B. Individual achievement
    C. Urban industrialization
    D. Marxist ideology

11. The author of this passage most likely believes that American labor unions:
    A. are more influenced by the American values of competition and materialism than their European counterparts.
    B. sacrifice status and wealth by adhering to religious values.
    C. seemed to reflect more European values during the 1930s than American values.
    D. are stronger, more widespread organizations than European trade unions.

12. Based on its use in the passage, the word "petrifaction" most nearly means:
    A. dissolution.
    B. conversion to stone.
    C. recognition.
    D. solidification.

PRACTICE ANSWERS

Passage 1 (Questions 1–6)

Sample Highlighting

P1. "wetland conversion"; **P2.** "wetland function" and "lost to"; **P3.** "obstacles" and "private initiatives"; **P4.** "alternative strategy" and "no general federal authority."

Sample Outlining

P1. History of wetlands conversion

P2. Benefits of wetlands, ways loss could have been minimized

P3. Current problems with conservation efforts and ways of minimizing further loss

P4. Current solution (save all threatened wetlands) hasn't worked; proposes new regional management strategy

1. **D**

The word *assumes* signifies to us that we are approaching an Inference question type. Inference questions ask us to find the necessary connecting information. With a paragraph reference, there is clear direction on where to begin the Investigate step. In paragraph 1 we're told "Wetland conversion began upon the arrival of European immigrants with their traditional antipathy to wetlands and with the will and technology to dry them out... when an estimated 450,000 acres per year were lost, primarily to agriculture." For answer choice **(A)**, costs were discussed in paragraph 3 but not paragraph 1. In answer choice **(B)** Wetland *conversion* began with the arrival of European immigrants, not wetland *conservation*. Answer choice **(C)** is an Out Of Scope answer choice. While the passage is about the wetlands and they may be a source of tourism income, that topic is never discussed within the passage. This leaves answer choice **(D)** as the correct answer, and further, this is a reasonable inference, as we are told that wetlands were converted to agricultural property.

2. **C**

For this Scattered Detail question type, we will have to find three answer choices that are mentioned in the passage and eliminate them. For **(A)**, the "technological innovations" of European settlers are mentioned in paragraph 1. "Development enterprises" are listed at the end of paragraph 2, which eliminates **(B)**. **(C)**, however, cannot be found in the passage. No mention is made of *increased rezoning* of wetlands, so this must be the correct answer. Finally, "conversion of wetlands for agricultur[e]" is listed at the end of the first paragraph, which removes **(D)**.

3. **D**

Typing this question is pretty straightforward, as you are given a word from the last paragraph and asked what it means in the passage, making this a Definition-in-Context question. The best thing to do with a question like this is to reread the targeted sentence in the passage, as well as the surrounding sentences. The sentence the word appears in is "There is no general federal authority to conduct such planning for wetland system protection." This sentence alone doesn't give a full answer, but the sentence before gives the necessary context by mentioning "federal, state, and local interests," or the organizations that are in power to conduct planning which matches **(D)**. Answer choices **(A)**, **(B)**, and **(C)** are all definitions for authority, but are not the definition that the author was referencing in this passage, which makes them all Out of Scope answers.

4. **B**

This Inference question is asking for an inference the author made without a passage reference. This would be a good question to leave for your second pass through the questions. Answer choice **(A)** is a Faulty Use of Detail choice from paragraph 4 that does not answer the question being asked. In paragraph 3, the passage tells us "At best, existing wetland laws and programs only slow the rate of loss," which Matches answer choice **(B)**. Answer choice **(C)** is an Extreme answer choice; unless an author is extreme, which this author is not, an Extreme answer choice will not be correct. The author thinks that the best way to fix this problem is to apply a new regional management strategy, making **(D)** Out of Scope.

5. **C**

The Roman numeral format and the verb *state* tell us that this is a Scattered Detail question. "Beneficial functions of the wetlands" are mentioned in the first part of paragraph 2. Relevant to this question are the details that "wetlands inhibit downstream flooding, prevent erosion along coasts and rivers,...and support scores of endangered birds, mammals, amphibians, plants, and fishes." Statements I and III are listed in this paragraph, whereas *renewable forestry options*, Statement II, are never mentioned.

6. **A**

This is a masked Main Idea question; asking what the author is a *proponent of* is another way of asking what the author likes. The author's goal in the passage is to *describe problems with current wetland conservation efforts and to propose a new strategy*. This prediction matches well with **(A)**. In the final paragraph, the author discusses the flaws of the current permit system, so **(B)** can be eliminated as an Opposite. **(C)** can also be eliminated for similar reasons because the author is a proponent of the preservation of—not *development of*—the wetlands. Finally, the author states in the third paragraph that "government incentives to induce wetland conservation through private initiatives are limited and poorly funded," so **(D)** can also be eliminated.

Passage 2 (Questions 7–12)

Sample Highlighting

P1. "strong materialistic bent"; **P2.** "egalitarian character"; **P3.** "effect contrary to"; **P4.** "prevented". **P5.** "American unions are"; **P6.** "reflect stagnancy and"; **P7.** "reflecting the basic."

Sample Outlining

P1. Individualism and materialism = American traits

P2. American traits come from egalitarianism, ex: Tocqueville

P3. European travelers' perspective on American system

P4. Egalitarianism and individualism prevent class consciousness

P5. Comparison of American and foreign labor unions

P6. Early predictions of American trade unions by Marxists

P7. American labor movement's values reflect general American values

7. **C**

This is a Definition-in-Context question asking for the author's definition of "strong materialistic bent." Immediately after the referenced sentence, the author mentions "the worship of the dollar, the desire to make a profit, the effort to get ahead through the accumulation of possessions" as an elaboration of the topic of materialism. Thus, materialism—a characteristically American trait—focuses on acquiring wealth, which matches **(C)**. Because materialism is attributed to Americans, and not *aristocrats* or *democracy* in general, **(A)** and **(B)** can be eliminated. Finally, the author never mentions *criticism* from American religion at all, so we can eliminate **(D)** as Out of Scope.

8. **A**

This is a Scattered Detail question asking for statements that are *NOT* true based on the passage. Statement I runs counter to the author's main point, most forcefully expressed in the first two paragraphs, that "individual achievement, not class solidarity," is a characteristically American trait. Statement II refers to the differences between American and foreign labor unions, which are listed in paragraph 5. Here, the author argues that American unions are more "narrowly self-interested"—in other words, they focus more specifically on labor issues than foreign unions. Statement II must therefore be true, and the answer must be **(A)**. Statement III is also true based on the beginning of paragraph 7: "*the American labor movement, ... like American religious institutions, may be perceived as reflecting the basic values of the larger society.*"

9. **D**

This is another Scattered Detail question because of the phrase *According to the passage* and the word *EXCEPT*. Secular and religious values, **(A)** and **(B)**, are addressed in paragraph 3, as both have "facilitated the 'triumph of American capitalism.'" American capitalism appears to have affected the goals of unions, according to paragraph 5, so these answer choices can be eliminated. "Urban industrial civilization" is listed as influencing unions in paragraph 7, so we can eliminate **(C)**. Foreign labor movements are contrasted with the American labor movement in paragraph 5, but no influence is described here. The closest match in the passage would be the Marxist prediction that "the American working class would finally follow in the footsteps of its European brethren," but this prediction actually failed to pan out. Therefore, **(D)** is correct.

10. **B**

For this Detail question, we can look for something that fits in with the two values the author mentioned as part of the traditional American value system: individual achievement and egalitarianism. **(B)** matches perfectly and is the correct answer. **(A)** and **(D)** are Opposites because "class solidarity" and "Marxist ideology" are more likely considered values of foreign democracies according to paragraph 5. Finally, while **(C)** is listed in the final paragraph as influencing labor movements, it is not part of the American traditional value system as described throughout the passage, so it can be eliminated.

11. **A**

This is an Inference question of the Implication subtype. Here, we should take into account the author's tone to determine his or her opinion. The author makes the case that American labor unions are more strongly influenced by traditional American values, such as individual achievement and egalitarianism, rather than more European values. This prediction is very close to **(A)**; "competition and materialism" are also mentioned as American characteristics in paragraphs 2 and 1, respectively. **(C)** is the Opposite of this prediction, so it can be eliminated. **(B)** is Out of Scope because *sacrific[ing] status and wealth* is not discussed as a religious value here. Finally, **(D)** is also Out of Scope; the author never talks about the relative *strength* of American labor unions as compared to European trade unions.

12. **D**

This is another Definition-in-Context question, so let's locate the relevant sentence: "Such changes in the structure of class relations seemed ... to reflect stagnancy and limitations on social mobility—the decline of opportunity and the petrifaction of class lines." There's parallelism here in the sentence's construction: "stagnancy" refers to *the decline of opportunity*, whereas "limitations on social mobility" refers to *the petrifaction of class lines*. Putting this together, if social mobility is limited, then class lines must be harder to break through, which matches with **(D)**. **(A)** is an Opposite, as *dissolution* of class lines would enable social mobility. Mere *recognition* of these lines, as in **(C)**, would not necessarily limit social mobility, thus eliminating this answer choice. Finally, **(B)** is far too literal—*petrifaction* can also mean the conversion of organic matter into stone, but an intangible concept like *class lines* would not actually be turned into stone.

CHAPTER 10

Question Types II: *Reasoning Within the Text* Questions

CHAPTER 10

Question Types II: *Reasoning Within the Text* Questions

In This Chapter

Introduction

LEARNING OBJECTIVES

After Chapter 10, you will be able to:

- Identify Function, Strengthen–Weaken, and Other *Reasoning Within the Text* questions
- Solve *Reasoning Within the Text* questions with focused strategies
- Apply the major principles of argument and logical structure to MCAT questions

In this chapter, we'll continue the treatment of question types first employed with *Foundations of Comprehension*, now examining the two types of *Reasoning Within the Text* questions that Kaplan has identified: Function questions, which ask about why the author included a piece of information or argument in the passage, and Strengthen–Weaken (Within the Passage) questions, which predominantly concern the ways in which arguments are backed by evidence and undermined by refutations. This chapter adopts the same general approach as Chapter 9: after identifying what makes the questions distinctive and offering several common question stems, we discuss strategies for each question type, illustrating them with a few Applied Examples.

Reasoning Within the Text questions account for approximately 30 percent of what you'll encounter on Test Day, according to both the AAMC's official statements and Kaplan's own extensive research of released AAMC material. In general, Function questions are roughly as common as Detail and Strengthen–Weaken (Beyond the Passage) questions, but they are less common than Apply questions. Strengthen–Weaken (Within the Passage) questions are more rare than Function questions.

Note: The question types, as well as the Kaplan Method for CARS Passages, Kaplan Method for CARS Questions, and Wrong Answer Pathologies, are included as tear-out sheets in the back of this book.

10.1 Function Questions

One of the reasons that the Kaplan Method for CARS Passages emphasizes reading for perspective (trying to understand the author's attitude and intentions) rather than just for detail is that the **Function question** type specifically asks about what the author is trying to *do* in the passage. Unlike a Main Idea question, which might ask about the overall goal of the passage, a Function question will ask about the purpose of only a portion of the passage, often in the context of the author's larger argument. Further, these questions will require some level of critical thinking about why a piece of information was included, meaning they involve an additional level of complexity beyond *Foundations of Comprehension* questions.

Sample Question Stems

- What is the author's apparent purpose in stating [quotation]?
- The author mentions [topic] in [paragraph reference] in order to:
- Which of the following is the most probable reason for the author's inclusion of a quotation from [person]?
- The author's reference to [concept] in [paragraph reference] is most likely supposed to show:
- When the author says [claim], she is emphasizing that:
- Which of the following is the example from [paragraph reference] most likely intended to suggest?
- The author compares [one concept] to [another concept] because:
- The author's principal motive for discussing [alternative position] is to explain that:

What should be readily apparent in the phrasing of Function questions is frequent mention of the author and the use of direct references to the text—especially through paragraph references. Language like *purpose*, *motive*, and *intention* indicate a Function question, as do phrases that end with *in order to* and *because*. Use these key phrases and traits to Type the question quickly. For Function questions, the Rephrase step tends to be fairly straightforward, as these questions usually state the task to be accomplished.

Strategy

Function questions are usually specific to a certain portion of the passage that can be directly referenced, making them a good option to solve now rather than later. Referring back to your Distill step will be key when working with a Function question. If you were reading for perspective by looking for Author keywords (which give a glimpse from the author's point of view), you may have already considered the information you need to answer such a question in your Distill step; it might even be highlighted or in your outline, if you chose either of those approaches. Keep in mind that, generally speaking, Function questions work in a nested way. In other words, the passage as a whole has a purpose, and each paragraph within it has a subordinate function that is distinct but that contributes to the larger whole. Each paragraph can in turn be broken down into sentences, each of which has its own particular role to play in the paragraph—and even sentences can be broken down into particular words or phrases.

Because a Function question will generally ask for the purpose of no more than a paragraph, to Investigate this question type you should recall or refer back to your Distill step for the specific paragraph (and perhaps also consider the author's overall goal in the passage). Then, if buzzwords in the question stem direct you to specific sentences, reread those portions of the paragraph if needed and think about how they fit into the purpose of the paragraph and the passage's general purpose. If you chose to Distill by Interrogating, you may find you don't even have to return to the passage, and can instead directly predict an answer. Formulate a statement of the function, and then start to look for an answer that matches.

Remember, if you can't find a perfect match, you can eliminate choices that are inconsistent with the passage at a higher level. For example, the purpose of a paragraph will not be at odds with the author's goal for the passage as a whole unless that paragraph represents a counterargument. Even when authors bring up information that conflicts with their main arguments, they commonly do so for the sake of shooting it down—answering or countering a refutation they introduced in a prior portion of the passage.

Function Questions—Applied Example

The most prevalent argument against doctor-assisted suicide relies upon a distinction between *passive* and *active* euthanasia—in essence, the difference between killing someone and letting that person die. On this account, a physician is restricted by her Hippocratic oath to do no harm and thus cannot act in ways that would inflict the ultimate harm, death. In contrast, failing to resuscitate an individual who is dying is permitted because this would be only an instance of refraining from help and not a willful cause of harm. The common objection to this distinction, that it is vague and therefore difficult to apply, does not carry much weight. After all, applying ethical principles of *any sort* to the complexities of the world is an enterprise fraught with imprecision.

Rather, the fundamental problem with the distinction is that it is not an ethically relevant one, readily apparent in the following thought experiment. Imagine a terminally ill patient hooked up to an unusual sort of life support device, one that only functioned to prevent a separate "suicide machine" from administering a lethal injection so long as the doctor pressed a button on it once per day. Would there be any relevant difference between using the suicide machine directly and not using the prevention device? The intention of the doctor would be the same (fulfilling the patient's wish to die), and the effect would be the same (an injection causing the patient's death). The only variance here is the means by which the effect comes about, and this is not an ethical difference but merely a technical one.

Example:

1. The author's apparent intention in discussing the "suicide machine" in paragraph 2 is to:

   A. support his thesis using an imaginative exercise.
   B. question the idea that vagueness is ethically relevant.
   C. explain the operation of a piece of medical equipment.
   D. propose a new method for performing euthanasia.

Solution: The first step in the method is to Type this question as a Function question. This can be done by recognizing the "intention . . . is to" portion of the question stem as asking why a specific piece of information was included. Your approach to Function questions will differ depending on which approach you took to the passage. However, they will all rely on one thing: knowing why the author mentioned the fact from the question stem. As this question stem has a direct task already listed, the Rephrase step is likely not necessary.

The question stem directly states a location and quoted term, meaning that whether you outlined, highlighted, or interrogated, the location of the relevant information in the passage is readily available. However, we don't actually need to dive back into the passage to answer this question. To investigate, recall the purpose of paragraph two or refer back to your outline or highlighting to find that purpose. We know that the author used the "suicide machine" thought experiment to support his argument, so this will suffice as a prediction. Going back to the text itself shouldn't be necessary unless the answer choices take us somewhere unexpected. Fortunately, that is not the case with this question, and we can see that **(A)** gives us precisely what we need. The time spent distilling the passage has been more than paid back with a quick correct response.

Among the wrong answers, **(B)** is wrong for bringing in *vagueness*, when the point of the thought experiment is to question whether the *passive/active distinction* is ethically relevant—not vagueness. **(C)** might be considered a Faulty Use of Detail because the operation of this machine is explained. However, the machine is being described not for its own sake—it's imaginary, after all—but simply to illustrate a point. The final incorrect option, **(D)**, would be a product of taking the thought experiment too literally.

10.2 Strengthen–Weaken (Within the Passage) Questions

Strengthen–Weaken questions span two of AAMC's delineated categories, but both types generally concern the logical relationships between conclusions and the evidence that *strengthens* them or the refutations that *weaken* them. Note that the only substantial difference between Strengthen–Weaken (Within the Passage) questions and Strengthen–Weaken (Beyond the Passage) questions is that the former stick to the passage as written while the latter will bring in some new element, usually appearing in the question stem, though on occasion only in the answer choices.

Sample Question Stems

- The author's suggestion that [claim] is supported in the passage by:
- For which of the following statements from the passage does the author provide the most support?
- The author states in [paragraph reference] that [claim]. This most strengthens the author's contention that:
- Which of the following objections considered in the passage most WEAKENS the author's thesis?
- Which of the following is a claim that the author makes without providing evidence?
- The view of [person] is challenged in the passage by:
- How does other information from the passage relate to the claim that [quotation]?
- What significance does the assertion that [claim] have for the author's argument?

As this list suggests, these types of questions often contain references indicating that the answers will be taken directly from the text and are heavy on words indicating connections between claims like *relate*, *support*, and *challenge*. Some question stems may be ambiguous about whether the support or challenge you're looking for will be coming from within the text or outside of it, in which case you'll also want to bear in mind the strategy for Strengthen–Weaken (Beyond the Passage) questions, detailed in Chapter 11.

Strategy

Every Strengthen–Weaken question has three pieces: **two claims** and the **connection** between them. You will always be given at least one of these elements, and your task will be to find the other(s), so begin your Investigate step by identifying where each piece can be found: either directly in the stem itself, somewhere in the passage, or in the answer choices.

If the *connection* is revealed in the question stem, it will typically be some variation of strengthen (support) or weaken (challenge), as the name of these questions suggests. However, when the connection does not occur until the answer choices, such as when a stem uses vague words like *relevance*, *significance*, or *impact*, the claims occasionally have some other relationship, such as identity (meaning the same thing) or even irrelevance. Once you know whether your task is to Strengthen, Weaken, or find some yet-to-be-discovered relevance, the next step of your investigation will be to research the status of any *claims* quoted or otherwise referenced in the question stem. The number of steps you must go through as you investigate will vary based on the number of claims in the stem.

If no other claims are mentioned, such as in a question like *Which of the following passage assertions is the LEAST supported?*, you should probably save the question for later and then use the process of elimination in your Investigate step.

KEY CONCEPT

Evidence is used to support a conclusion through a one-way relationship. A refutation is used as a counterargument against a conclusion through a one-way relationship.

If the question stem refers to two claims, then the task must be to find the nature of the relationship they share, so think about whether one claim supports the other, remembering that evidence makes a conclusion more likely to be true and refutations make conclusions less probable.

In most cases, though, you'll be presented with only one claim in the question stem, so plan to investigate the given statement as a conclusion, piece of evidence, or a refutation. Start your research with the relevant sentence, but check the surrounding text for language suggestive of a relationship. When you are researching the passage, Logic keywords are just about as important for Strengthen–Weaken questions as they are for Inference questions. If the question stem specified the relationship, pay special attention to that one; otherwise, keep an eye out for any logical connections made to the claim, using those to set expectations.

If your initial plan of attack proves unsuccessful, try process of elimination, crossing out any answer choice that does not establish the correct kind of relationship. Do not forget that support is unidirectional: if the "arrow" points the wrong way, it cannot be the right choice. So, for instance, if you are asked to find a claim that *supports* the author's thesis, a potential wrong answer is an implication that could be drawn if you assumed the thesis was true—in other words, a conclusion that the thesis itself *supported.*

Strengthen–Weaken (Within) Questions—Applied Example

The most prevalent argument against doctor-assisted suicide relies upon a distinction between *passive* and *active* euthanasia—in essence, the difference between killing someone and letting that person die. On this account, a physician is restricted by her Hippocratic oath to do no harm and thus cannot act in ways that would inflict the ultimate harm, death. In contrast, failing to resuscitate an individual who is dying is permitted because this would be only an instance of refraining from help and not a willful cause of harm. The common objection to this distinction, that it is vague and therefore difficult to apply, does not carry much weight. After all, applying ethical principles of *any sort* to the complexities of the world is an enterprise fraught with imprecision.

Rather, the fundamental problem with the distinction is that it is not an ethically relevant one, readily apparent in the following thought experiment. Imagine a terminally ill patient hooked up to an unusual sort of life support device, one that only functioned to prevent a separate "suicide machine" from administering a lethal injection so long as the doctor pressed a button on it once per day. Would there be any relevant difference between using the suicide machine directly and not using the prevention device? The intention of the doctor would be the same (fulfilling the patient's wish to die), and the effect would be the same (an injection causing the patient's death). The only variance here is the means by which the effect comes about, and this is not an ethical difference but merely a technical one.

Example:

2. Which of the following roles is played in the passage by the claim that the difference between killing and letting die is ethically relevant?

    I. It is contradicted by the assertion that the distinction between active and passive euthanasia is only technical.
    II. It bolsters the contention that applying ethical principles precisely is difficult.
    III. It underlies the most common argument against physician-assisted suicide.

    A. III only
    B. I and II only
    C. I and III only
    D. I, II, and III

Solution: The first thing to notice in your Type step is that this question is a Roman numeral question. A glance at the answer choices also tells you that this is a Strengthen–Weaken question based on the use of the words *contradicted, bolstered,* and *underlies.* The question asks about *roles* that are *played in the passage* by a statement given in the stem, so you can identify this as a Strengthen–Weaken (Within the Passage) question. However, it's a Roman numeral question, and these are usually more time-consuming. Your best bet is to save this for the end of the question set.

When you do solve this question, rephrase the base stem of the question first and leave the Roman numerals aside for the time being. Then, Investigate to find the claim referenced in the question stem. As is often the case in more complex Strengthen–Weaken (Within the Passage) questions, there is no single sentence that contains all the words in the assertion; rather, parts of it are spread throughout the text. The two lines that are most important are the opening sentences of each paragraph, the first of which refers to "the difference between killing someone and letting that person die" and the second of which maintains that "the distinction . . . is not an ethically relevant one." To Rephrase, we could ask ourselves *What is the connection between the two paragraphs?*

As you investigate, you can see that the claim that the distinction is ethically relevant must be what is "relie[d] upon" (Evidence keyword) by the so-called "most prevalent argument against doctor-assisted suicide." In other words, the claim that the distinction is ethically relevant plays a supporting role in that argument. Second, it is clear that the second paragraph is denying this claim. These initial observations already offer a sense of two roles that the distinction plays, a fairly thorough prediction.

MCAT EXPERTISE

CARS authors often use multiple *terms* (words or phrases) to describe the same *concept*, or underlying idea. On Test Day, pay special attention to the ways in which authors use terms, especially when you see Opposition keywords. When you come across dualisms, you can draw a set of columns adjacent to your map on your note board and jot down what words the author uses for each side of the contrast. This can serve as a handy reference for any synonymous language you might encounter in the questions and answer choices. So, for the ethics passage, you would put *killing* and *active euthanasia* in one column and *letting die* and *passive euthanasia* in the other.

At this point, you can look at the answer choices to see how the Roman numerals are distributed. We generally recommend starting with the most common numeral or, alternatively, whichever seems easiest for you. Statements I and III both appear three times, so start with the shorter of them. Statement III suggests the claim *underlies* the most common argument, which is precisely as predicted. Therefore, Statement III must be true and **(B)** can be crossed off.

Turning to Statement I, you'll note the mention of "the distinction between active and passive euthanasia," which you were told in the first sentence was "in essence, the difference between killing someone and letting that person die." This is consistent with the expectation set earlier that the second paragraph challenges the assertion that the claim is ethically relevant. The final sentence confirms it: "this is not an ethical difference but merely a technical one." The "not" tells you that this is the contradiction that Statement I suggests, so it must also be true, eliminating **(A)**.

There are still two answer choices remaining, so you will have to deal with Statement II. The contention that it mentions did not figure into our prediction, so check the text to find the reference, which is located at the end of the first paragraph. How does this assertion relate to the original claim that the distinction is ethically relevant? The clue is the keyword that precedes the assertion: *After all*. Even though it may sound like a Conclusion keyword on the surface, it's actually an Evidence keyword, which means that this assertion about *applying ethical principles* is in truth used to support something else. Specifically, this statement bolsters the author's belief that the "common objection" to the distinction carries little weight. This is not the relationship suggested by Statement II, which says that the claim that the distinction is ethically relevant supports how hard it is to apply ethical principles precisely. Thus, Statement II is false. Only Statements I and III are true, making **(C)** the only match, and thus the correct answer.

Example:

3. On the basis of the author's discussion, which of the following items from the passage LEAST challenges the argument for the prohibition of active euthanasia?

   A. The thought experiment involving two suicide machines from the second paragraph

   B. The assertion that the distinction between passive and active euthanasia is too difficult to apply

   C. The argument that the distinction between passive and active euthanasia is only technical

   D. The claim that the effect and the intention are the same regardless of the type of euthanasia

Solution: Question 3 is somewhat tricky to untangle, with its multiple negative terms, but you can use the phrases *challenges* and *from the passage* to Type this question as a Strengthen–Weaken (Within the Passage) question—specifically, Weaken. However, the *LEAST* means a **Scattered** format, one in which you'll probably have to test all of the answer choices. Save this for later if possible.

When rephrasing, begin by clarifying the *argument for the prohibition of active euthanasia*. Although the order after the dash in the first sentence is switched, it should be clear from the subsequent sentence that *active euthanasia* refers to the act of *killing*, which is supposedly forbidden because of the Hippocratic oath that the physician takes. This argument prohibiting active euthanasia is, in fact, the *most prevalent argument* from the first line. The correct answer, then, will be the one that challenges this argument the least.

The Scattered form suggests process of elimination as the best investigation method, but before resorting to that, it doesn't hurt to see whether the author actually talks about a challenge that he or she regards poorly because that could be the very answer you're seeking. Indeed, the author does say in paragraph 1 that "the common objection . . . does not carry much weight." The objection referenced is that it's difficult to apply the distinction between killing and letting someone die, so this can serve as your prediction.

Looking at the answers, you can see that this prediction matches **(B)**. However, you should be cautious with this sort of question. While the passage says that the objection carries little weight, it does not say that it carries no weight at all, meaning that if there were an answer choice that had no effect or even supported the argument, that would be *even less* of a challenge. As it turns out, the remaining answer choices are all aspects of the counterargument made in the second paragraph, and all do indeed challenge the original argument. Now you can be confident that **(B)** is the correct answer.

10.3 Other *Reasoning Within the Text* Questions

There are a few rarer types of questions that do not neatly fall into either the Inference or Strengthen–Weaken (Within the Passage) categories, but that definitely concern passage reasoning and that do show up in some CARS sections. These can take many different forms and all are rare, so we'll just focus our discussion on three typical tasks.

Clarification

Questions that ask about **clarification** concern a relationship that is very similar to support, as it is also a one-way relationship. One assertion clarifies another if the two share roughly the same meaning, but the "clarifying" part is typically more specific or exact. Because the clarifying language tends to be more precise, its truth value is easier to assess, and thus you should think of "clarifying" statements as supporting evidence for "clarified" conclusions. Approach them more or less as you would a Strengthen–Weaken (Within the Passage) question, except keep in mind that the meanings should be roughly synonymous.

With the ethics passage, for instance, you could see a Clarification question like *Which of the following clarifies the author's statement that the common argument against physician-assisted suicide rests upon the distinction between passive and active euthanasia?* The answer would most likely come from one of the sentences that followed, which explained the difference between the two more concretely, including the reasons why one is supposedly permitted and the other is not. In addition to words like *clarify*, words like *explain* and *reflect* are used in questions to indicate this kind of relationship.

KEY CONCEPT

In a Clarification question, look for an answer choice that is nearly synonymous with the given claim, only that is more specific or exact.

Weakness

Weakness questions are somewhat related to Inference questions, but they concern *implicit weaknesses* and *reasonable objections* to arguments discussed in the passage. Instead of the Denial Test, the best way to investigate is often via the process of elimination by directly examining the effect that answer choices have on the argument in question. The correct answer will have the most significant negative impact on the argument, perhaps even contradicting it altogether.

KEY CONCEPT

Answering a Weakness question is just like using the Denial Test, discussed in Chapter 9 for Inference questions. The difference is that the correct answer choice will be detrimental to the arguments in the passage *without* being negated.

One example of a Weakness question for a passage we've seen would be *Which of the following is the greatest inherent weakness in the author's use of a thought experiment to support the main argument?* This is a more complex type of *Reasoning Within the Text*, and it is one among a number of rarer questions that require you to appraise the strength of the author's reasoning. The answer to this example might be the fact that thought experiments force the author to rely upon readers' imagination and intuition, which may not always result in the same conclusion as the author intended.

Paradox

Finally, by a **paradox**, we mean an *apparent* logical contradiction, a set typically consisting of two assertions that seem inconsistent, but only at first glance. These will usually include two distinct claims from the text, phrased in a way to make them sound conflicting, followed by a question like *How would the author resolve this dilemma?* or *How might the passage account for this discrepancy?* Sometimes one of the claims will be a new element, which would technically make such questions *Reasoning Beyond the Text*, although they should still be approached with the same strategy in this case.

The correct answer to a Paradox question must be *consistent* with both of the claims given in the question stem. If possible, it should also not conflict with anything that the author says elsewhere in the passage. Thus, to resolve paradoxes, you should use process of elimination, marking out any answer choice that is inconsistent with one or both of the claims (or with the passage as a whole).

KEY CONCEPT

A paradox is a set of two claims that appear to be inconsistent on the surface. The correct answer in a Paradox question will be consistent with both of the claims, and it will usually attempt to explain the surface inconsistencies between the two claims.

Conclusion

Although often variable in appearance, *Reasoning Within the Text* questions test only a few essential skills: identifying inherent weaknesses in arguments; identifying the function of parts of the argument; understanding relationships of consistency and conflict; and recognizing the connections in passages between conclusions, evidence, and refutations. Regardless of how challenging these questions may seem to you now, you have the ability to improve your reasoning skills! The solution is to practice using Logic keywords to identify support relationships, and applying the other strategies discussed in this chapter as you Type, Rephrase, and Investigate questions. These tactics will also be useful when working on the final class of questions, *Reasoning Beyond the Text*, the subject of the upcoming chapter.

Concept and Strategy Summary

Function Questions

- **Type: Function questions** ask about what the author is trying to *do* with a piece of information in the passage.
    - These questions are similar to Main Idea questions, although they focus on the purpose of only one portion of the passage (usually one sentence or one paragraph).
    - Function questions tend to use words like *purpose, motive,* or *intention,* or phrases like *in order to* or *because*
- **Rephrase:** Function questions will always have a similar Rephrase task, which will result in something similar to *What purpose does [this detail] serve in the passage?*
- **Investigate:** If buzzwords in the question stem direct you to specific sentences, recall or reread those portions, thinking about how they fit into the purpose of the paragraph and the overall passage.
- **Match:** Your expectations should match the right answer. If there is no clear match, or if you cannot perform any of the earlier steps of the Kaplan Method for CARS Questions, use the process of elimination, removing any answer that conflicts with the author's main argument or the paragraph's purpose.

Strengthen–Weaken (Within the Passage) Questions

- **Type: Strengthen–Weaken (Within the Passage) questions** concern the logical relationship between conclusions and the evidence that strengthens them or the refutations that weaken them.
    - These questions often contain words like *relate, support, challenge, relevance, significance,* or *impact.*
    - These questions are closely related to Strengthen–Weaken (Beyond the Passage) questions, which simply bring in a new piece of information rather than using information directly from the passage.
- **Rephrase:** Your Rephrase should center around the task of identifying the two pieces of information and the connection between them; you will usually be given at least one of these elements and will have to find the other(s).
    - Identify where each piece of the argument can be found: in the question stem, in the passage, or in the answer choices.
    - If no claims are given in the question stem, triage the question and answer it by process of elimination later.
    - If one claim is given in the question stem, determine if it is a conclusion, a piece of evidence, or a refutation.
    - If two claims are given in the question stem, identify the relationship between them.

- **Investigate:** Research the relevant text to determine the missing claim or the connection between the claims. Use Logic keywords to help assemble the argument.
- **Match:** Your expectations should match the right answer. If there is no clear match, or if you cannot perform any of the earlier steps of the Kaplan Method for CARS Questions, use the process of elimination.

Other Reasoning Within the Text Questions

- **Clarification questions** ask for statements that are roughly synonymous, but the clarifying statement tends to be supporting evidence for the conclusion because it is more specific or exact.
  - These questions often contain words like *clarify*, *explain*, or *reflect*.
  - Approach these questions as you would Strengthen–Weaken (Within the Passage) questions, except that the meanings of the two claims should be roughly synonymous.
- **Weakness questions** ask for implicit refutations to arguments discussed in the passage.
  - These questions often contain words like *implicit weaknesses* or *reasonable objections*.
  - Approach these questions using the Denial Test as you would for Inference questions, except that the correct answer will be the most detrimental to the argument made in the passage *without* being negated.
- **Paradox questions** ask for the resolution of an apparent logical contradiction.
  - These questions often contain words like *paradox*, *dilemma*, or *discrepancy*.
  - Approach these questions through the process of elimination, crossing out any answer choice that is inconsistent with one or both of the claims of the paradox or with the passage as a whole.

Worked Example

Use the Worked Example below, in tandem with the subsequent practice passages, to internalize and apply the strategies described in this chapter. The Worked Example matches the specifications and style of a typical MCAT *Critical Analysis and Reasoning Skills* (CARS) passage.

The paragraphs of this passage may seem lengthy, but most of that volume is dedicated to examples to support the conclusions that are provided at the start of each paragraph. As each paragraph appears to center around one theme with examples, Outlining should be extremely effective. To work this passage efficiently, just make sure not to fall into the trap of trying to note all the supporting details. The language and wording are not overly complex, making it relatively easy to spot those arguments and extract a short summary for the outline.

OUTLINING METHOD

Passage	Expert Thinking
Certain contemporary forms of literary criticism draw on modern sociology and political science to understand literary works. There has been a conservative reaction to these schools of criticism, accusing them of imposing modern ideas on old texts. For example, some would consider it an implausible claim that Shakespeare's *The Tempest* can be interpreted as a play about "colonialism" and "imperialism"; after all, these terms were not even in use when Shakespeare wrote the play. These concepts must therefore be modern ones, and it is anachronistic to suppose that Shakespeare had them in mind. Besides, as Ben Jonson wrote, Shakespeare "was not of an age, but for all time," and it trivializes his genius to suppose that he had in mind the fashionable concerns of any one period. The conservative reading of *The Tempest* sees it as a play about "universal" themes like estrangement and reconciliation.	What argument does the author introduce in the first paragraph? *The author states that modern literary criticism uses modern ideas on old texts, and that there has been an accusatory reaction to this from the conservative side.* Is there evidence to back up this argument? *Yes. The author provides* The Tempest *as an example. However, we do not need to write down any details on that in our outline.* Does the author take a stance on the issue? *No keywords point to tone at this point in the passage.* P1: Conserv: modern crit uses modern bias, evidence
But writers do live in specific societies and are affected by the cultures of the times and places in which they live. The establishment of colonies—the building of empires—was an issue of keen concern in England in the early 17th century. It was a matter of national prestige and also a potential source of private wealth. All the great powers of Europe were competing for the wealth of the East and West Indies. The rich hoped to add to their fortunes; the poor hoped to begin their lives anew in the New World. Richard Hakluyt's *Voyages*, a series of published accounts of European explorations in Asia, Africa, and the Americas, was one of the most successful publishing ventures of Elizabethan England. Moreover, although the words *colonialism* and *imperialism* had not been coined yet, the ideas they connote already existed, in the sense that some Europeans perceived ethical problems relating to empire-building. The Spanish priest Bartolomé de las Casas had already condemned the cruelty of the Spanish regime in Mexico and the Caribbean, and the French essayist Michel de Montaigne had already compared the "cannibals" favorably with decadent Europeans.	This paragraph starts with a contrast key word ("But"). What does that indicate is coming next? *This paragraph will center around the author disagreeing with the view from the previous paragraph.* What is the author's argument? *The author argues that while certain modern words were not used in the time of older literary texts, the ideas behind those texts already existed.* Is there support for the author's claims? *Yes. Multiple examples are provided. We do not need to try to analyze or remember all the details, as our outline will send us back to this paragraph if needed.* P2: Auth ≠ conserv, mod ideas already existed, evidence

OUTLINING METHOD	
Passage	**Expert Thinking**
Now let us look again at *The Tempest*. Here is a play about a European family ruling a remote island by superior European technology (magic, learned from books) and the enforced labor of the native population. When another group of Europeans arrives on the island, one of them imagines an ideal commonwealth in terms derived, as scholars have long recognized, from Montaigne's essay about the native people of Brazil. The prostrate Caliban reminds Trinculo of a "dead Indian" who might be exhibited in England for crowds willing to pay to see an exotic "monster." And scholars have long recognized that the story of *The Tempest* is suggested in part by accounts of the *Sea Venture*, shipwrecked in Bermuda in 1609 on the way to the Virginia colonies.	What new ideas or themes are introduced in this paragraph? *The first sentence of this paragraph tells us that the author is coming back to* The Tempest *example. Since we do not have any Contrast keywords, it is safe to assume that the author will try to apply the argument from the previous paragraph to* The Tempest, *which conservatives used as an example in the opening paragraph. There is no need to write down specific details here.* P3: Auth: mod ideas in The Tempest
With all this in mind, are we really to believe that neither Shakespeare nor anyone who saw the play in London in 1611 was reminded of the colonial enterprise that England was then undertaking in America? Who is making the implausible claim?	How can we identify the author's conclusion? *The author used rhetorical questions to indicate a concluding thought.* What is the author's conclusion? *Those who saw the play in London were reminded of England's actions in America.* P4: Auth: conserv making implaus claim

Question	Analysis
1. The author mentions "Richard Hakluyt" primarily in order to:	This is a Function question which is asking how the author uses a particular detail in the structure of the passage. "Richard Hakluyt" was mentioned in paragraph 2, which should be the focus of our Investigate step.
A. serve as an example of authors who wrote about topics while lacking specific language to describe them.	Much of this paragraph consisted of examples of authors who were writing about topics that didn't have words for them yet. Answer choice **(A)** is correct.
B. as an example of a popular contemporary of Shakespeare.	While Shakespeare does appear in paragraph 1, the author doesn't suggest that Hakluyt and Shakespeare are contemporaries, making **(B)** a Faulty Use of Detail answer choice.
C. serve as an example of authors who invented new ideologies.	**(C)** is an Extreme answer choice; Hakluyt didn't invent colonialism, he just described it in his writing.
D. to serve as a counter point to Bartolomé de las Casas.	**(D)** is an Opposite answer choice: Bartolomé de las Casas and Hakluyt were used in the same paragraph as examples of people who condemned colonialism and imperialism.

Question	Analysis
2. Which of the following is used in the passage to support the conservative interpretation of Shakespeare?	This is a Strengthen–Weaken (Within the Passage) question because it asks for the evidence used for a given conclusion. We know the conservative critics see Shakespeare as a timeless, "universal" playwright and not one whose themes are about only a specific time. The conservative argument in paragraph 1 hinges on the claim that the terms "colonialism" and "imperialism" did not exist yet, and therefore it would be "anachronistic to suppose that Shakespeare had them in mind."
A. Imperialism and colonialism are anachronistic terms for the 17th century.	**(A)** fits perfectly with the prediction and is the correct answer.
B. *The Tempest* has its roots in a story regarding English ships headed for America. **C.** The specific society a writer lives in is essential to understanding his themes. **D.** European technology in the 17th century was far superior to other technology.	**(B)**, **(C)**, and **(D)** can be eliminated quickly because they do not come from paragraph 1, the only place where the conservatives' view is given any support. These answer choices come from paragraphs 2 and 3, which are not used to support the conservative view.

Question	Analysis
3. How would the author of this passage resolve the apparent paradox that it is an implausible claim that Shakespeare's *The Tempest* can be interpreted as a play about "colonialism" and "imperialism"?	This is an Other *Reasoning Within the Text* question. The "implausible claim" is discussed in Paragraph 1, but the author's response to that claim is discussed in the rest of the passage. By looking at our outline, we can see that the author thought those modern ideas already existed and gave examples to support that theory. In fact, the author focuses most of the passage on critiquing the argument from the first paragraph, centering around his assertion that the ideas already existed even if the words did not.
A. Ideas can only occur when they are able to be described by specific language.	**(A)** sounds like the conservative critics and is inconsistent with the author's central argument. This choice should be eliminated.
B. All ideologies are present in all stories and are ubiquitous regardless of time period.	The words *all* and *ubiquitous* indicate that **(B)** is an Extreme answer choice and should be approached with caution. For an Extreme answer choice to be correct, the author must hold an extreme viewpoint, which this author does not.
C. Even though the words had not been invented, the ideas behind them existed in the time when Shakespeare wrote *The Tempest*.	This answer choice is very close to our prediction from the Investigate step and uses moderate language. Therefore, **(C)** is the correct answer choice.
D. Thematic ideas are not present in older stories in which the author was living in a different time period.	**(D)** implies that older stories do not have thematic elements at all. This idea is inconsistent with the author's argument that "colonialism" and "imperialism" are present in *The Tempest*.

Question	Analysis
4. The author mentions the East and West Indies for what purpose (paragraph 2)?	While this question uses the word *purpose* and it may feel like a Main Idea question at first blush, it's actually asking about the function of the East and West Indies in building the author's argument. Approach this question by locating the reference in the passage and either recalling or reading the sentence before and after it. It's stated in the passage that "all the great powers of Europe were competing for the[ir] wealth" and "the building of empires—was an issue of keen concern in England in the early 17th century."
A. As an example of one of the most successful publishing ventures of Elizabethan England	**(A)** is a Faulty Use of Detail answer choice. The most successful publishing venture of Elizabethan England mentioned in the passage was "Hakluyt's *Voyages*, a series of published accounts of European explorations in Asia, Africa, and the Americas."
B. As an example of empire building that Shakespeare participated in as part of his research for *The Tempest*	There is no reference to Shakespeare witnessing firsthand the colonialism that is referenced in *The Tempest*; **(B)** is an Out Of Scope answer choice.
C. As an example of empire building that existed in when Shakespeare was writing *The Tempest*	**(C)** is closest to our investigation of the paragraph surrounding the mention of the East and West Indies and is the correct choice.
D. To show that everywhere visited by Europeans was colonized	**(D)** This is an Extreme answer choice; while the Europeans were involved in colonialism and imperialism, it is never implied that everywhere the Europeans traveled was a place they colonized.

Question	Analysis
5. The fact that the terms "colonialism" and "imperialism" were not coined yet in Shakespeare's time has what effect on the author's argument?	This is another Strengthen–Weaken (Within the Passage) question. The author's idea throughout the passage is that even though the words *colonialism* and *imperialism* did not exist, these themes were still present in *The Tempest*. Therefore, the fact that these words "were not coined yet" is consistent with—but does not strengthen or weaken—the author's argument.
A. It strengthens the author's argument about Shakespeare's works. **B.** It weakens the author's argument about *The Tempest*. **C.** It strengthens the author's argument about *Voyages*.	**(A)**, **(B)**, and **(C)** can be eliminated immediately because they state that the author's argument would be *strengthen[ed]* or *weaken[ed]*.
D. It doesn't affect the author's argument.	The prediction—that the author's argument is neither strengthened nor weakened—matches **(D)**.

PRACTICE QUESTIONS

Passage 1 (Questions 1–6)

GO ONLINE

It would be difficult to overstate the complexity of the Japanese language. The system of writing (or more properly, systems) represents a fusion of almost entirely foreign characters and a spoken language so linguistically isolated that philologists have yet to discover a precursor. Not unlike many other ancient languages, Japan lacked any system of writing at all for much of its history. Making up for lost time, though, no fewer than three different systems of writing are now employed.

The first Japanese system of writing was not Japanese. The *kanji*, a group of logographic Chinese characters, each representing a word or idea, were adopted with minimal change around the seventh century. Few languages are so geographically close yet linguistically dissimilar. As a result, the Japanese adopted a modified Chinese pronunciation for each kanji (the *on-yomi*), while retaining the native Japanese spoken word that most closely fit each kanji's meaning (the *kun-yomi*). In modern Japanese, the *on-yomi* is used for certain kanji and the *kun-yomi* for others, with compound words often involving both. Further adding to the confusion, the Chinese language contains many words in which variations in tone alone indicate drastically different meanings. The adaptation of these words to Japanese pronunciation led to a number of homophones that has, without hyperbole, been called "embarrassing" and "alarming" by scholars of the language.

The *hiragana* syllabary was developed in the eighth century by court women, who were not permitted to study *kanji* because they were deemed unfit to master its complexities. In response, they developed a simplified, flowing form of the kanji that represented all the sounds in spoken Japanese. Hiragana is phonetic rather than logographic and is therefore far more accessible to a foreign learner than the kanji. Because Japanese is an open language, most consonants cannot be expressed by themselves. Hiragana is therefore not strictly an alphabet. *Katakana* came about around the same time as hiragana, also as an attempt to simplify the kanji. The sparse, angular characters correspond fairly closely to the hiragana and, as befitting their origin among Buddhist monks, have a look generally considered more masculine than hiragana, which was originally called *onnade*, or "women's hand." The katakana have essentially become the print counterparts of the "cursive" hiragana.

With so many systems jostling for position, each used more or less independently of the other, it would not be unreasonable to anticipate that a national movement toward systematization of the language would settle on a single one. A national movement was in fact started after World War II: a radical idea encouraging the use of all three systems together. A glance at any Tokyo newspaper will reveal kanji used to represent most standard actions and ideas, hiragana to indicate grammatical inflections and tenses, and katakana to represent adopted foreign and technical words, as well as to indicate emphasis. The use of the three systems has become sufficiently standardized in this way that deviations often lend a piece of writing strong connotations. A piece written entirely in katakana, for example, may be disconcerting to a modern reader and may have a vaguely pre–World War II military air to it, much as a piece written all in capital letters with telegraph punctuation might in English. While such a complex system has made the language's learning curve high for native speakers and foreigners alike, it has also contributed to a stunning richness of expression such that any list of world's great works of art a hundred years from now will have to be written partially in kanji, hiragana, and katakana.

1. The author mentions Chinese *kanji* (paragraph 2) for what purpose?
   A. To show the large number of homophones in the language is due to the closeness of Japanese and Chinese pronunciation.
   B. To illustrate that the presence of homophones in a language can be considered embarrassing.
   C. To show it as a precursor to the Japanese method of writing that caused some linguistic quirks.
   D. To show that homophones are dependent on variations in tone.

2. The author's primary purpose in the passage is to:
   A. argue that the Japanese language is overly complex.
   B. describe the origins of the Japanese language's complexity.
   C. propose a simplification in how Japanese is written.
   D. trace the origins of logographic writing systems.

3. According to the passage, which of the following pieces of Japanese literature would NOT likely be written entirely in katakana?
   A. a modern Japanese novel
   B. a list of adopted foreign words
   C. a 9th-century Buddhist text
   D. an early 20th-century general's log

4. Based on the author's description, open languages generally contain:
   A. borrowed systems of writing and speaking from many different sources.
   B. intrinsic acceptance of change and reform.
   C. syllables that end in vowels.
   D. few consonant sounds.

5. As can be inferred from the passage, the group of Buddhist monks who developed katakana:
   A. was predominantly or entirely male.
   B. used hiragana as a model.
   C. was considered unfit to master the complexities of kanji.
   D. was closely involved with the military of the time.

6. Which of the following is a claim the author makes without providing evidence?
   A. The first Japanese system of writing was not Japanese.
   B. The use of the three systems has become sufficiently standardized in a way that deviations often lend a piece of writing strong connotations.
   C. Japanese lacked any system of writing at all for much of its history.
   D. The katakana have essentially become the print counterparts of the "cursive" hiragana.

Passage 2 (Questions 7–12)

The palette of sights and sounds that reach the conscious mind are not neutral perceptions that people then evaluate: they come with a value already tacked onto them by the brain's processing mechanisms. Tests show that these evaluations are immediate and unconscious and applied even to things people have never encountered before, like nonsense words: *juvalamu* is intensely pleasing and *bargulum* moderately so, but *chakaka* is loathed by English speakers. These conclusions come from psychologists who have developed a test for measuring the likes and dislikes created in the moment of perceiving a word, sound, or picture. The findings, if confirmed, have possibly unsettling implications for people's ability to think and behave objectively. This is all part of preconscious processing, the mind's perception and organization of information that goes on before it reaches awareness—these judgments are lightning fast in the first moment of contact between the world and the mind.

Some scientists disagree with the claim that virtually every perception carries with it an automatic judgment, though they, too, find that such evaluations are made in many circumstances. This cohort posits a narrower scope of stimuli that elicit the response. That is, these scientists believe that people don't have automatic attitudes for everything, but rather for areas of interest.

In responding to a stimulus, a signal most likely travels first to the verbal cortex, then through white matter tracts to the amygdala, where the effect is added, and then back to the occipital lobe through the same or similar pathways. The circuitry involved can do all of this in a matter of a hundred milliseconds or so, long before the individual experiences any conscious awareness of the word. This creates an initial predisposition that gets things off on a positive or negative footing. These reactions have the power to largely determine the course of a social interaction by defining the psychological reality of the situation from the start.

Although perhaps counterintuitive, the "quick and dirty" judgment tends to be more predictive of how people actually behave than is their conscious reflection on the topic. This may represent a new, more subtle tool for research on people's attitudes, allowing scientists to assess what people feel without their having any idea of what exactly is being tested. One could detect socially sensitive attitudes people are reluctant to admit, like racial and ethnic biases, because these automatic judgments occur outside of a person's awareness, as part of an initial perception. They are trusted in the same way senses are trusted, not realizing that seemingly neutral first perceptions are already biased.

Conclusions from both camps are based on a method that allows them to detect subtle evaluations made within the first 250 milliseconds—a quarter of a second—of the perception of words. The measurement of liking can be made outside the person's awareness because if the first word is presented in less than a quarter of a second, the reaction to it never registers in consciousness, though it can still be read.

7. The author's description of reactions to words like *juvalamu*, *bargulum*, and *chakaka* relies on which of the following assumptions?
   - **A.** These words can be pronounced in less than 250 milliseconds.
   - **B.** The meaning of a word is not necessary for an emotional response to it.
   - **C.** Familiarity with a word can cloud judgment of one's reaction to the word.
   - **D.** An individual could repeat back the words after hearing them.

8. Which of the following, if true, would serve to most strengthen the argument of an opponent to the author?
   - **A.** Many of our actions are influenced by perceptions unknown to our consciousness.
   - **B.** In Swahili, *juvalamu* and *chakaka* mean "enjoyable" and "severe pain," respectively.
   - **C.** People's actions are most regulated by conscious thought patterns rather than unknown feelings.
   - **D.** Humans perceive their surroundings subjectively.

9. The view of the cohort of scientists mentioned in paragraph 2 is most challenged in the passage by:
   - **A.** the description of the neural circuits involved in responding to a stimulus.
   - **B.** the claim that the "quick and dirty judgment" is more predictive of behavior than conscious reflection.
   - **C.** the referenced study of reactions to nonsense words in English speakers.
   - **D.** the hypothesis that measurements can be made outside of a person's awareness.

10. Based on the passage, information retrieved from these types of perception experiments could best be used by psychologists to:
    - **A.** help patients with language barriers.
    - **B.** map out the pathological thought patterns in a murderer's mind.
    - **C.** identify hidden attitudes that cause two individuals to repeatedly clash on various issues.
    - **D.** determine why one sibling has math skills while the other excels in literature analysis.

11. Based on information in the passage, in the author's view, which of the following statements is NOT true?
    - **A.** Information regarding external stimuli is processed so quickly that it does not become part of our conscious awareness.
    - **B.** Automatic judgments occur on stimuli with which there is great familiarity.
    - **C.** Automatic judgments have little effect on a person's mood.
    - **D.** Ethnic biases may be influenced by attitudes of which we are unaware.

12. If given the chance to expand on the points put forth in the passage, the author would most likely argue:
    - **A.** to use this type of experimentation to map the pathway through which neurological signals travel.
    - **B.** that the evidence presented in the passage is inconclusive and directs psychologists in no specific direction.
    - **C.** that automatic judgments have little or no effect on important behavior patterns.
    - **D.** to continue with further experimentation, seeking to identify the roots of problems found in human relationships.

PRACTICE ANSWERS

Passage 1 (Questions 1–6)

Sample Highlighting

P1. "linguistically isolated"; **P2.** "Chinese character"; **P3.** "Hiragana is phonetic" and "simplify the kanji"; **P4.** "sufficiently standardized" and "richness."

Sample Outlining

P1. Japanese language and writing complex: at least 3 systems

P2. Chinese (logographic) introduced = kanji; on-yomi *vs.* kun-yomi

P3. Syllabic/phonetic: hiragana (feminine, script) *vs.* katakana (masculine, print)

P4. Post-WWII: All three combined and systematized, different uses for each

1. **C**

For this Function question, start with where *kanji* are mentioned. Kanji appears in paragraph 2, where the author says that the adaptation of Chinese words with variations in tone to Japanese resulted in lots of homophones. The function of including kanji is to introduce the idea that Japanese came from the Chinese and doesn't have as many variations in tone as Chinese does. **(C)** fits this prediction. While **(B)** may be a true statement, it is not related to kanji and is a Faulty Use of Detail answer choice. **(D)** is an Opposite; it is the loss of "variations in tone" as words moved from Japanese to Chinese that led to the formation of homophones. As for **(A)**, the author specifically stated that "Few languages are so geographically close yet linguistically dissimilar," so we know that pronunciations must be very different.

2. **B**

This is a Main Idea question, so predict using the author's overall goal established during your Distill step: *to examine the origins and complexity of three Japanese writing systems.* Only **(B)** involves both the *origins* and *complexity* of the language. Notice that we can use a vertical scan of the first words of the answers to eliminate **(A)** and **(C)** because the author is neutral and does not make any strong *argu[ments]* or *propos[als]*. As for **(D)**, the author discussed the origins of Japanese only—not multiple *logographic writing systems*—and, even then, this answer choice is too narrow as it addresses the author's purpose only in paragraph 2.

3. **A**

This is an Apply question asking for an example of a text that would *NOT likely be written entirely in katakana.* Where does the author discuss the uses of writing in katakana? Referring back to the passage (or recalling if you used the Interrogate method for distilling the passage), we find that it is introduced in paragraph 3 and that the modern approach of using all three systems together—including katakana—is described in paragraph 4. There, the author writes that katakana is used to "represent adopted foreign and technical words, as well as to indicate emphasis" and that a piece written entirely in katakana "may be disconcerting to a modern reader and would have a pre–World War II military air" to it. Based on this information, **(B)** and **(D)** can immediately be eliminated. **(C)** can also be eliminated based on the description of the origins of *katakana* in paragraph 3: both hiragana and katakana appear around the *eighth century*, and katakana specifically "origin[ated] among Buddhist monks." The answer must therefore be **(A)**, which makes sense: a modern piece of literature would be expected to combine all three writing systems.

4. **C**

For this Inference question of the Implication subtype, start with where *open languages* are mentioned. Paragraph 2 states that "because Japanese is an open language, most consonants cannot be expressed by themselves." It also points out that hiragana (and, by extension, katakana) is "not strictly an alphabet," but rather a "syllabary." Taking these pieces of information together, we can determine that open languages must express consonants together with vowels and that the language is built on these consonant–vowel combinations (syllables). **(C)** reflects this idea, highlighting the syllabic nature of the language. While the author does not specifically state that vowels end syllables in Japanese, the two Japanese terms given in the paragraph—hiragana and katakana—both demonstrate this pattern. **(D)** is a Distortion because although the author does say that consonants are not often used by themselves, there is no mention that they are few in number overall. While **(A)** describes the Japanese language, it does not reflect the author's use of the more general term "open languages." Finally, **(B)** is a literal use of the word *open* and does not fit the context described by the author.

5. **A**

The word *inferred* shows that this is an Inference question. The Buddhist monks who developed katakana are highlighted in paragraph 3. Let's review the main points: katakana was created by "Buddhist monks" and looks "more masculine than hiragana." The author also notes that hiragana was developed by women and was known as "women's hand." Given the contrast between katakana and hiragana on the basis of gender, we can infer that katakana looks masculine because it was developed by men. **(A)** must therefore accurately describe this group of Buddhist monks. While **(C)** might look tempting, we know only that the women who developed hiragana were "considered unfit to master the complexities of kanji." The author never stated anything similar about the monks who developed katakana, so we cannot make that inference. In the last paragraph, the author points out that katakana now has a "vague... military air" to it for the modern reader, but that does not mean that the monks who created katakana were *closely involved with the military of the time.*

6. **C**

This is a Strengthen–Weaken question asking for a claim in the passage that lacks evidence. Note that all four answer choices are sentences taken verbatim from the passage. For this question, we will have to address each answer choice as we look for evidence that supports the claim. The "first Japanese system of writing," **(A)**, was addressed in paragraph 2. Immediately following this sentence, the description of "kanji" is given—which is that "first... system of writing." Therefore, this answer choice can be eliminated. The use of "three systems" and the effects of "deviations" from the standardized approach are detailed in paragraph 4; the subsequent "piece written entirely in katakana" is evidence to support this claim, eliminating **(B)**. **(C)** is mentioned at the end of paragraph 1, but that's all the information we get about the Japanese language before writing systems were developed. This answer is therefore correct. Finally, **(D)** is supported by the sentence that immediately precedes it, which describes the "sparse, angular characters" of katakana as counterparts to the hiragana.

Passage 2 (Questions 7–12)

Sample Highlighting

P1. "evaluations"; **P2.** "narrower scope"; **P3.** "initial predisposition"; **P4.** "more subtle tool" and "automatic judgement"; **P5.** "detect subtle."

Sample Outlining

P1. Value assigned to perceptions before reaching awareness (preconscious processing)

P2. Some scientists think scope is narrower: areas of interest

P3. Brain pathway of preconscious processing

P4. Applications of the theory, ex: socially sensitive attitudes and biases

P5. Methods of measuring

7. **B**

This is an Inference question of the Assumption subtype focusing on the author's example in the first paragraph. In this "nonsense words" example, words devoid of any importance in the English language appear to trigger an emotional response. This might be surprising—usually we would think that any emotional response to language would be based on the meaning of the word. However, if individuals still have responses to these nonsense words, knowing the definition of the word must not be required. This assumption matches closely with **(B)**, making it the correct answer. **(A)** is a Faulty Use of Detail; while the last paragraph states that a word would have to be given in less than 250 milliseconds to avoid conscious perception, the example in the first paragraph does not require that the word never reaches conscious perception. **(C)** is a Distortion. While this answer choice may very well be a true statement, the example described in the first paragraph does not address familiar words at all—only nonsense words; thus, we cannot infer anything about familiar words for this example. Finally, **(D)** is Out of Scope; whether or not the individual can repeat back the words has no clear bearing on the individual's feelings about the word.

8. **C**

The words *if true* and *strengthen* indicate that this is a Strengthen–Weaken (Beyond the Passage) question, which is similar to the Strengthen–Weaken (Within the Passage) question type discussed in this chapter, except it brings in new information. The question asks to *strengthen the argument of an opponent to the author*, which would logically be the same as weakening the author's conclusion. The author's main conclusion in the passage is that many impressions are formed preconsciously, so any answer choice that goes against this premise would be correct. **(C)** says exactly the opposite of what the author argues, claiming that it is *conscious thought*—not *unknown feelings*—that dictates behavior. **(A)** and **(D)** both fit cleanly with the author's argument and therefore would not weaken his or her conclusion. **(B)** is Out of Scope as the author does not address the responses to these nonsense words in any populations besides English speakers.

9. **C**

This Strengthen–Weaken (Within the Passage) question type requires a bit of decoding. The *cohort of scientists mentioned in paragraph 2* believes that automatic attitudes do not occur in response to "everything, but rather…areas of interest." To challenge this claim, we would need some evidence of an automatic attitude formed in response to something unlikely to be a person's area of interest. We find a match to this prediction in **(C)**; nonsense words are unlikely to be an area of interest for most people, and yet an automatic response was still generated. The other answer choices all bring in other components of the passage, but the "neural circuits, predict[ability] of behavior," and methods for "measurements" all have nothing to do with the "areas of interest" claim.

10. **C**

This is an Apply question asking for an example of the possible uses for the *information retrieved from the... perception experiments* described in the passage. Possible uses of this information are explored in paragraph 4. Our prediction, then, is an answer that identifies "attitudes and biases"; this matches best with **(C)**. While the example of racial and ethnic biases is mentioned in the passage, other "attitudes people are reluctant to admit" are also considered. The other answer choices are all Out of Scope, as they bring in *language barriers, pathological thought patterns,* and differences in cognitive abilities between *siblings*—none of which is even hinted at in the passage.

11. **C**

Here we've got another Inference question, but this one is a Scattered Inference question due to the *NOT* in the question stem. Keep the author's main conclusion in mind before looking at the answer choices—that incoming information has value placed on it before we're even aware of it. Three answers will agree with the author's argument, and one should work against it. In this case, **(C)** is unusually easy to spot: it contradicts the basic conclusion of the experiment the author cites in paragraph 1. In this experiment, automatic judgments placed on nonsense words appeared to generate an emotional response. **(A)** and **(D)** both focus on the author's point that we are unaware of the processing that is happening. **(B)** might also be acceptable to the author; just because *automatic judgments occur on stimuli* with which one is unfamiliar doesn't mean that these judgments won't occur on stimuli that are familiar.

12. **D**

For this Strengthen–Weaken (Within the Passage) question in which we must continue an argument already present in the passage, consider the word *argue*: whatever the author argues will have to match the ideas he or she already presented in the passage. First, knock out any answer choices that weaken the author's stance; this allows us to eliminate **(B)** and **(C)**, as they would ruin the author's central thesis. Then, **(D)** is a logical extension of what the author describes for the majority of the passage, so it's the best fit. While the author does mention the "pathway" in **(A)**, this idea is a small point in the passage; the primary focus is on psychology, not neuroanatomy.

CHAPTER 11

Question Types III: *Reasoning Beyond the Text* Questions

CHAPTER 11

Question Types III: *Reasoning Beyond the Text* Questions

In This Chapter

Introduction

LEARNING OBJECTIVES

After Chapter 11, you will be able to:

- Identify Apply, Strengthen–Weaken, and other *Reasoning Beyond the Text* questions
- Solve *Reasoning Beyond the Text* questions with strategies specific to each question type
- Differentiate between Probable Hypothesis, Alternative Explanation, and Passage Alteration questions

As a physician, you'll quickly realize that patients rarely present with the exact mix of signs and symptoms you may have spent countless hours memorizing in medical school. Additionally, when patients describe their symptoms, the language they use rarely coincides with the sophisticated terminology found in textbooks and journals. As a result, a large part of medical training involves interacting directly with patients in order to better understand how to assimilate the knowledge learned in class with the practicalities of real life. Medical schools are interested in students who can take information they have elicited from a patient and go a step further with it, not just matching symptoms to a diagnosis, but rather, identifying future diagnostic steps and applying information learned from prior patients to new ones. This is why this skill is so heavily tested in the *Critical Analysis and Reasoning Skills* (CARS) section. Like an ICU, the MCAT tests whether you are able to take what you have learned and apply it to a novel situation with speed and precision.

In this chapter, we'll examine the Apply and Strengthen–Weaken (Beyond the Passage) question types. As in the previous question types chapters, we'll look at some common question stems, specific strategies, and a few worked examples for each question type. We will conclude with a brief discussion of the rarer kinds of *Reasoning Beyond the Text* questions.

The AAMC reports that 40 percent of the questions in the CARS section should be classified as *Reasoning Beyond the Text.* It further divides this categorization into questions that require you to apply or extrapolate ideas from the passage to a new context (Apply questions) and those that require you to evaluate the effect new information would have if it were incorporated into the passage (Strengthen–Weaken [Beyond the Passage]). The fundamental difference between the two question types is one of direction: Apply questions ask you to determine how the passage relates to a new situation, while Strengthen–Weaken (Beyond the Passage) questions ask you to determine how the new information might affect the passage. While the AAMC suggests that the *Reasoning Beyond the Text* category is split evenly between these question types, our intensive study of released AAMC materials has shown that Apply questions tend to be slightly more common than Strengthen–Weaken (Beyond the Passage) questions. Note that it is possible, though rare, to get a question that doesn't fall into either question type but still qualifies as *Reasoning Beyond the Text.*

KEY CONCEPT

Apply and Strengthen–Weaken (Beyond the Text) test your deductive reasoning. In both cases, a new situation is provided. Apply questions focus on how the passage relates to the new information (through a Response, Outcome, or Example). Strengthen–Weaken (Beyond the Passage) questions focus on how the new information impacts the passage.

Note: The Question Types, as well as the Kaplan Method for CARS Passages, Kaplan Method for CARS Questions, and Wrong Answer Pathologies, are included as tear-out sheets in the back of this book.

11.1 Apply Questions

Reasoning Beyond the Text questions always contain information in the question or answer choices (or both) that is not stated or suggested by the passage. *Reasoning Beyond* questions commonly begin with words like *suppose*, *assume*, and *imagine*, followed by an elaborate scenario, ending with a question connecting the new content to the author or passage. Apply questions will ask you to take information from the passage and apply it to a new situation. Strengthen–Weaken (Beyond the Passage) questions will test the opposite relationship, providing you with new information and then asking you to determine how it impacts the passage.

Apply questions are one of the most common of the CARS question types covered on the MCAT. In Apply questions, the text is used as a starting point that you must then apply to a new context. There are three common tasks, each constituting roughly one-third of the Apply question pool, that you will be asked to carry out. We categorize them as *Response*, *Outcome*, and *Example* Apply questions based on the words commonly seen in their respective question stems.

Sample Question Stems

- Consider the following: [new info]. The author would most likely respond to this by claiming:
- With which of the following claims would [the author or an alternative viewpoint from the passage] be LEAST likely to agree?
- Suppose that [details of new scenario]. Based on the passage, what would the author most likely advise in this case?
- Imagine [new info]. Which of the following, according to information presented in the passage, is the most reasonable outcome?
- If the passage's author is correct, the most likely consequence of [new situation] would be:
- Assume that [new info]. One could reasonably expect, on the basis of the passage, that:
- Which of the following best exemplifies the author's notion of [quotation from the passage]?
- Which of the following phenomena would the author most likely characterize as a [concept]?
- [New info]. The author would most likely classify this as:

At least a third of Apply question stems are similar to the first three samples above, concerning how the author (or, less frequently, another individual discussed in the passage) would respond to a particular situation. These questions commonly ask for the author's likely *response* or *reply*, a claim that the author would be *most likely to agree with*, or the statement *least consistent with* one of the views discussed.

Other Apply question stems take the form of the next three samples, investigating the most probable *outcome*, *result*, *expectation*, or *consequence* in a situation that is in some way analogous to one discussed in the passage. In other words, these questions provide you with a cause and ask about the likeliest effect, based on the passage.

Most of the remaining third of Apply question stems resemble the final three cases, asking for *examples* or *instances* of ideas discussed in the passage. Usually, the concept or term will be given, and your task is to find an item from a specified context (or from the "real world") that best exemplifies it as it was used in the passage. Question stems like the last one are rarer and more difficult, starting with an outside case and asking you how the author would categorize it: *What is this an example of?*

MCAT EXPERTISE

A lengthy question stem preceded by words like *suppose*, *imagine*, or *assume* is often a good indication of a *Reasoning Beyond the Text* question. From there, jump ahead to the last line before the question mark or colon at the end of the question stem to Type the question as either Apply or Strengthen–Weaken (Beyond) and identify the specific task being asked of you. Then, read through the new information and Rephrase carefully, always watching out for analogies to and similarities with the passage text.

Strategy

Once you've identified that new information is being provided in the question stem, you may find it helpful to jump to the end of the question stem to see what the question is really asking. This particular technique can be very helpful in your Rephrase step. Once you've picked up on the key language needed to determine the task, analyze the new information in the question stem closely for hints that connect it back to the passage. Then, go back and reread the relevant portions of the passage to make your prediction. When rephrasing the stem, keep in mind the question type and the task associated with it.

Response

For example, if the stem asks *How would the author respond to . . . ?* or *Which of the following claims would the author be most likely to endorse?*, your task is to get inside the author's head. The correct answer to a Response question should be consistent with the author's beliefs, which are typically reflected in the passage through the use of Author keywords, originally discussed in Chapter 5 of *MCAT CARS Review*. If you find yourself with a prediction that does not match directly with an answer choice, begin by eliminating any answers that are logically inconsistent with the author's assertions. If you are asked about a viewpoint other than the author's, utilize a similar strategy by putting yourself in the mind-set of the alternative perspective and recognizing what that person believes.

Outcome

If a question asks you about the probable outcomes of a scenario, look for words in the passage that indicate cause-and-effect relationships (Logic keywords). Identify any causes in the passage that are analogous to what is presented in the question stem and use their corresponding effects as the basis for your prediction.

Example

When you are called upon to identify examples, begin by finding the relevant text from the passage. Specifically focus on text that provides definitions, explanations, or the author's own examples of the concept in question. Take note of necessary conditions (which MUST occur in all instances of the concept) and sufficient conditions (which are enough on their own to make an instance qualify as that concept). Sufficient conditions are easier to match to, prediction-wise, but necessary conditions are useful when using an elimination strategy as they can be easily ruled out.

Apply Questions—Applied Example

There is no shortage of evidence for the existence of systemic biases in ordinary human reasoning. For instance, Kahneman and Tversky in their groundbreaking 1974 work proposed the existence of a heuristic—an error-prone shortcut in reasoning—known as "anchoring." In one of their most notable experiments, participants were exposed to the spin of a roulette wheel (specially rigged to land randomly on one of only two possible results) before being asked to guess what percentage of United Nations member states were African. The half of the sample who had the roulette wheel stop at 65 guessed, on average, that 45 percent of the UN was African, while those with a result of 10 guessed only 25 percent, demonstrating that prior presentation of a random number otherwise unconnected to a quantitative judgment can still influence that judgment.

The anchoring effect has been observed on repeated other occasions, such as in Dan Ariely's experiment that used digits in Social Security numbers as an anchor for bids at an auction, and in the 1996 study by Wilson et al. that showed even awareness of the existence of anchoring bias is insufficient to mitigate its effects. The advertising industry has long been aware of this bias, the rationale for its frequent practice of featuring an "original" price before showing a "sale" price that is invariably reduced. Of course, anchoring is hardly alone among the defective tendencies in human reasoning; other systemic biases have also been experimentally identified, including loss aversion, the availability heuristic, and optimism bias.

Example:

1. Suppose a consumer who is looking for an inexpensive replacement for her outmoded refrigerator is enticed by a local retailer's ads for a discount sale promising savings of 50 percent or greater on all appliances. The author would probably warn the consumer that:

    A. sales are scams designed to exploit the consuming public.
    B. the pre-markdown prices are most likely set artificially high.
    C. heavily discounted merchandise is likely damaged or stolen.
    D. making a rational decision about what to buy is impossible.

Solution: The question starts with "suppose" so we are likely dealing with a *Reasoning Beyond* question. Given the long question stem, you might consider skipping the new information in the stem, to determine the task required. Your Rephrase step should result in needing to determine what the author would warn the consumer, which is a type of response. As a result, as you work through the question, you want to be thinking about views that the author holds. The ads for the sale should draw you to paragraph 2, in which the author states: "The advertising industry has long been aware of this bias, the rationale for its frequent practice of featuring an 'original' price before showing a 'sale' price that is invariably reduced." Although the author is not explicit, the use of quotation marks here is a case of "scare quotes," suggesting that the so-called "original" price is just there to make the "sale" price seem lower. Thus, it's reasonable to infer that the author would warn the consumer about the anchoring effect intended with pre-sale prices.

The closest match to this prediction is **(B)**. **(A)** can be ruled out as a Distortion because the language is too strong. Advice or admonitions that an author would provide should both be consistent with what the author says in the passage and also similar in tone. **(D)** is also a Distortion because of the word *impossible*. Finally, **(C)** is Out of Scope. The author suggests that discounts are offered to manipulate buyers into believing they have found a better deal, not to trick them into buying products that turn out to be faulty.

MCAT EXPERTISE

Although forming a prediction prior to looking at the answer choices is generally recommended, since *Reasoning Beyond the Text* questions contain new elements in the answers, sometimes looking at the first one or two answer choices can give you a better idea of the form that the correct answer will take. If the options diverge significantly from what you expected, go back and modify your original expectations before moving on to any remaining choices. Matching to the correct answer is generally less time-consuming than crossing out all three incorrect choices, so revising your prediction is usually a more optimal strategy than process of elimination.

Example:

2. Imagine that a psychologist specialized in the study of systemic reasoning biases. On the basis of the information presented, this psychologist could most reasonably be expected to:

   A. have a higher likelihood of misjudging numerical quantities when not given an anchor.
   B. make significantly fewer mistakes in reasoning than those ignorant of anchoring bias.
   C. be equally as susceptible to errors resulting from the anchoring effect as anyone else.
   D. avoid entirely the logical fallacies that ordinary human beings commit systematically.

Solution: The question stem is relatively short, so there's no need to jump to the end before reading the part after *Imagine*. Because we're searching for something *this psychologist could most reasonably be expected to do*, you can Type this as an Outcome question. Ask yourself whether the passage suggests anything about what happens to psychologists with greater knowledge of these biases in reasoning. Since there is no explicit reference, and the question stem is not heavy on other clues, expect to briefly look through the answer choices for additional clues on what to search for in the passage as the first part of your Investigate step.

The very first possibility suggests a consequence that would occur without the presentation of an anchor; however, the passage is only concerned about what happens when an anchor is present. It does not tell us anything about how accurate people's judgments of quantities are without the anchoring effect. We therefore have no basis to determine what effect having no anchor would have. Eliminate **(A)** as Out of Scope.

(B) suggests that such a psychologist would make fewer mistakes because of her knowledge of the anchoring effect. Is there anything in the passage that would warrant this conclusion? The second paragraph states that Wilson and his colleagues "showed even awareness of the existence of anchoring bias is insufficient to mitigate its effects." Not only does this rule out **(B)** as an Opposite, but it also gives you an idea of what you can expect in a correct answer—that a psychologist is just as likely as anybody else to fall into these errors due to anchoring.

This revision to your Investigation pays off when you read **(C)**, which is almost an exact match for the new prediction. On Test Day, you would select this choice and move on to the next question without paying much attention to **(D)**—which could be ruled out for being a Distortion or an Opposite, constituting an even more extreme version of **(B)**.

Example:

3. Which of the following would the author be LEAST likely to consider a case of anchoring bias?

    A. An unusually high opening bid at an annual charity auction leads to a sizable increase over previous years in total proceeds collected.

    B. The sequel to a popular film is deemed a failure because it could not quite beat the record-smashing box office receipts of the original.

    C. A shipping website receives reports of greater levels of customer satisfaction after starting deliberately to overestimate delivery times.

    D. Poor initial sales figures for a new video game console motivate its manufacturer to reduce significantly the system's suggested retail price.

Solution: Despite asking for the *LEAST likely* case, we should still be able to Type this as an Example Apply question. It is hard to know what to predict based on the limited information in the question stem, but you can still set expectations about the correct answer by investigating the concept mentioned. The anchoring effect is the primary subject of this short passage, but the author only provides examples of the phenomenon without giving it an explicit definition. It's difficult to say precisely what would be sufficient to constitute anchoring bias, but we can isolate some necessary conditions. In each of the passage examples, a baseline numerical expectation, or "anchor," is set (either at random, as in the case of the experiments mentioned, or deliberately high in order to manipulate purchasers), which then skews the judgments people make about quantities. Any answer choice that satisfies these prerequisites should be eliminated because you are asked to find the *LEAST likely* example. Your Match will be found by process of elimination.

(A) is clearly a case of the anchoring effect; the passage even made mention of an experiment that used bids at an auction as the dependent variable. You can reason that the lofty opening bid must have caused other participants to heighten their appraisals of the items for sale, which in turn led to larger final sale prices and increased total proceeds.

Even though **(B)** is unlike anything found directly in the passage, it follows the model we anticipated for wrong answers. In this case, the original film is serving as the anchor that biases judgments of its sequel. Stating that the new movie *could not quite beat the record-smashing box office receipts of the original* suggests that it still generated a lot of revenue, meaning that the assessment of failure was probably in error. Because the anchoring effect is said to be a systemic bias, this implication of error should quell any remaining doubts about eliminating **(B)**.

Turning to **(C)**, we find another case that departs from the passage. Here, the anchor would be the estimated time of arrival for a particular shipment. If the site deliberately overestimates shipping times, its customers will consistently have to wait less time than they are told to expect, which will lead many to think that they are receiving excellent service. This situation is precisely analogous to the example of the original price marked down for sale, and hence it should also be eliminated.

Thus, **(D)** must be correct. While similar to one of the examples discussed on a superficial level, the monetary value of the sales (which would undoubtedly be orders of magnitude *higher* than the original price of one individual console) does not *bias* the manufacturer to lower the cost per system. Rather, this is a case of a rational response to an economic problem: when demand is too low, reduce the price.

Now, to be clear, this reduction in price could *lead to* a case of anchoring bias if, say, consumers started to purchase the console in greater quantities, believing it now to be a better deal. But the answer choice does not focus on that effect nor on any of the effects of cutting the cost, rather only mentioning its cause. And so, **(D)** is indubitably the one *LEAST likely* to count as anchoring bias for the author.

11.2 Strengthen–Weaken (Beyond the Passage) Questions

Like Strengthen–Weaken (Within the Passage) questions, **Strengthen–Weaken (Beyond the Passage) questions** explore evidence–conclusion relationships. Two notable distinctions between the two question types are that Strengthen–Weaken (Beyond the Passage) questions (1) contain at least one claim not from the passage that is unique to the question stem or answer choices and (2) treat the passage as flexible, in that they can be modified by outside forces.

Sample Question Stems

- Suppose [new info]. This new information:
- Which of the following statements, if true, would most bolster the author's argument about [topic]?
- Assume that [new info]. This assumption weakens the author's claim that:
- [New info] would most strongly support the view of:
- Recent research on [topic] suggests [new info]. Which of the following assertions from the passage is most logically consistent with these results?
- Some theorists have argued that [new info]. Based on the discussion in the passage, which of the following would present the greatest CHALLENGE to their argument?
- [New info]. In conjunction with information presented in the passage, it would be most reasonable to conclude that:

- Imagine that [new info]. What impact would this have on the arguments made in the passage?
- Which of the following study findings would most seriously undermine the author's thesis?

If the question includes new information and asks about logical relationships using words like *support*, *challenge*, and *consistency*, you can safely Type the question as a Strengthen–Weaken (Beyond the Passage) question. In some cases, the new information may be hidden in the answer choices, so watch out for clues that suggest the correct answer will come from outside the passage, such as words like *would* and *could*.

Strategy

As with Strengthen–Weaken (Within the Passage) questions, your primary task is to identify the three relevant parts: the conclusion, the evidence or refutation, and the nature of the connection (strengthen, weaken, or some unspecified relevance). Begin by determining which component (or, rarely, which two components) you are looking for, which you can target by rephrasing the question stem. Rather than reading all the new details during your first pass, jump to what immediately precedes the question mark or colon in order to figure out what the question is asking. Then, reread the question stem, keeping an eye out for any hints of analogy. Then, return back to the passage to confirm the relationship as the last step of your Investigation. Remember that Logic keywords from the passage can help you identify relevant evidence–conclusion relationships. For example, if a new experimental finding described in the question stem is similar to a study in the passage that was used to support the author's thesis, a good prediction is that this new finding will *strengthen* the author's thesis.

Note that the correct answers to Strengthen–Weaken (Beyond the Passage) questions rarely match predictions exactly the way they did for Strengthen–Weaken (Within the Passage) questions. As a result, it is more important to focus on relationships in your predictions, rather than specific wording, especially since answer choices are usually only incidentally related to the text.

BRIDGE

Strengthen–Weaken (Beyond the Passage) questions are extremely similar to Strengthen–Weaken (Within the Passage) questions, except that the former bring in new information, while the latter ask about arguments contained in the passage. Thus, a similar strategic approach can be used for both question types. Make sure to review Strengthen–Weaken (Within the Passage) questions, discussed in Chapter 10 of *MCAT CARS Review*, in tandem with this discussion of Strengthen–Weaken (Beyond the Passage) questions.

Strengthen–Weaken (Beyond the Passage) Questions—Applied Example

One of the first examples of the ascendance of abstraction in 20th-century art is the Dada movement, which Lowenthal dubbed "the groundwork to abstract art and sound poetry, a starting point for performance art, a prelude to postmodernism, an influence on pop art . . . and the movement that laid the foundation for surrealism." Dadaism was ultimately premised on a philosophical rejection of the dominant culture, which is to say the dominating culture of colonialist Europe. Not content with the violent exploitation of other peoples, Europe's ruling factions once again turned inward, reigniting provincial disputes into the conflagration that came to be known by the Eurocentric epithet "World War I"—the European subcontinent apparently being the only part of the world that mattered.

The absurd destructiveness of the Great War was a natural prelude to the creative absurdity of Dada. Is it any wonder that the rejection of reason made manifest by senseless atrocities should lead to the embrace of irrationality and disorder among the West's subaltern artistic communities? Marcel Janco, one of the first Dadaists, cited this rationale: "We had lost confidence in our culture. Everything had to be demolished. We would begin again after the *tabula rasa*." Thus, we find the overturning of what was once considered art: a urinal becomes the *Fountain* after Marcel Duchamp signs it "R. Mutt" in 1917, the nonsense syllables of Hugo Ball and Kurt Schwitters transform into "sound poems," and dancers in cardboard cubist costumes accompanied by foghorns and typewriters metamorphosize into the ballet *Parade*. Unsurprisingly, many commentators, including founding members, have described Dada as an "anti-art" movement. Notwithstanding such a designation, Dadaism has left a lasting imprint on modern Western art.

Example:

4. According to some estimates, prior to the beginning of World War I in 1914, more than four-fifths of the world's landmass was controlled by European nations or former colonies such as the United States. If this figure is accurate, what effect does it have on the passage?

   A. It bolsters the author's suggestion that European colonialism was an overbearing force.
   B. It weakens the author's assertion that World War I was instigated by provincial disputes.
   C. It strengthens the author's claim that Europe is the only place in the world that mattered.
   D. It challenges the author's insinuation that European rulers ignored the rest of the globe.

Solution: With such a long question stem, you'll want to skip right to the question itself, which asks for the *effect* on the passage. Type this as a Strengthen–Weaken (Beyond the Passage) question and that your task will be to determine the relevance of this new data. As you Rephrase the question stem, think about how the information presented either supports or challenges statements from the passage. The evidence provided has nothing to do directly with the Dada movement, which doesn't start until after the commencement of the war, but it does pertain to the author's discussion of Europe and World War I at the end of the first paragraph. So, your Investigate step will require rereading that portion of the text to see whether anything there would be impacted. The figure cited supports the author's opinionated characterization of *the dominating culture of colonialist Europe*.

(A) matches this prediction perfectly. *Overbearing* and *dominating* are synonyms, so even the language in the answer choice closely matches that in the passage. On Test Day, select this answer and then move on to the next question without reading the wrong answers.

For our purposes, however, it's worth reviewing where the others go wrong. While **(B)** may point to a claim that the author actually makes, the new information provided in the question stem does not pose a threat to it, making it a Faulty Use of Detail. The catalyst for World War I could still have been provincial disputes, regardless of how much land each country controlled. **(C)** is wrong for two reasons: first, it's Out of Scope because this is not something the author endorses; the author sarcastically states "the European subcontinent apparently being the only part of the world that mattered" when pointing out that "World" War I actually took place exclusively in Europe. Also, even if the author did have this view, the statement is a value judgment—a matter of opinion—which cannot be directly affected by geographic facts. Finally, **(D)** is also Out of Scope. Though it does contain a factual claim that would be challenged by the question stem, the author never insinuated that *European rulers ignored the rest of the globe*. Rather, the author refers to rulers "turn[ing] inward" after "violent exploitation of other peoples," suggesting they were actually gazing outward before the start of the war, the time period noted in the question stem.

Example:

5. Which of the following, if true, would most threaten what the author says in the final sentence of the second paragraph?

    A. A large majority of members of the general public, when asked to identify the most important work of art of the 1900s, fail to mention an example from the Dada movement.

    B. Other prominent 20th-century artistic movements, such as Surrealism and Pop Art, were also commonly described as "anti-art" by their most influential participants.

    C. The consensus among art historians today is that Dada was merely a brief departure from the principal themes in European art that evolved during the 20th century.

    D. Some of the founding members of the Dada movement were sympathetic to the radical view that, far from being anti-art, Dada was the purest form of art imaginable.

Solution: The hypothetical *if true* is strong evidence that the answer choices will be new elements and that this is indeed a Strengthen–Weaken (Beyond the Passage) question. The word *threaten* further implies that the task is to Weaken the author's claim.

MCAT EXPERTISE

Appeals to authority or expert opinion are common in CARS passages, particularly those involving disciplines like the arts or literature, in which value judgments and other opinions play a prominent role. As you read, take note of the field you are reading about and the types of argumentation the author chooses to use. Quoting experts may provide decent support for a passage on an artistic movement, but such testimony will carry far less weight in more empirical social sciences like psychology or economics. Nonexpert opinions tend to carry even less weight: while authors may occasionally draw on popular opinion to support arguments, actual surveys of public opinion are seldom seen outside of a small number of cases, confined primarily to political science.

After the Rephrase step, the first part of your investigation should involve returning to paragraph 2 and reading the last line, as referenced in the question stem, in addition to reading the preceding one for context. Though second to last sentence suggests that some people think of Dada as "anti-art," the last one contests this by pointing out Dada's "lasting imprint on modern Western art." Thus, we are most likely looking for a refutation of the author's idea that Dada was influential.

It's also possible that the correct answer challenges the author's rejection of the term *anti-art*. So, if we cancel out the confusing double negatives, the correct answer could also support the idea that Dada is anti-art.

Our first prediction matches **(C)** since the scholars' idea of a "brief departure" clashes directly with the "lasting imprint" from the text. An appeal to expert opinion is an acceptable form of evidence for most arts and literary passages. Also, since the author name drops and cites quotations frequently in the passage, consensus among experts should carry even more weight in this case.

While **(A)** does not support what the author says, it also does not pose a large challenge. Popular opinion does not necessarily reflect whether a work of art is influential or not. Additionally, leaving a "lasting imprint" is not identical to being the singularly most important movement of the century. Thus, the threat presented by **(A)** is weak at best. **(B)** is consistent with the passage, strengthening the point that simply calling a movement "anti-art" does not necessarily make it so. **(D)** has no impact on the last sentence of paragraph two. Even if some of the "founding members" of Dadaism did not think of the movement as being "anti-art," this answer choice targets the penultimate sentence—not the last one.

MCAT EXPERTISE

According to our research of released AAMC material, *Reasoning Beyond the Text* questions that are neither Apply nor Strengthen–Weaken (Beyond the Passage) questions are rare, numbering only one to two, on average, per CARS section.

11.3 Other *Reasoning Beyond the Text* Questions

Some questions that require you to think beyond the passage won't necessarily fall under the categories of Strengthen–Weaken (Beyond the Passage) or Apply question. Since these types of questions do not occur frequently, we will limit our discussion to three examples that have appeared on past MCAT exams.

Probable Hypothesis

In many ways, **Probable Hypothesis questions** are similar to Apply questions, but instead of asking about the outcomes of new situations, they ask about the likely causes. After presenting the new details, these stems will ask for *a probable hypothesis, the likely cause,* or *the most reasonable explanation based on the passage.* Though working backward from a given effect to its probable cause may seem more challenging, you will still use Logic keywords that reveal analogous cause–effect relationships in the passage to form your predictions. If no match to your prediction can be found, eliminate any answers that contradict claims the author states or suggests elsewhere in the passage.

Alternative Explanation

Alternative Explanation questions are also interested in potential causes. They begin by providing a phenomenon that may be pulled directly from the passage but then ask you to provide a cause that is not provided in the passage and may be dissimilar from anything previously discussed. Since it is almost impossible to form a focused prediction, expect to evaluate each answer choice and eliminate those that would not produce the result in the question stem. If you get stuck between multiple answers that seem equally likely to serve as the cause, eliminate those that would most conflict with other parts of the passage. While a correct alternative explanation won't be one the author has already provided, it should also not significantly contradict what the author has already stated.

Passage Alteration

One other less common *Reasoning Beyond the Text* question type will ask about changes that the author could make to the passage to make it consistent with new information. These are appropriately called **Passage Alteration questions**. In many ways, these are like the rare instances of *Reasoning Within the Text* questions that require resolving paradoxes, except that they will include some new information that contradicts what the author says or implies. The correct answer to these questions will typically be the one that produces the desired effect with the *least* amount of modification to ideas originally presented in the passage.

Conclusion

And so, this brings us to the end of our discussion of question types. If you are still confused about which name corresponds to which type, don't worry! The common stems, tailored strategies, and Worked Examples you have seen thus far are designed to serve only as your first exposure to the intricacies of the Kaplan Method for CARS questions, the question types and tasks, and the Wrong Answer Pathologies. Continue practicing until the Method becomes second nature for you. In the end, it will be less important to be able to name the question type than to know how to approach it. The best way to improve your performance with CARS questions is practice accompanied by effective review—the subject of our final chapter.

Concept and Strategy Summary

Apply Questions

- **Type: Apply questions** require you to take the information given in the passage and extrapolate it to a new context. These questions often begin with words like *suppose*, *consider*, or *imagine*. Apply questions typically ask for one of three tasks:
  - They may ask for the author's **Response** to a situation, using words like *response*, *reply*, *most likely to agree with*, or *least consistent with*.
  - They may ask for the most probable **Outcome** in a situation, using words like *outcome*, *result*, *expectation*, or *consequence*.
  - They may ask for an **Example** of an idea discussed in the passage, using words like *example* or *instance*.
- **Rephrase:** If the question stem is long, jump to the end to determine what it's asking. Read any information given in the question stem closely, looking for hints that connect it to the passage.
- **Investigate:** Reread the relevant text if needed, keeping in mind the specific type of Apply question involved.
  - For Response questions, determine the author's key beliefs, which are generally indicated in the passage by Author keywords.
  - For Outcome questions, pay attention to cause–effect relationships in the passage, which are generally indicated in the passage by Logic keywords.
  - For Example questions, look for passage excerpts that provide definitions, explanations, or the author's own example and note any necessary or sufficient conditions.
- **Match:** Match your expectations with the right answer. If there is no clear match, or if you cannot perform any of the earlier steps of the Kaplan Method for CARS Questions, use process of elimination.
  - Eliminate any answer choices that are inconsistent with the author's views, especially for Response questions.
  - Eliminate any answer choice that does not contain necessary conditions (which must occur in all instances of a concept), especially for Example questions.

Strengthen–Weaken (Beyond the Passage) Questions

- **Type: Strengthen–Weaken (Beyond the Passage) questions** focus on the logical relationship between conclusions and the evidence or refutations that strengthens or weaken them, respectively.
  - These questions are closely related to Strengthen–Weaken (Within the Passage) questions, which use information directly from the passage, rather than bringing in new information.
  - These questions often contain words like *relate*, *support*, *challenge*, *relevance*, *significance*, or *impact*. Words like *could* or *would* often help differentiate them from Strengthen–Weaken (Within the Passage) questions.
- **Rephrase:** Determine the two claims and the connection between them; you will usually be given at least one of these elements and be tasked with finding the other(s).
  - Identify where each piece of the argument can be found: in the question stem, in the passage, or in the answer choices.
  - If no claims are given in the question stem, plan to triage it for later and answer it using process of elimination.
  - If one claim is given in the question stem, determine if it is a conclusion, a piece of evidence, or a refutation.
  - If two claims are given in the question stem, identify the relationship between them.
- **Investigate:** Research the relevant text to determine the missing claim or the connection between them. Use Logic keywords to help assemble the argument.
- **Match:** Match your expectations with the right answer. If there is no clear match, or if you cannot perform any of the earlier steps of the Kaplan Method for CARS Questions, use process of elimination.

Other Reasoning Beyond the Text Questions

- **Probable Hypothesis questions** ask for causes of new situations presented in the question stem.
  - These questions often contain words like *probable hypothesis*, *likely cause*, or *most reasonable explanation.*
  - Approach these questions like you would Apply questions, but look specifically for analogous cause–effect relationships in the passage.
- **Alternative Explanation questions** ask for causes that differ from the ones given in the passage while still providing an explanation for a phenomenon.
  - These questions often contain words like *alternative explanation*, *other cause*, or *different reason.*
  - Approach these questions by eliminating answer choices that would not lead to the effect in the question stem. If you get stuck between multiple answers, eliminate those that conflict most significantly with the passage.
- **Passage Alteration questions** ask for changes the author could make to the passage to make it consistent with new information.
  - These questions often contain words like *alter*, *change*, or *update.*
  - Approach these questions by looking for the answer that produces the desired effect with the least amount of modification to the ideas in the passage.

Worked Example

Use the Worked Example below, in tandem with the subsequent practice passages, to internalize and apply the strategies described in this chapter. The Worked Example matches the specifications and style of a typical MCAT *Critical Analysis and Reasoning Skills* (CARS) passage.

An initial glance at this passage reveals that there are multiple types of art described in the body paragraphs. Interrogating this passage is a strong choice, as drawing connections between each of the mentioned art types and their underlying purpose in building the author's argument will be key to answering test questions. Further, this passage is not too long and has a structure that appears to lend itself well to chunking. Outlining would also be a strong choice, given that there are multiple terms with descriptive detail for each, and an outline is a simple way of keeping track of the location of a large number of details. In any case, we can predict during our passage Preview that the structure and high detail content of this passage will necessitate some referral back into the passage for questions regardless of Distill method chosen.

INTERROGATING METHOD

Passage	Expert Thinking
Summer to winter to summer yet again, morning to night and then dawn once more. All things in life seem to cycle, and so, too, do trends in art. Styles, of course, do coexist and always have. Life is rarely as neatly divided between night and day as we might wish. But throughout history, art has followed one main avenue and then reversed direction time and time again, thus producing the Classicism *vs.* Romanticism (or Expressionism) dichotomy.	S1–4. Why does the author talk about such different concepts as life and art? *The author is drawing a parallel between life and art, as both life and art are apparently cyclical (in art, this refers to "style"). The author then brings up life being not neatly divided; we can predict from the start of the next sentence ("But") that art IS evenly divided, and the author is going to talk about that.* S5. Why does the author bring up Classicism and Romanticism? What do you think the author will discuss next? *These are the two main streams that art follows according to the author. Art following one path, then reversing direction seems to be a connection to the cyclical art trend noted in earlier sentences. We can anticipate that the author is going to elaborate more on the dichotomy.*
Classicism in art primarily refers to clean, cool imagery. In the High Classical period in ancient Greece, idealized sculpted figures of young men and women were perfect in proportion, the picture of health and vitality. The subsequent ancient Greek Hellenistic art swept in a more expressive era in which figures depicted actual people, with an emphasis on their individuality. Eschewing the sleek lines of the Classical period, sculptors lent their images a sense of weight so that clothing and hair looked a bit waterlogged. However, this additional substance often produced a sense of expressive motion: the goddess Nike (Victory) of Samothrace races forward as her windswept drapery creates wet wings behind her. The rational distance of the earlier Classical sculptures has given way to expressions that convey a more passionate, romantic essence.	S1–7. Why is High Classical art contrasted with Greek Hellenistic art? *High Classical art is defined by ideal sculptures with perfect proportions, whereas Greek Hellenistic art is subsequent to High Classical art, and is more expressive and weighted. The two are presented in this paragraph as examples of the two bigger art types in paragraph1: High Classical art is the first type (Classicism), and Greek Hellenistic art is the second type (Romanticism).*

INTERROGATING METHOD

Passage	Expert Thinking
Not surprisingly, the French Neoclassicists, from about 1750 to 1850, looked to the Classical age for inspiration. Painting and sculpture contained the same refined, dignified qualities as the earlier work, although employing contemporary subjects. Portraits of both aristocrats and commoners reveal the late 18th- to early 19th-century "re-vision" of Classical times in everything from fashion to furniture and architecture. It was the Romanticists, though, who put the soul back into art. Their technique was looser, emitting the sense that the artist's hand had just lifted off the canvas or finished chiseling the stone. Brooding compositions described exotic locales in the Middle and Far East. Heroic stories detailed contemporary shipwrecks, battles, and civilian revolutions. Neoclassical works tasted of buttered toast where romantic pieces tasted of hot spice.	S1–8. Why is Neoclassical art contrasted with the Romanticists? *The refined Neoclassical art, which is similar to the Classical art in paragraph 2, is contrasted with the looser but exotic Romantic art. The contrast serves to highlight the dichotomy brought up in paragraph 1, were art starts in one direction (Classical) then reverses (Romanticism). We can now see a clear trend: the author is trying to provide recurring examples of the cyclic nature of art.*
Classicism evolved into two camps during the 20th century. A realism trend continued, in which artists depicted the world along the lines of human perception. The Regionalists in the early part of the century reflected life in America's backcountry. Grant Wood's 1930 painting "American Gothic" presents a no-nonsense farm couple staring the viewer straight in the eye. They exude the basic goodness and solidity of their nature. They stand together for eternity, more as emblems of an age and ideal than true individuals. Interestingly, the same cool distance resulted later on in the abstraction of Minimalism, beginning around the early 1960s. Minimalist artists created no figures or references to the outside world. Instead, the sharp edges of their geometric shapes, unmixed colors, and lack of visible brush or carving stroke embodied the same distilled, classical calm.	S1–9. How does this paragraph differ from paragraphs 2 and 3? Why does the author bring up Regionalists and Minimalists? *Paragraphs 2 and 3 focused on the Classical* vs. *Romanticism dichotomy, but in paragraph 4, the author is only talking about two streams of Classicism in the 20th century. Regionalist, focusing on ideal goodness, and Minimalists, focusing on abstract geometric shapes, are both Classical streams that embody the same cool distance. This diverges from the cyclic trend that has been emphasized since P1.*

INTERROGATING METHOD

Passage	Expert Thinking
Between these two periods, America birthed Abstract Expressionism, its most fervent art form. Painters abandoned realistic, figurative images and thrust their inner emotions or the invisible vibrations of the universe onto canvas. Virtuoso brushwork and color flash across flat surfaces with a magnetism and energy unknown before. These huge compositions take your breath away. Abstract Expressionism, the nation's first unique art movement, exudes all the brashness of a young upstart, even as the more classically oriented works that bracket this movement recall an esteemed, stately heritage.	S1–5. Why does the author bring up Abstract Expressionism as occurring between the two styles from P4? *Abstract Expressionism is not Classicism, and it is stated to be unique. However, from its description, we can see links to Romanticism, and given that this new style arises between the two Classical trends in P4, we are again returning to the theme of the cyclic nature of art.*
The time between the cycle from cool to expressive and ideal to romantic has shortened of late. The current art scene hosts art from both sides now. Perhaps the future will bring us more artists who freely sew the two ends of the continuum together.	S1–3. Why does the author see the future art movement possibly combining both dichotomies? *The time between cycles of Classicism and Romanticism seems progressively shorter and shorter until the present movement has a mix of both, and the author hopes for more mixes of the two dichotomies in the future.* *Overall, the author has built an argument that the nature of art styling is cyclical and alternates between versions of Classicism and Romanticism.*

Question	Analysis
1. Assume that most people at the time said the work of the Regionalists was emotionally overwhelming. What effect would this have on the author's arguments?	Because this question brings in new information and asks about its *effect*, this is a Strengthen–Weaken (Beyond the Passage) question. Regionalism supposedly evolved from Classicism, but *emotionally overwhelming* is a much better descriptor for Abstract Expressionism, as described in paragraph 5: these artists "thrust their inner emotions ... onto canvas, creating huge compositions [that] take your breath away." Because the new information in the question stem seems to go against the author, investigate the answer choices that include the word *weaken*.
A. It would strengthen the assertion that the time period between styles is currently shortening.	**(A)** and **(B)** can be eliminated immediately because they say *strengthen*.
B. It would strengthen the claim that Classicism primarily refers to clean, cool imagery.	
C. It would weaken the claim that art styles can coexist.	Because the new information implies that Regionalists have some expressive attributes, the claim that Classicism and Romanticism can coexist is *strengthened*, not weakened—making **(C)** an Opposite answer.
D. It would weaken the claim that Regionalism was a type of classicism.	The new evidence suggests that Regionalism could have been misclassified as Classicism when it has expressive aspects, confirming **(D)** as the correct answer.

Question	Analysis
2. In 1801, a French Neoclassicist announced that "I seek to infuse the modern era with the historical weight of a great past." On the evidence of the passage, he could have best achieved his goal by producing:	This is an Apply question of the Example subtype. Given the Roman numerals, it is a Scattered Apply question. As a French Neoclassicist, this artist would aim to have the "refined, dignified qualities" of classicism, "although employing contemporary subjects." The quote in the stem also indicates that a reference to something *historical* or the *past* should also appear in the answer choice.
I. a statue of Napoleon Bonaparte in which he is costumed plainly as an ancient Greek emperor.	In Statement I, we see Napoleon is the contemporary subject matter, but the artist is inspired by ancient times and uses a plain costume, reminiscent of Classical style. This fits the criteria, so Statement I must appear in the correct answer.
II. a full-length portrait of Romantic novelist Victor Hugo resembling one of his heroic characters.	In Statement II, representing Hugo as a hero makes no connection to the past—"refined" and "dignified" or otherwise. This falls short of what we need, so Statement II must not appear in the correct answer.
III. a painting of Queen Marie Antoinette in which she is indistinguishable from the courtiers who surround her.	Statement III similarly does not make a connection to the past, and therefore must not appear in the correct answer. Further, to make the Queen blend in with her court would be the opposite of making her appear "refined" and "dignified."
A. I only	**(A)** contains the correct statement and is the right answer.
B. III only	**(B)**, **(C)**, and **(D)** contain incorrect statements.
C. I and II only	
D. II and III only	

Question	Analysis
3. According to the passage, which of the following musical experiences is most analogous to Minimalism?	This is an Apply question asking for a musical example that is similar to Minimalism in art. Minimalism is described at the end of paragraph 4, and all of the characteristics of this style are said to "embody . . . distilled, classical calm" and is a form of Classicism that involves abstraction and ideals like calm.
A. Agitated classical music in a large concert hall	**(A)** intentionally uses the word *classical* as a Faulty Use of Detail, but the word *agitated* indicates that this music is anything but calm. Eliminate this answer choice.
B. Electronic elevator music playing quietly	Elevators are typically calm places—the phrase *elevator music* even tends to have the connotation of innocuous or boring, implying that this music is quite calm. This makes **(B)** the correct answer.
C. Hard rock music blasting through speakers	We can rule out **(C)** because *hard rock* and *blasting* indicate that this music anything but calm.
D. A repetitive tape loop of country music	A repetitive tape loop would likely be irksome and not particularly calm, making **(D)** incorrect as well. This answer choice is a good trap if you are familiar with music history, as many Minimalist composers did indeed employ tape loops in their music; however, this answer is not supported by the passage and is therefore incorrect.

Question	Analysis
4. With which statement would the author most likely DISAGREE?	This is an Inference question of the Implication subtype; we are looking for something that the author would not agree with. It is hard to form a solid prediction for this question, so analyze each answer choice and eliminate the ones that are consistent with the author's opinions or that are Out of Scope.
A. Late 18th- to early 19th-century French Romantic art had a sense of personality that Neoclassicism lacked.	**(A)** clearly addresses the dichotomy we are getting comfortable with: classical = cool, romantic = expressive. The author would agree that *Romantic art* has more personality than a Classical movement, so eliminate this answer choice.
B. At any given time, it can be difficult to pinpoint a strong dichotomy between prevailing styles and previous ones.	**(B)** reflects what the author tells us in the first paragraph: that "styles, of course, do coexist" and that "life is rarely as neatly divided ... as we might wish." This answer can be eliminated as well.
C. French Romanticism lacked an immediacy that was apparent in Neoclassical painting.	**(C)** goes against the author's description of Classicism as having a coolness and distance to it, whereas Romanticism has more expressive, emotional impact. Romanticism should have the immediacy that Classicism lacks—this answer choice is an Opposite, meaning the author would *DISAGREE* with the statement, making **(C)** the correct choice.
D. French Romanticism was the polar opposite of Neoclassicism.	This passage repeatedly draws a polar contrast between Classicism and Romanticism. **(D)** is a valid inference based on the passage and therefore can be eliminated.

Question	Analysis
5. Suppose that most late 18th-century French drawings are exotic and exciting. Which passage assertion would be most WEAKENED by this conclusion?	This is another Strengthen–Weaken (Beyond the Passage) question that provides new evidence and asks us to identify the conclusion that is most *WEAKENED* by it. The *late 18th-century* is a reference to the time period given at the beginning of the third paragraph, "1750 to 1850." This is the French Neoclassicist era—a Classical period—but *exotic* and *exciting* are descriptors of Romanticism. This implies that many artists were not actually using Classical themes at the time.
A. Classicism was more popular than Romanticism in late 18th-century France.	**(A)** uses comparative language, but how *popular* one style is in comparison to another is not addressed at all in the passage and so this choice can be eliminated.
B. The Neoclassicists' inspiration had pervasive effects.	This claim is made in paragraph 3: according to the author, Neoclassicism impacted "everything from fashion to furniture and architecture." However, because we were surprised to hear about Romanticism in a Classical period, it no longer seems that the Neoclassicists were as "pervasive" as the author described. Therefore, **(B)** is the correct answer.
C. Inner emotions can be imbued into artwork.	**(C)** is certainly a valid inference, but it is not affected in any way by the question stem's information.
D. Neoclassical works are less dramatic than works of the Classical age.	**(D)** also makes a comparison that was never mentioned explicitly in the passage—the author never addresses how *dramatic* one Classical style is compared to another Classical style.

Question	Analysis
6. Suppose that romantic artists and Classically oriented artists began borrowing heavily from one another. This finding would support the view that:	This is a Strengthen–Weaken (Beyond the Passage) question, based on the words *Suppose* and *would support the view*. The new evidence in the question stem is consistent with the final point the author makes—that these two styles may eventually be "sew[n] ... together." To carry out your Investigation, scan for an answer choice that is consistent with the passage goal.
A. Romanticism and Classicism are independent movements.	**(A)** contradicts the author's point in the last paragraph. Further, the question stem shows a case where the two movements are converging—not that they are *independent* of each other. Eliminate this answer choice.
B. the past inevitably influences the future.	**(B)** uses extreme language and was never overtly claimed by the author. While the author provides some examples of the "past ... influenc[ing] the future," such as French Neoclassicism, the word *inevitably* is far too strong, as many movements that are not strongly influenced by the past, such as Abstract Expressionism, are also described.
C. the cycle from cool to expressive art styles is shortening.	**(C)** might be tempting because it is a claim the author makes. However, "borrowing heavily" from each other at one point in time would mean there wouldn't be two distinct styles that are cycling.
D. artists might be beginning to make less of a distinction between these two schools of art.	**(D)** uses the Moderating keyword *might* and correctly states the author's main point at the end of the last paragraph. That's our match.

Question	Analysis
7. Which of the following, if true, would constitute a reason Neoclassicists looked to Classical Greece for inspiration?	For this Strengthen–Weaken (Beyond the Passage) question, we are looking for a reason Neoclassicists would look to the High Classicism style in ancient Greece for inspiration. It is challenging to make a prediction, but we should look at the answers with an eye toward the themes the author has identified as part of the Classical mode.
A. Nineteenth-century French nobility admired the ancient period because it was one in which even ordinary citizens acquired important art.	**(A)** does not address the idea of Classicism at all; further, there is no obvious reason why *nobility* would like an art style that *even ordinary citizens* could acquire. Eliminate this answer choice.
B. Neoclassical painters and architects were impressed by the wide range of human feelings that Classical Greek sculptors captured in their marble works.	**(B)** discusses *feelings*, which the passage indicates are actually the hallmark of the opposite art style: Expressionism. Eliminate this Opposite answer choice.
C. Nineteenth-century French elite idealized the cultural expressions of the ancient past as conveying regal grandeur, devoid of intense emotions.	**(C)** references the absence of emotions, which matches our understanding of Classical values. In addition, it makes sense that the *French elite* would be attracted to art that *convey[ed] regal grandeur*. This is the correct answer choice.
D. Neoclassicists found that appealing to patrons' interest in antiquity allowed them to move their style toward the Expressionist ideal.	**(D)** implies that Neoclassicists desired to move their style out of the Classical mode and into the Expressionist mode, but there is no evidence in the passage that these artists desired to do so.

PRACTICE QUESTIONS

Passage 1 (Questions 1–5)

GO ONLINE

Post-structuralist literary criticism was developed largely in reaction to Saussurian linguistic theory, which first expressed the relationship between words and the concepts they denote. In Saussurian linguistics, an actual word is referred to as a "signifier"—the "sound image" made by the word "train," for instance, constitutes a signifier. At the same time, the idea evoked by the signifier is termed a "signified." Saussure argued that the structural relationship between a signifier and a signified constituted a "linguistic sign." He saw language as made up entirely of such signs, or structural relationships, and argued that the relationship that constituted these signs was actually arbitrary and based on common usage rather than on some necessary link. He did believe, however, that certain "signifiers" (words) could be permanently linked to specific "signifieds" (concepts) in order to create stable, predictable relationships that evoked constant meanings.

In contrast to Saussurian linguistics, the post-structuralist view contends that there exists no system of describing ourselves, or of communicating with one another, which does not somehow use our indigenous language systems. To post-structuralists, language defines our identities and is required if we are to maintain those identities. In this view of language, any signifier always signifies another signifier. Definitions and meanings always take the form of metaphors: one term can only be defined as being another term. To change the meaning of a term, one must only change the metaphor through which that term is defined. Meaning shifts from one signifier to another, and because of this, no act of signification is ever fully closed or fully complete.

Because the post-structuralists do not view necessary connections as composing permanent linguistic sign relationships, they reject the idea of absolute meaning. Because language constantly shifts along a chain of meaning, "absolute" meanings cannot exist. Language in this view can never be viewed as entirely stable. Whereas Saussure believed that linguistic sign relationships could create stable, consistent meanings between terms and the images they evoke, post-structuralists argue that meaning can be established only through discourse. Thus, meaning is never absolute, immutable, or concrete because it is always dependent on the differing and constantly shifting discourse in which language terms operate.

It is here that the divergence of these two schools of thought becomes readily apparent, insofar as they concern themselves with fields beyond linguistics. The implications of the nearly irrefutable, albeit bleak, reasoning that is so fundamental to post-structural thought reach far beyond the confines of linguistics. The idea of conceptual instability is a manifestation of the existential phenomenology that heavily influenced the work of innumerable scholars in disparate fields through the 20th century. Where structuralism was grounded in linguistics and made inroads into the human sciences, the very ideas that presuppose post-structuralism are rooted more ambitiously in the central discussion of human nature. It is for this reason that existentialism and post-structuralism continue to flourish and have inserted themselves into our enduring understanding of what it is to be human while structuralism and Saussurian linguistics hold a devoted place in scarcely few discussions beyond linguistic relativism.

1. Which of the following best adheres to the post-structuralist view of meaning as presented in the passage?
   - **A.** The words "signifier" and "signified" are expressed in different ways across several languages.
   - **B.** Each member of a literary club puts forth a unique interpretation of a fairy tale based on his or her academic background.
   - **C.** The relationship between a father and his son shifts dramatically during the course of a novel.
   - **D.** All ten members of a focus group derive the same meaning from the preview of a new situation comedy.

2. In addressing a class, a professor describes the mind first as a blank slate to be written on and later as a garden to be cultivated. In light of the information in the passage, the professor's method best represents:
   - **A.** the immutable nature of meaning.
   - **B.** a Saussurian relationship between words and the concepts they denote.
   - **C.** the arbitrary use of language in academia.
   - **D.** a post-structuralist change in the meaning of a term.

3. "Broad, open-ended study is frivolous and indulgent. It is through deliberate, focused study that knowledge advances." How does this statement affect the author's argument in paragraph 4?
   - **A.** It weakens the author's argument.
   - **B.** It strengthens the author's argument.
   - **C.** It neither strengthens nor weakens the author's argument.
   - **D.** It could both strengthen and weaken different parts of the author's argument.

4. A study finds that humans taught a fabricated word and its definition immediately form an association between that word and its definition. After repeated exposure to the word in different contexts, the same people consistently offered different definitions for the word than the one they were initially given. This strengthens:
   - **A.** post-structural theory because the initial signifier did not maintain its association with the initial signified.
   - **B.** post-structural theory because the initial signifier was most likely understood through a metaphor that changed.
   - **C.** Saussurian linguistic theory because of the initial association of signifier and signified.
   - **D.** neither theory, either theory, or both theories; more information on the underlying mechanism that precipitated the change is needed.

5. Which of the following examples is LEAST analogous to the Saussurian understanding of linguistic relativism?
   - **A.** A dog's response to the "sit" command could not be changed upon retraining, though its response to the "stop" command could be changed.
   - **B.** A study reveals that words learned in any language stimulate the same neurons regardless of attempts to retrain the meaning of these words over several years.
   - **C.** The definition of the word "cool" has evolved drastically over the years depending upon social context and common usage.
   - **D.** A study revealed that the presentation of a word to a listener immediately resulted in a localized brain region becoming activated before the listener could offer the word's definition.

Passage 2 (Questions 6–10)

Where once they were surrounded and protected by vast wilderness, many of America's national parks are now adversely affected by activities occurring outside their boundaries. The National Park Organic Act established the national park system and empowered the Secretary of the Interior to manage activities within the parks. According to this act, however, conditions outside park boundaries are not subject to regulation by the National Park Service unless they involve the direct use of park resources.

Several approaches to protecting the national parks from external degradation have been proposed, such as focusing on enacting federal legislation granting the National Park Service broader powers over lands adjacent to the national parks. Legislation addressing external threats to the national parks twice passed the House of Representatives but died without action in the Senate. Giving the states bordering the parks a significant and meaningful role in developing federal park management policy was also brought to the table as a possible remedy, though this solution is rife with its own problems.

Because the livelihood of many citizens is linked to the management of national parks, local politicians often encourage state involvement in federal planning. In spite of this, current state legislation has been an ineffective legal solution to the dangers facing park wildlife and the parks themselves. For instance, state legislatures have not effectively addressed the fundamental policy issue of whether states should be responsible for protecting park wildlife. State land use and environmental statutes, moreover, are often not intended to solve problems of this nature. Timber harvesting, ranching, and energy exploration compete with wildlife within the local ecosystem while priorities among different land uses are not generally established by current legislation. Additionally, often no mechanism exists to coordinate planning by the state environmental regulatory agencies, thus limiting the impact of legislation aimed at protecting park wildlife and the larger park ecosystem.

Even if these deficiencies can be overcome, state participation must be consistent with existing federal legislation. States lack jurisdiction within national parks themselves, and therefore state solutions cannot reach activities inside the parks, thus limiting state action to the land adjacent to the national parks. Most of this land falls under federal oversight by default due to ownership laws. Under the supremacy clause, federal laws and regulations supersede state action if state law conflicts with federal legislation, if Congress precludes local regulation, or if federal regulation is so pervasive that no room remains for state control. With the assumption that federal regulations leave open the possibility of state control, state participation in policy making must be harmonized with existing federal legislation.

The residents of states bordering national parks are affected by park management policies as well. They in turn affect the success of those policies. This interrelationship must be considered in responding to the external threats problem. Local participation is necessary in deciding how to protect park wildlife. Local interests should not, however, dictate national policy, nor should they be used as pretext to ignore the threats to national parks or the regions that surround them.

6. Suppose that a state government's environmental policies were contributing to the gradual extinction of a species in a national park within its territory. How would this information affect the author's argument?
   - **A.** It would strengthen the author's argument.
   - **B.** It would weaken the author's argument.
   - **C.** It would neither strengthen nor weaken the author's argument.
   - **D.** It would strengthen the author's argument only if it were shown that the National Park Service was not trying to save this species.

7. The author of the passage would most strongly believe that which of the following legislative issues where federal and state policy are in conflict can be effectively overcome?
   - **A.** A state seeking to legalize capital punishment does not entirely comply with federal regulations but is allowed to revise their proposal.
   - **B.** A state seeking to protect a species, including its habitat within a major national park, appeals to the supremacy clause to pass its legislation in its entirety.
   - **C.** A national park seeking legislation pertaining to the misuse of nearby municipal lands appeals first to the state legislature, which proposes a thorough review of federal precedent.
   - **D.** A state seeking to legalize physician-assisted suicide has given thorough consideration to the relevant federal statutes and existing legal precedent in drafting the proposed legislation.

8. In light of the viewpoints presented throughout the passage, the author would most likely support which of the following actions by state governments?
   - **A.** Efforts to buy back land adjacent to national parks from the federal government
   - **B.** Efforts to force the federal government to give up control of national parks to states
   - **C.** Efforts to curb activities that are harmful to national park wildlife
   - **D.** Efforts to reduce the power of the Secretary of the Interior

9. Which of the following situations is analogous to the challenge facing national parks, as outlined by the author?
   - **I.** A man trying to complete his work in his own office is having difficulty doing so due to the noise emanating from the office next to his.
   - **II.** A state-run park cannot influence the process of the construction of a hydroelectric dam outside the park on a river that the park is responsible for conserving.
   - **III.** A municipality can regulate the release of toxic pollutants outside its boundaries only if it is funding the decontamination process.

   - **A.** I and II only
   - **B.** I and III only
   - **C.** II and III only
   - **D.** I, II, and III

10. Which of the following, if true, would most WEAKEN the author's main contention in Paragraph 4?
    - **A.** There are examples of states having overcome minor legislative discrepancies with the federal legislature after lengthy and extremely expensive appeal processes.
    - **B.** Under the National Park Organic Act, states are capable of appealing any decision supported by the supremacy clause that pertains to national parks.
    - **C.** No state has ever passed legislation that conflicts with federal regulations pertaining to national parks.
    - **D.** The National Park Service experiences a revision of its mandate every three years.

PRACTICE ANSWERS

Passage 1 (Questions 1–5)

Sample Highlighting

P1. "linguistic signs" and "permanently linked"; **P2.** "define our identity" and "metaphors"; **P3.** "reject" and "only through discourse"; **P4.** "far beyond" and "scarcely few."

Sample Outlining

P1. Saussure: arbitrary signifier (word) relation with signified (concept) = linguistic sign, can become concrete

P2. Post-structuralist: signifier defined only by other signifiers; understand through metaphors that can change

P3. Saussure = language can be stable, post-structuralist = language never stable, depends on discourse

P4. Post-structuralism endures because of broader ties to human nature; Saussurian theory restricted to linguistics

1. **B**

This question challenges us to apply our understanding of the post-structuralist theory to the answer choices in order to find an example. Paragraph 3 makes clear that the post-structuralists believe that "meaning is never absolute, immutable, or concrete." Looking for the choice that best fits with this notion brings us to **(B)**. The fact that the "unique interpretations" are due to differences in "academic backgrounds" supports the notion that exposure to different fields of discourse leads to the differences in interpretation, which is consistent with the language of paragraph 3. **(A)** cleverly uses words from the passage in a new context. The fact that words are expressed differently in different languages does not affect the absolute meaning of these words; they're expressed in various ways, as one would expect in different languages, but that doesn't mean they are defined differently. In **(C)**, the relationship between father and son may shift, but we'd have to see some variation in interpreting the *meaning* of this shift to get us into post-structuralist territory. As for **(D)**, a lack of consensus would seem more consistent with the post-structuralist view, although it's conceivable that the ten focus-groupers arrived at their opinions through discourse with each other. However, because such a state of affairs isn't indicated, we can't assume it.

2. **D**

In Application questions like this one, we must determine how the new example relates to what's in the passage. In passages where the goal is *to explain the differences between* two things, that usually means figuring out what camp the new situation falls into. The major task here is figuring out what the professor is doing, and how it relates to the theories in the passage. When the "professor describes the mind ... as a blank slate, or as a garden," she is using metaphors, which brings us into the post-structuralist camp. Moreover, she's shifting the metaphor, which relates precisely to the penultimate sentence of paragraph 2: "To change the meaning of a term, one must only change the metaphor through which that term is defined." Therefore, the answer is **(D)**. **(A)** and **(B)** are Opposites because they presuppose constant meaning, whether it is described as immutable or Saussurian—Saussure posits "stable, predictable relationships that evoke constant meanings," according to the end of the first paragraph. **(C)** is Out of Scope as there is no mention of the use of language in academia in the passage.

3. **C**

This question provides new information and asks how it impacts the passage, making this a Strengthen–Weaken (Beyond the Passage) question. The new quotation suggests that widely applicable fields of study, such as post-structuralism as described in this paragraph, are frivolous and that narrowly focused fields, such as Saussurian linguistics, advance knowledge. However, let's consider the author's argument in this paragraph. The author focuses on the differences in scope between the two camps and why post-structuralism is more enduring. He or she makes no mention of the advancement of knowledge or frivolity. The author may imply some partiality to post-structuralism, but there is no argument made regarding the focus of the new quotation. Thus, **(C)** is correct as this statement has no bearing on the author's argument.

4. **D**

This Strengthen–Weaken (Beyond the Passage) question challenges us to identify not only whether the new information *strengthens* post-structuralism or Saussurian theory, but also why. The first thing to recognize is that the study's results could support either theory. Even though Saussurian theory suggests that words "could be permanently linked" to concepts, it also recognizes that the relationship between a word and the concept it represents is "actually arbitrary and based on common usage rather than on some necessary link." We also know that post-structuralism definitely supports changing definitions. Thus, we have to look at the reasoning in the answer choices to determine the correct answer. **(A)** may sound tempting because the reasoning sounds as if it's disproving Saussurian theory, but remember that Saussurian theory does account for "signified" changing depending on "common usage." Also, Saussurian theory being proven incorrect wouldn't necessarily prove post-structuralism correct. **(B)** requires assumptions we just can't make because there is not enough support in the passage for it being a "metaphor that changed" that caused the change in meaning. **(C)** is also tempting because it mentions Saussurian theory and one of its tenets, although *the initial association of signifier and signified* does not really add any new information to strengthen the author's description. More importantly, this answer choice doesn't explain the all-important second half of the new information: why the association changed. Therefore, the answer must be **(D)**. This answer choice identifies the real problem with the other choices: we don't know the mechanism behind the change in association. Both theories account for a change in associations, so the mechanism by which the association changed needs to be described before we can draw any conclusions about which theory the study supports.

5. **B**

The challenge in this Application question is to transfer our knowledge of Saussurian linguistics to find three appropriate analogies or one choice that does not fit. This description appears in paragraph 1 of the passage, where the author makes it clear that "signifiers" are tied to "signifieds" in order to produce a sign, or linguistic understanding. This relationship can be, but is not necessarily, permanent. **(A)** adheres to this description well. The meaning can be permanently linked or changed, depending on the circumstances. **(B)** is Extreme in its categorization of semantic links as permanent. The passage said that links are "arbitrary and based on common usage rather than on some necessary link," so the categorization in this answer choice of all links being concrete is Extreme, and thus correct. **(C)** fits in perfectly with the previous quotation that links are "based on common usage." Finally, **(D)** is also a perfect explanation of Saussurian theory. The signifier is presented, it triggers brain activity, and the definition (signified) is provided.

Passage 2 (Questions 6–10)

Sample Highlighting

P1. "adversely affected" and "not subject"; **P2.** "broader powers"; **P3.** "not effectively addressed"; **P4.** "limiting states action" and "harmonized"; **P5.** "should not."

Sample Outlining

P1. Problem: national parks affected by surrounding lands, but Park Service has little control

P2. Solutions tried: more power over neighboring lands (didn't pass), empower states to help

P3. Problems with state legislation as a solution

P4. State legislation must be consistent with federal policy

P5. Elements of local participation needed but should not be overriding (warnings)

6. **A**

Here, the question stem asks us to reason how a novel situation *affect[s] the author's argument*, making this a Strengthen–Weaken (Beyond the Passage) question. First, we must consider what the author thinks about the effect of state governments on national parks. There's evidence throughout the passage, and especially in paragraph 3, that the author thinks that state governments' "land use and environmental statutes . . . compete with wildlife within the local ecosystems." Therefore, the new example of state policies harming a species in a national park supports the author's argument. **(A)** is correct. **(B)** and **(C)** are, of course, necessarily incorrect by this logic. **(D)** throws the curveball of offering the same position—that this information *strengthens the author's argument*—but with a conditional qualifier: that *only if* statement. Whether or not the National Park Service is trying to save this species is irrelevant because the state policy still serves as an example of a state statute affecting wildlife in the ecosystem.

7. **D**

For this Apply question, we are asked to find which scenario would elicit a given Response from the author: which "legislative . . . conflict can be effectively overcome." The author's argument can be found in paragraph 4, where it is argued that "state participation in policy making must be harmonized with existing federal legislation." This prediction should be sufficient to attack the answer choices. **(A)** offers a case of a state *not entirely comply[ing] with federal regulations*, which is the Opposite of what we're looking for. **(B)** again implies a case of a state at odds with the federal government, but the state *appeals to the supremacy clause* this time. Referring back to paragraph 4, we see that the supremacy clause dictates that federal statutes take precedence and thus the state will not even be successful by appealing to the supremacy clause. **(C)** gives the scenario of a federal service appealing to the state government about *municipal lands*. Not only do we know nothing about where municipal governments fall within this scheme, but also the passage gives no information about why the federal government would be appealing to the state government about something neither have any particular control over. If anything, the federal government would have much more say given the passage's description of the supremacy clause. **(D)** proposes a scenario where the state is seeking to align itself with federal requirements. This is the best match for our prediction and is the correct answer.

8. **C**

This is another Apply question of the Response subtype. We are looking for an answer *the author would most likely support*. First, let's consider what the author wants state governments to do. In general, the author is in favor of cooperation with the federal government and a greater effort to protect wildlife within the parks. Of these two priorities, only **(C)** offers a course of action that matches. **(A)** is a Faulty Use of Detail; states already own the *land adjacent to national parks*, so there would be no need to *buy back* this land. Both **(B)** and **(D)** would serve to limit the power of the federal government, which is never stated or implied by the passage.

9. **B**

For this Apply question, we must first identify what the author has deemed a "challenge facing national parks." The opening paragraph states that "national parks are . . . adversely affected by activities occurring outside their boundaries" and that "conditions outside park boundaries are not subject to regulation . . . unless they involve the direct use of park resources." These are the challenges that make way for the rest of the author's discussion. Therefore, we are looking for examples similar to these claims. Statement I is analogous to the above scenario because the *man trying to complete his work* is being adversely affected by conditions outside of his own space, much like the parks. **(C)** can be therefore eliminated. Statement II again offers a scenario where an entity is unable to control conditions outside its own space that would affect it. However, this time, the park should be able to influence proceedings according to the scenario described in the passage as "park resources" are committed to conserving the river. Therefore, Statement II is incorrect, ruling out **(A)** and **(D)**. We are only left with **(B)**, the correct answer. We already know Statement III to be correct, but upon reading, it becomes apparent that this choice presents an entity being adversely affected by forces outside its own borders that it can act upon only if it assumes financial responsibility, or the tying-up of its resources. This situation is analogous to the situation described in the first paragraph.

10. **B**

This Strengthen–Weaken (Beyond the Passage) question directs us back to paragraph 4 where we must first identify the author's *main contention*. The author argues in this paragraph that "state participation in policy making must be harmonized with existing federal legislation." Thus, we need to identify the answer choice that most strongly suggests that state policy making needn't necessarily be harmonized with federal legislation to weaken this statement. **(A)** does indeed suggest that states can overcome differences with federal legislation, however, only for "minor ... discrepancies and after lengthy and extremely expensive appeal processes." Does this weaken the contention that states must be consistent with the federal government? Slightly—it demonstrates that it is possible to achieve victories. Still, it is a very weak refutation of the author because only minor victories can be achieved at a great cost. Further, it doesn't mention whether any of these minor victories were won in pursuit of caring for the national parks. **(B)** offers the point that states can legally appeal the supremacy clause in *any* case pertaining to national parks, implying both major and minor cases. This would have a much more profound effect on the author's argument, weakening it. While this answer choice is the correct one, it is worthwhile to eliminate the others because it would be possible for another answer choice to have a more significant weakening effect. **(C)** actually strengthens the author's contention that states need to be consistent with the federal legislature—otherwise, they'll never achieve anything. Finally, **(D)** has no apparent effect on the author's argument. We don't have enough information to determine if this *revision of [the] mandate* would result in more state power and less need to be consistent with the federal government, so we can eliminate this answer.

CHAPTER 12

Effective Review of CARS

CHAPTER 12

Effective Review of CARS

In This Chapter

Introduction

LEARNING OBJECTIVES

After Chapter 12, you will be able to:

- Troubleshoot common errors on missed questions with Why I Missed It Sheets
- Apply post-phrasing analysis to difficult passages
- Manage and adjust your pacing within the CARS section to meet the 90-minute deadline
- Build endurance and vocabulary through targeted practice

This final chapter is a troubleshooting guide for raising your score. Experience with the CARS section is a prerequisite for the material in this chapter. So, before studying this chapter in detail, you must practice the strategies discussed throughout this book, especially the Kaplan Method for CARS Passages from Chapter 7 and the Kaplan Method for CARS Questions from Chapter 8. For the material in this chapter to be most effective, it's best if you have also taken a few timed and scored CARS sections, possibly on one or two Full-Length practice exams. Once you've built this familiarity with the CARS section—and with your personal pitfalls in the CARS section—you'll be ready for troubleshooting! We'll look at five proven ways to increase your *Critical Analysis and Reasoning Skills* (CARS) section score—including some methods you'll find useful for the science sections as well.

Hands down, the best way to improve is to learn from the mistakes you make on practice tests, which is why we begin our discussion with Kaplan's Why I Missed It Sheets (WIMIS). The post-phrasing strategy discussed in the following section builds on the WIMIS, helping you think more like the writers of the MCAT. After we discuss post-phrasing, we'll talk about a simple approach for managing your pacing on Test Day, and we'll discuss what you can do to build your test-taking stamina. We'll close with a discussion of one final advantage you can give yourself for CARS: a stronger vocabulary.

12.1 Learning from Your Mistakes

At this juncture, you have read quite a bit about how to approach CARS, you have had the opportunity to practice, and you may even have completed one or more Full-Length Exams. Now is the time to reflect on your performance and adjust your study plan based on your personal test-taking pathologies, which are identifiable patterns in your past errors that can help to guide your future CARS studying.

MCAT EXPERTISE

WIMIS are a way to create and maintain a pool of data on your own performance on the MCAT. Because they're structured as tables, spreadsheet tools can be very helpful for analyzing your WIMIS. Consider using a spreadsheet-creating program to build your WIMIS so you can add to them from multiple devices, sort them by different columns to more easily see trends, and organize them in different ways that best suit your current needs.

Why I Missed It Sheets

It can be overwhelming to try to discern what your strengths and areas of opportunity are by looking at an entire CARS section at once. Rather, take time to analyze each question you answered incorrectly. One way to help manage this process—and to keep a clear record of your performance over time—is to set up **Why I Missed It Sheets** (**WIMIS**). Create a document with at least four columns: question number, question type, topic (or discipline), and "Why I Missed It." In the first column, make a list compiling all of the items you answered incorrectly on the test. In the second column, identify the question type based on the Kaplan classifications we reviewed in the previous three chapters (or, if you're making a WIMIS for one of the sciences, include the question topic). The third column should include the topic (alternatively: academic disclipline, when reviewing science) for the passage. Lastly, in column four, take time to review your incorrect response and identify the specific reason why you got the question wrong. Table 12.1 demonstrates what WIMIS should look like.

Question Number	Question Type	Topic/ Discipline	Why I Missed It
Full-Length 1, #12	Detail	Music	Misread question stem. Misinterpreted the question stem, leading me to the wrong paragraph for information.
Full-Length 1, #21	Strengthen–Weaken (Within the Passage)	Art	Didn't understand question stem. Chose a piece of evidence that would strengthen the critics' point of view, but was looking to strengthen the author's point of view.
Full-Length 1, #38	Inference	Population Health	Too much conjecture. Chose an answer choice that *may* be true based on the information in the passage but does not have to be—Inference questions have answers that *must* be true based on the passage.
Full-Length 1, #53	Main Idea	Psychology	Scope too narrow. Answer choice was too narrow, only describing two of the paragraphs but not the entire passage.

Table 12.1 Sample Why I Missed It Sheet (WIMIS)

BRIDGE

Need some help identifying the question type for your Why I Missed It Sheets? Don't forget to review Chapters 9, 10, and 11 of *MCAT CARS Review* for thorough coverage of each question type.

The most important element of this table is the *Why I Missed It* column. Do not just categorize your test-taking mistakes with a generic *I didn't understand* or *careless mistake*. Such comments will not give you insight into why you missed the question. At the other extreme, you are never going to see this specific question again, so a hyper-specific description of the mistake you made on this specific question isn't going to help you in the future. Responses that are either too vague or too specific both result in the same problem: you're spending this time reviewing your performance, but you're not learning what to do better on the next test! The goal of WIMIS is to learn to identify questions in the future that are similar to questions you've missed in the past, and then to avoid the mistakes you made on such questions in the past. Your response in the *Why I Missed It* column should always keep this goal in mind.

To write good *Why I Missed It* responses, start by identifying exactly what you did wrong on the specific question you're reviewing, then generalize that error just a bit by recording only the mistake without including the unique details of that question or passage. The responses in the above table illustrate this technique. Take the first response as an example. This student identifies the specific mistake, "Misinterpreted the question stem ... leading me to the wrong paragraph." But this student doesn't provide unnecessary details about the question or passage. In the *Why I Missed It* column, this student also labels this mistake with the general label, "Misread question stem." In fact every one of this student's *Why I Missed It* entries follow this formula: general label, then specific details, all with little or no information unique to the passage or question. You don't have to follow this format exactly on every question, but it is a good default.

WIMIS make it easier to identify flaws that are becoming trends, which helps you better address these trends earlier in your prep. Consider the following pairs of "bad" and "good" examples:

> **Bad:** *I missed this because the answer choice declares that* Beowulf *was written by an aristocrat, but the passage said that the author was only "probably" a member of the royal court.*
> **Good:** *Misread question stem. I missed this because the answer choice stated something as fact that the author of the passage did not.*
>
> **Bad:** *I missed this because I thought the question was asking for the author's opinion on Mendelssohn's work, but it was actually asking for his opinion on the critics of Mendelssohn's work.*
> **Good:** *Didn't understand question stem. I missed this because I confused which opinion a question stem was asking for.*
>
> **Bad:** *I missed this because I recognized "unmoved mover" from the question stem, which is in paragraph 3, but the answer they were looking for actually comes from the description of the "Central Headquarters" sentence in paragraph 4.*
> **Good:** *Referred to passage incorrectly. I missed this because I failed to see how the paragraphs fit together (how an idea in one paragraph is further explained in subsequent paragraphs) before moving on to the questions.*

When reviewing your WIMIS, in addition to looking for trends in the *Why I Missed It* column, compare the question type column to the topic/discipline column to narrow the scope of your future study plan. Is there a recognizable pattern you can utilize to optimize your performance? For instance, maybe looking at the question type column, you observe that you are repeatedly missing *Reasoning Within the Text* questions. That's good information. But maybe by also comparing with the topic column, you observe that you most often miss this question type on anthropology passages specifically. Now, not only have you identified a question type for further practice, but also you know to be on guard next time you're facing a *Reasoning Within the Text* question on an anthropology passage. Use your WIMIS to plan which passage varieties you'll focus on in the future, to identify what part of the question or passage you're consistently spending too much time on (or too little), or to determine which chapters you need to reread in *MCAT CARS Review*. You want to make sure that you are moving forward with a plan that specifically supports your areas of opportunity.

In addition to the basic data we've discussed so far, you can optionally collect several additional pieces of data to further flesh out your WIMIS and make them even more valuable tools. You can add a column to record the time you spent on each passage and question, which can be valuable for discovering passage types, question types, or content or topic areas where you tend to fall behind. Note that when reviewing the Kaplan Review Page, the time associated with the first question in a question set will typically represent the time you spent Distilling the passage plus the time you spent answering that question. You could also add a column to indicate the type of wrong answer you chose, which can help you identify which wrong answer pathologies you find most tempting. Some students also find it helpful to actually split out the one *Why I Missed It* column into two columns: one column to capture a general label for the error and a second column to record more detail. Make your WIMIS your own by customizing them to capture data points that are most relevant to the most problematic parts of your prep.

REAL WORLD

In medical school, you'll attend morbidity and mortality (M&M) conferences, in which physicians review clinical and medical errors, patient complications, and other healthcare quality measures to investigate the root causes of problems in order to fix them directly. These conferences are not dissimilar to creating Why I Missed It Sheets—you can only improve if you know what you're doing wrong!

Troubleshooting Common Wrong Answer Trends

WIMIS are a great tool for uncovering trends among your wrong answers. But of course, diagnosing the problem is only the first step. We must now treat the problem. So, let's discuss three of the most common trends among wrong answers, and what to do about them.

If you repeatedly read questions incorrectly, you are not alone—MCAT students often misread or misinterpret questions in CARS. So, if misreading questions comes up as a trend in your WIMIS, you'll know to practice the CARS question strategy. Remember, establishing the question type is a critical component of the CARS question strategy. You might be less familiar with CARS-style content, but the tradeoff is that all CARS questions are subdivided into just a few predictable question types, each with its own predictable (and practiceable!) strategy. Rephrasing the question stem to ensure you fully understand the question is also essential. As you review your practice tests, reexamine the questions and the answers you selected that were incorrect. In hindsight, when looking at the question a second time, do you find yourself shaking your head because you simply misinterpreted what was being asked? This could indicate that you are reading the question stems too hastily and moving on to the answer choices before you even know *what* you're supposed to be answering. If this is the case, force yourself to take the extra time to paraphrase a difficult question stem so that you know precisely what the question is asking. Time spent paraphrasing is not wasted time. The added clarity will allow you to Investigate more quickly, which will help you Match an answer choice rapidly and accurately. In the next section, we will discuss a review technique called post-phrasing, which is a particularly helpful exercise for any student who commonly misreads the question.

If your WIMIS demonstrates a pattern in which you keep missing the same question types, then there is an easy way to gain time and confidence on this test: start to triage those questions. First, go back and reread the appropriate chapter in *MCAT CARS Review* so you can recognize these question types with greater ease. Chapters 9, 10, and 11 introduced the major question types and highlighted the common question stems that fall into each type. Remember that triaging a question to the end of the set is the opposite of admitting defeat. Very often, as you answer the other questions in the set, you learn more about the author's argument, which maximizes your chances on the triaged question. Here's a good way to think about it: triaging a question is less like running from your foe, and more warming up before going into battle. Remember also that each question type has an associated strategy that you can practice. Many students find that their least favorite question types early on becomes among their most reliable question types by Test Day. And this turnaround starts by identifying a problematic question type using WIMIS.

Finally, don't neglect identifying the Wrong Answer Pathologies of the incorrect answers that you chose. Simply knowing the Wrong Answer Pathologies that most often sway you from choosing the correct answer may be enough to help you avoid falling for the same traps in future tests. For example, if you know that you frequently choose Out of Scope answer choices in Main Idea questions, then you'll know to ask yourself *Is this answer truly within the confines of the passage, or does it bring in something else?* before finalizing your answer for Main Idea questions in the future.

BRIDGE

Knowing your personal test-taking pathologies is essential to improving your score. In addition to looking at the question types, topics or disciplines, and how you read question stems, don't forget to look at Kaplan's classifications of Wrong Answer Pathologies (Faulty Use of Detail, Out of Scope, Opposite, and Distortion), discussed in Chapter 8 of *MCAT CARS Review*.

12.2 Thinking Like the Test Maker: Post-Phrasing

Does this scenario sound familiar? You skim through the explanation for a question that you missed, shrug, and think, "I guess I get it." When you miss a practice question, you've actually made two mistakes. Not only were you tempted by a trap answer, but also you looked right at the right answer and thought, "Nah! Not that one!" To turn today's wrong answers into tomorrow's right answers, you need a proactive technique for reviewing questions. Post-phrasing is that technique. With **post-phrasing**, you go over both why incorrect answers are wrong and why correct answers are right. This strategy is especially helpful if you frequently experience the classic test-taking dilemma: *I can usually narrow it down to two answer choices, and then I always go for the wrong one.*

To post-phrase, open a question you answered incorrectly, and make a note of the correct answer. Now, you'll carefully dissect the question. Begin by identifying the Type of the question. Recall that each question Type is associated with a general strategy so, if necessary, briefly review the general strategy for that question Type. Next, paraphrase the question stem. To make it very concrete, say your paraphrase out loud, write it down, or type it out. By doing this, you are essentially practicing a formal version of the Rephrase step from the question method. Your focus in this step should be to carefully decode what the question stem is asking you to do.

Once you've Typed the question, reviewed the best strategies for addressing that question type, and Rephrased the question, Investigate the solution. When post-phrasing, write or type out your prediction for the correct response. While you wouldn't write down a prediction on Test Day, the post-phrasing exercise is designed to ensure you are making concrete predictions as part of your Investigation. It's essential to take the time to write the prediction down: many students find that a poorly focused prediction keeps them from finding a Match on MCAT questions. The more you practice writing out your predictions during the post-phrasing process, the more routine making predictions will become.

You may notice that the description of post-phrasing, thus far, basically describes a regimented and very deliberate use of the Kaplan Method for MCAT Questions. But at the end, the focus will not be on matching your prediction anymore; after all, you already know the correct answer because you're reviewing a question you answered incorrectly. Instead, turn your attention to the incorrect answers. Determine why each incorrect answer is unsuitable. Does it fail to match your prediction? Does it have a common Wrong Answer Pathology? Are there other subtle flaws in logic or reasoning that misrepresent the author's arguments? Pay special attention to the wrong answer you actually chose. Why was this answer tempting? Why is it ultimately wrong? By taking the time to write or type out an explanation of the flaw for each incorrect answer, you'll become better at identifying flaws in future wrong answers. Finally, after all this, *now* take a look at Kaplan's explanation and check how well your reasoning matches up.

To summarize, remember this saying: *Practice alone doesn't make perfect. Perfect practice makes perfect.* Post-phrasing is that perfect practice. You are taking a question that you did not answer correctly, and you are modeling the steps, actions, and strategies that you *should have used* to get that question right, which reinforces those behaviors for next time. And lastly, after such a thorough review of a question, what's the best next step? Summarize the results of this detailed, post-phrasing investigation in your WIMIS of course. You don't have to post-phrase every question you missed on every practice exam. Instead, save post-phrasing for the questions that really deserve a thorough examination: the ones where you were stumped, where you were stuck between two answers, and the ones where you made a silly mistake that you want to forever avoid in the future.

Alternatively, a variant of post-phrasing can be used as an exercise to improve your recognition of Wrong Answer Pathologies in passages you haven't seen yet. Consider practicing with a few passages following the steps below:

1. Find a fresh CARS passage from your Kaplan resources, as well as its explanations.
2. Go through and select the correct answers for the passage. Don't read the explanations, just record the correct answers and set them aside for now.
3. Outline the passage as you normally would.

4. When you work on the questions, do so in a fundamentally different way: as above, the goal here is not to find the right answers (as you have already identified them) but to reason why each incorrect answer is wrong. Read the right answer to be sure you understand it, but spend more time identifying the types of wrong answers presented by the other choices.
5. Look at the explanations for each incorrect answer choice and compare them with your notes. If there are any major omissions or other discrepancies in your explanations, be sure to note them prominently. The more you practice, though, the closer you should find the correspondence between your post-phrasing notes and Kaplan's explanations.

Remember, on the MCAT the right answer is not just the "most correct" answer. Each question has one right answer and three answers with objective flaws, meaning that the right answer is objectively correct because it is the only answer without flaws. Through regular post-phrasing practice, you will begin to see the patterns emerge. What seem now like very subtle flaws will soon start to seem very pronounced. As a result, you will find yourself recognizing the correct answer, the one that doesn't suffer from any flaws, and thinking *this one just feels right*. Furthermore, taking the time to post-phrase will expand your awareness of how the test is written and what the expectations are for MCAT questions. Through post-phrasing practice, you will refine your thought process and your approach to passages and questions, making you better prepared for Test Day.

BRIDGE

In addition to Wrong Answer Pathologies, take time to review the Signs of a Healthy Answer presented in Chapter 8 of *MCAT CARS Review*. These include appropriate scope, agreement with the author, and "hedging" language that creates weaker claims.

12.3 Improving Your Timing

Proper management of the clock can make a significant difference to your score in any section, but the timing constraints in the CARS section differ from those of the other sections. Consider a few basic facts about the CARS section:

- It lasts 90 minutes.
- There are 9 passages.
- There are 53 questions.

It's a simple matter of arithmetic to see that 9 passages in 90 minutes allows you 10 minutes for each passage. Because all of the points actually come from answering the questions, though, you'll want to ensure that you allot enough time for the questions. Passages vary in difficulty, meaning that some will certainly take longer to read than others, but as a rule of thumb you should aim to complete the Preview, Read, and Distill steps for the passage in about 4–5 minutes. Your timing will vary based on the Distill approach you Choose, but even the most intense Interrogation of a very challenging passage should take no longer than 6 minutes. Subtracting those combined ~40 minutes for reading leaves you with 50 minutes for the 53 questions, which works out to almost exactly one minute per question. Keep in mind that the number of questions associated with a passage can vary from 5 to 7, meaning that a more precise expectation for completing a passage and its questions is somewhere between 9–11 minutes.

MCAT EXPERTISE

While it is most common to have five to seven questions associated with a given passage, there have been instances of passages with only four or as many as eight questions. This should not significantly impact how you approach these passages, although—all else being equal—a passage with eight questions will give you more points for the same amount of reading.

Pacing Guidelines

Now, trying to ensure that each passage takes 4–5 minutes and each question a little less than one minute would actually be counterproductive: not only would you lose a substantial amount of time checking the clock, you'd likely heighten your anxiety each time you came across a question that takes a little extra time to get through. A better approach to managing your time is to check your timing only at a limited number of predetermined points during the section. We recommend checking the clock after every other passage. Assuming that you take 10 minutes to read a passage and answer its questions, you should ideally be no more than 20 minutes further into the section each time you check the clock. We recommend jotting down how much time remains when you start the third, fifth, seventh, and ninth passages (it should be at least 70, 50, 30, and 10 minutes, respectively). If your timing appears to go off-course, then during your review of the test, note which passages (and potentially Distill approaches) seem to have caused this trouble. Could you have triaged these time-consuming passages or distilled them differently, and answered more questions correctly in the limited time you had left?

Hitting these recommended time points does not leave you much of a cushion for going back to previous passages and questions, but this approach is by far the best way to maximize your score. Given that each CARS passage contains between 500 and 600 words, jumping between passages is far less realistic in CARS than in any of the science sections, where passages can be less than half that length. While it is recommended that you skip passages that you decide will be too time-consuming in your Preview step, we do not recommend that you leave a passage unfinished with the intention of returning later. Once you decide to work on a passage, commit to finishing it.

Managing Question Timing

The timing guidelines given so far are useful for managing the CARS section. But mastering the split between reading each passage and answering its questions can still be tricky. In your online resources, we record the time spent working on each question so you can get a sense of which questions take you longer than others. (Note that this mechanism requires you to click on the question when you begin working on it to accurately record this information.) Keep in mind that the time we record for the first question of each passage will include the time you spend critically Reading and Distilling the passage in the beginning. Thus, you should strive to have the first question of each set answered within around 5 minutes (or 300 seconds), and every subsequent question should be close to the one-minute mark, calculated as an average. If you've chosen to Distill primarily by Highlighting, you should expect to see that you answered your first question more quickly, and then spent more time on each question. Alternately, if you mostly choose to Distill by Interrogating, you should expect to see a time of 6 or even 7 minutes for the first question, with much shorter times for the subsequent questions in each set. If you use a mix of methods, or if you primarily use the Outlining method, 5 minutes average for the passage and first question should be your target.

MCAT EXPERTISE

If the seconds ticking down causes you too much anxiety, an alternative way of working on your timing is to use the timer on your phone or some other timepiece to "time up"—that is, to count upwards from zero to see how much time it's taking you to read the passage and to complete each question. Timers with "lap" functions can be especially useful because these can record how long each question takes. Generally, this method works best with only one or two passages at a time. Keep in mind, however, that an alternate timer cannot be used on Test Day!

Once you have a better sense of which types of passages and questions take you longer to complete, you should use this information in conjunction with what you've learned from each WIMIS to guide your *now* or *later* decisions in the first step of each method (Preview for passages and Type for questions). Beginning the section with the passages that are easiest for you will allow you to get ahead on the timing curve. Similarly, once you've committed to a passage, saving the toughest questions for the end of that set will give you an opportunity to gain additional familiarity with the passage as you work on its more manageable questions. Because questions can repeat the same theme, you may even find the answer to a challenging question while working on an easier question.

12.4 Building Endurance

Preparing for the MCAT is like preparing for a marathon. You cannot run 1 or 2 miles a day for two months and then expect to be successful in a 26.2-mile marathon. Runners ramp up to the 26.2 miles, routinely running long distances in preparation for marathon day.

How often are you studying and for how long? If you are only studying in 30- or 60-minute increments, you might find when you sit down to complete a Full-Length Exam that you have not built up your endurance for that type of test. Not only do you need to study and practice regularly, but you also need to simulate the Test Day experience, which requires concentrating for 6 hours and 15 minutes of testing time. Gear up for practice tests—and Test Day itself—by periodically holding long study and practice sessions. When working on CARS, try studying in 90-minute increments to mimic the amount of time you'll have for this section. Long study periods can also be helpful for the sciences, as you will have 95 minutes for each science section on Test Day.

12.5 Enhancing Your Vocabulary

The AAMC claims that there's no outside content required for the CARS section of the MCAT, but in truth this is not 100 percent accurate—you may come across a question that has an element of common knowledge (the number of days in a week, for example). Furthermore, while very few questions hinge on knowing the definition of a piece of jargon from outside knowledge, a strong vocabulary on Test Day can be a tremendous asset. If nothing else, you can become more comfortable with academic writing in the various disciplines that the AAMC includes in the CARS section. Do this by familiarizing yourself with plenty of examples.

BRIDGE

While you will not be expected to know the definition of a piece of jargon from outside knowledge, Definition-in-Context questions hinge on your ability to determine the meaning of a word or phrase from the rest of the passage. These questions are discussed in Chapter 9 of *MCAT CARS Review*.

The Kaplan and AAMC Full-Length Exams are, of course, the best place to go to get samples of CARS-style passages and questions. But students who want to go the extra mile also have the option of practicing with outside reading.

Reading Plan

The more often you read, the stronger your reading comprehension skills become, and the faster you evolve into a more efficient reader. This practice will help you build up a wider array of words that you recognize by sight, meaning that you'll spend less time deciphering the text. Believe it or not, reading is a skill that you can practice and improve, just like any skill. So, to prepare for CARS, read on a consistent basis. This practice this will increase your reading speed so that you can maneuver swiftly through passages.

Set up a regimented reading schedule for at least 20 minutes a day, which is just enough time to get through two passages according to the the timing guidelines for the CARS section. Also, keep an array of reading materials handy for when you have time available. This mini-library could consist of a list of links kept on a smartphone or other wireless-enabled device. It is surprising how much time we spend waiting—for a friend at a coffee shop, for a group to go out in the evening, for a professor to arrive to class, or for a meeting. These are all examples of times that could be used productively to sharpen reading skills to prepare for CARS.

When selecting materials, try to simulate the variety of passages you'll encounter on Test Day with a blend of texts from both the humanities and the social sciences. Use your WIMIS to determine which types of passages cause you the greatest confusion or frustration, the ones that slow you down the most on your Full-Length Exams. Focus on reading those problematic types of passages in particular; the more you read these difficult texts, the more familiar you will become with their jargon and with other conventions, and the less intimidating these passages will start to seem. Consider using one of the following online services to find academic journal articles in the disciplines that give you the most trouble:

- JSTOR (jstor.org)
- Oxford Journals (academic.oup.com/journals)
- Google Scholar (scholar.google.com)
- Project MUSE (muse.jhu.edu)
- The Directory of Open Access Journals (doaj.org)
- Sage Journals (journals.sagepub.com)

Whenever you come across a word that you don't recognize, stop and take a moment to look that word up and write down its definition *in your own words*. Keeping a list of these new terms and looking over them periodically will go a long way toward building your vocabulary. As your vocabulary expands, you will be able to either recognize words or swiftly infer their meaning, which will increase your efficiency both while Reading and Distilling the passage and while tackling question stems and answer choices.

In addition to learning the meanings of new words, reading humanities and social sciences articles will help you become acquainted with the major themes and concepts that frequently appear on the CARS section. Familiarity with a concept can allow you to glean more from a passage when that concept is mentioned, enabling you to read more quickly if the text simply repeats what you already know. Be careful, however, not to bring in any ideas that the passage does not include when answering its questions—there's a reason Out of Scope is a common Wrong Answer Pathology! Because there are only nine passages in CARS, the likelihood that you will have read about the exact scenario presented in a CARS passage is not high, but it is very likely that you will have read about similar topics.

Finally, any form of academic reading forces you to think critically about the ideas the author is presenting. Just as when you practice with passages in CARS, try distilling a few paragraphs of an academic article. Focus on how the author structures the argument: what conclusion does he or she want the audience to reach? What evidence is used to support that conclusion? Are there any flaws in the author's logic? Focused, regular reading will help prepare you for CARS by bolstering your reading comprehension and reasoning skills, as well as your comfort with academic texts and the challenges they bring.

Conclusion

Medical schools want to admit students with strong reasoning skills because higher-order thinking is necessary for both appropriately diagnosing patients and conducting groundbreaking research. When given a constellation of symptoms and concerns, a physician needs to generate a differential diagnosis—a list of the potential ailments described by the symptoms. From this list, the medical team must rule out unlikely diagnoses and provide evidence for the most likely diagnosis. What starts as a list of a hundred possible causes of headache—from migraines and tension-type headaches to intracranial bleeds and brain tumors—is reduced to the one most likely cause by asking appropriate questions (*When did the headache start? Where do you feel the headache? Have you ever had anything similar before?*), by performing a thorough physical exam (cranial nerve function, eye exam, looking for evidence of trauma), and by running appropriate laboratory and imaging tests (head CT, MRI, inflammatory markers).

Patients want to trust the expertise of their doctors, and you want to ensure that you are able to deliver the best treatment to your patients. The skills in CARS that enable you to determine the author's perspective, to distinguish her voice from others in the passage, to predict a response to a question, and to match your predictions (while eliminating incorrect answers) are the same skills that will serve you well as a physician. In the future, you may not be expected to think critically about dance theory, musicology, archaeology, and linguistics, but you will have to synthesize disparate pieces of information, consider assumptions about patient care, and respond appropriately to all parts of your patients' questions—both what they say and what they leave for you to infer. As pointed out in this chapter, the skills tested in the *Critical Analysis and Reasoning Skills* section can always be improved through pointed and actionable review of your past performance. Congratulations on reaching the end of *MCAT Critical Analysis and Reasoning Skills Review.* Though this book now comes to an end, for you this is merely a beginning—good luck on the MCAT, and in all your endeavors in medicine!

Concept and Strategy Summary

Learning from Your Mistakes

- Create **Why I Missed It Sheets** (**WIMIS**) to look for your test-taking pathology patterns.
  - Make a table with at least 4 columns: question number, question type, topic/discipline, and "Why I Missed It."
  - In the Why I Missed It column, describe the error you made in the question as specifically as possible (with respect to the thought pattern), but in a way that allows you to extrapolate what you learn from that question to future questions.
  - Look for patterns in your WIMIS.
- If you misread questions, be sure to slow down and rephrase the question stem to make sure you know what question you are actually trying to answer.
- Reread the relevant chapters in *MCAT Critical Analysis and Reasoning Skills Review* as needed.

Thinking Like the Test Maker: Post-Phrasing

- Find a CARS passage from your Kaplan resources, as well as its explanations. This could be a passage you have already read or a new one.
- Go through and indicate all the correct answers for the passage. Don't read the explanations—just indicate the correct answers and set aside for now.
- Distill the passage as you normally would.
- When you work on the questions, do so in a fundamentally different way: the goal is not to find the right answers (as you have already identified them). Rather, the goal is to reason why each incorrect answer is wrong. Read the right answer to be sure you understand it, but spend more time identifying the types of wrong answers presented by the other choices.
- Look at the explanations for each incorrect answer choice and compare them with your notes. If there are any major omissions or other discrepancies in your explanations, be sure to note them prominently.

Improving Your Timing

- Aim to read a passage in about 4 minutes; aim to answer each question in about one minute.
- Each passage and its questions together should take somewhere between 9 and 11 minutes.
- Check the clock after every other passage and its questions. Each passage pair should take about 20 minutes.
  - At the beginning of the section, you have 90 minutes left.
  - After two passages, you have about 70 minutes left.

- After four passages, you have about 50 minutes left.
- After six passages, you have about 30 minutes left.
- After eight passages, you have about 10 minutes left.
- After nine passages, you have 0 minutes left and are finished with the section.

Building Endurance

- Increase stamina by studying in 90-minute increments when possible (equal to the amount of time for the CARS section).
- Periodically, simulate the Test Day experience with study and practice for 6 hours and 15 minutes (equal to the amount of testing time).

Enhancing Your Vocabulary

- Read academic texts for at least 20 minutes a day to sharpen reading skills in preparation for CARS.
- Choose articles on topics that cause you trouble on Full-Length Exams (as revealed by your WIMIS).
- Practice Distilling (Highlighting, Outlining, and Interrogating) using these articles.

Notes

Notes

Notes

Notes

Notes

Notes

Notes

Notes

Notes

Notes

Notes

Notes

Notes

Notes

Notes

Notes

Notes

Notes

Notes

Notes

Notes

Notes

Notes

Notes

Notes

Notes